Parking Structures

Planning, Design, Construction, Maintenance, and Repair

Parking Structures

Planning, Design, Construction, Maintenance, and Repair

ANTHONY P. CHREST
MARY S. SMITH
SAM BHUYAN

VAN NOSTRAND REINHOLD
New York

Copyright © 1989 by Van Nostrand Reinhold

Library of Congress Catalog Card Number 89-30621
ISBN 0-442-20655-0

Printed in the United States of America

Van Nostrand Reinhold
115 Fifth Avenue
New York, New York 10003

Van Nostrand Reinhold International Company Limited
11 New Fetter Lane
London EC4P 4EE, England

Van Nostrand Reinhold
480 La Trobe Street
Melbourne, Victoria 3000, Australia

Macmillan of Canada
Division of Canada Publishing Corporation
164 Commander Boulevard
Agincourt, Ontario M1S 3C7, Canada

16 15 14 13 12 11 10 9 8 7 6 5 4 3 2 1

Library of Congress Cataloging-in-Publication Data

Chrest, Anthony P.
    Parking structures: planning, design, construction,
  maintenance, and repair / Anthony P. Chrest, Mary S. Smith, Sam
  Bhuyan.
      p.   cm.—(Structural engineering series)
  Bibliography: p.
  Includes index.
  ISBN 0-442-20655-0
    1. Parking garages—Design and construction—Handbooks, manuals,
  etc. I. Smith, Mary S., 1952–   . II. Bhuyan, Sam. III. Title.
  IV. Series.
  TL175.C48   1989
  690′.538—dc19
                                                   89-30621
                                                   CIP

Other Structures-Related Books

ICE INTERACTION WITH OFFSHORE STRUCTURES by A.B. Cammaert and
  D.B. Muggeridge
FOUNDATION ENGINEERING HANDBOOK by Hans Winterkorn and H.Y. Fang
FOUNDATION ENGINEERING FOR DIFFICULT SUBSOIL CONDITIONS by
  Leonardo Zeevaert
ADVANCED DAM ENGINEERING edited by Robert Jansen
TUNNEL ENGINEERING HANDBOOK by John Bickel and T.R. Kuesel
SURVEYING HANDBOOK edited by Russell Brinker and Roy Minnick

To our families

Claire and David Chrest
Doug, Brad, Geoff, and Emily Smith
Gail, Samuel, Robbie, and Stephanie Bhuyan

Preface

This book contains up-to-date information on planning, design, maintenance, and repair of multilevel parking structures. Based on our firm's experience in designing over 600 built structures, it provides owners, designers, and builders with information which they can put to immediate practical use.

My fellow authors, Mary Smith and Sam Bhuyan, are acknowledged experts in their respective fields. Much of the subject matter in Mrs. Smith's chapters on functional design, access design, and safety and security has never before been published. The same is true for many of Mr. Bhuyan's topics in the chapters on maintenance and repair, which have never appeared in book form. I hope my own experience in design, administration, specification preparation, and construction in the United States, Australia, and New Zealand will help with the practical application of the advice expressed in this handbook. To our knowledge, no one book of this scope, including chapters on structural systems, durability design, specifications, and construction, exists today.

As editing author, I gratefully acknowledge the efforts of my fellow authors in preparing quality manuscripts, along with their cooperation in all stages of the production of this book.

ANTHONY P. CHREST

Kalamazoo, Michigan

Acknowledgments

Our heartfelt thanks to the people of Walker, all of whom have helped make this book possible, but especially to Thomas A. Butcher and Richard T. Klatt for their help with the functional standards; Judie Lenz, Patricia A. Prentice, Sandra Scott, and Judith A. Williamson for manuscript preparation; Howard Linders and Frank Transue for their understanding and support. Also, our sincere appreciation for the improvements suggested by Technical Reviewers James E. Staif and Ronald G. Van Der Meid, and for the input received for Chapter 5 from Thomas J. Downs, Harald G. Greve, Norman G. Jacobson, Jr., Howard R. May, and William S. Phelan. Finally, our heartfelt thanks to our manuscript editor, Carol Orstein, for her quiet, helpful competence and patience.

Authors

ANTHONY P. CHREST is Senior Vice President and Corporate Chief Engineer at Walker Parking Consultants/Engineers, Inc. A registered structural engineer and professional engineer in several states, he has worked with parking structures for eighteen years. He coauthored the *Parking Garage Maintenance Manual*, articles in professional journals, and is a member of several professional organizations, including American Concrete Institute (ACI) Committee 362 on parking structures. He has spoken nationally at ACI seminars and has served as an expert witness across the United States.

MARY S. SMITH is Vice President and Director of the Transportation Group at Walker Parking Consultants/Engineers, Inc. She joined Walker in 1975 and is a registered professional engineer. Mrs. Smith is generally acknowledged as one of the country's leading experts on functional design for parking facilities. Mrs. Smith has spoken at the national conventions of the National Parking Association and the Urban Land Institute. As a member of the prestigious Parking Consultants Council of the National Parking Association, she has led the development of standards for the design of parking for the handicapped and of parking geometrics.

SAM BHUYAN is Vice President and Director of the Technical Resources Group at Walker Parking Consultants/Engineers, Inc. Mr. Bhuyan is a registered professional engineer and is directly involved in providing parking facility restoration services. These services include evaluation of structures, preparation of condition appraisals and construction documents, and project representative services for restoration projects. Mr. Bhuyan is also involved in the evaluation, selection, and implementation of repair materials, and methods and corrosion-protection systems for parking structures. He has authored technical articles in the *Transportation Research Record*, the *Concrete Construction* and the *Parking Professional* magazines.

Contents

Chapter 1
Introduction

ANTHONY P. CHREST

1.1 BACKGROUND

Parking structures can be found all over North America. They serve office buildings, shopping centers, banks, universities, and hospitals, and may be located in urban or suburban areas.

Parking structures have been designed and built for decades. Why then, the need for this book?

Parking structure design is more difficult than is immediately apparent, possibly leading to deficiencies in the finished building. Yet this need not be so. It is hoped that the direction and advice given in this book will raise awareness of the complexities of parking structures and lead to their improved design, construction, maintenance, and repair.

Parking structures can be deceptively difficult to plan, design, and construct. Aside from consideration to the impact on traffic in the surrounding streets, attention must be given to entrances and exits, revenue control, internal traffic and pedestrian circulation, patron security, openness requirements, structure durability, maintainability, and other matters which are not usually encountered in urban building types.

As a result, even experienced designers and builders can be caught by practices which they have used before, but which will not work in parking structures. Owners, too, may make decisions based on their previous experience, but that experience may not be applicable to parking structures. Much of this advice will also apply to surface parking lots as well.

1

1.2   PURPOSE

The purpose of this book is to explain some of the more common peculiarities of parking structures which set them apart from other building types, followed by advice on dealing with these unique features.

1.3   PARKING STRUCTURE PECULIARITIES

It is important to know that building codes recognize two types of parking structures—open and closed. Open parking structures do not require mechanical ventilation or sprinklers as do closed structures; therefore, they are less expensive.

For a building code to admit a parking structure as open, it must meet specific requirements in that code. For instance, the Uniform Building Code states, in part,

> For the purpose of this section, an open parking garage is a structure of Type I or II construction which is open on two or more sides totalling not less than 40 percent of the building perimeter and which is used exclusively for parking or storage of private pleasure cars. For a side to be considered open, the total area of openings distributed along the side shall be not less than 50 percent of the exterior area of the side of each tier.

Other codes have similar requirements, differing only in degree. A parking structure not meeting the code requirements for openness will be considered closed. A closed parking structure carries more stringent requirements for ventilation and fire protection, especially if automobile service will be inside the structure. The term used in most building codes for parking structures which are not open is *garage*.

It is always important, then, in dealing with building departments and code authorities to identify your project properly. In this book, unless we specifically state otherwise, we are usually discussing *open* parking structures, though most of the material applies to garages and surface lots as well.

Chapter 2 deals with internal circulation. In years past, a 300–500-car parking structure was considered an average size project, and a 1000-car deck, huge. These days, structures of 1000–3000 cars are not uncommon and decks as large as 12,000 cars are being planned. With the advent of larger capacity structures, the old rule-of-thumb methods for determining the number of entrance and exit lanes, and internal circulation routes to permit traffic to flow smoothly, are no longer adequate. Further, parking related dimensions—stall width and length, parking angle and parking module (the clear dimension between opposite walls of a parking bay)—are often limited by local ordinance. The local rules may often be outdated, leading to uneconomical parking structures.

Parking-related dimensions will differ for different patron types. At one end of the spectrum is a deck used for office parking only. Since the office worker will tend to park in the same spot all day, and probably in the same spot or close to it every day, he will quickly become used to the circulation pattern within a structure. Also the office worker will not require a wide parking stall. At the other end of the spectrum is a deck used for shopping center parking. Shoppers may be unfamiliar with the structure, will park for shorter periods, and will need wider parking stalls to load and unload passengers and packages. Chapter 3 deals with access issues, such as revenue control systems and designing entrances and exits for peak activity levels. As parking facilities have gotten larger, these issues have also become more significant.

Parking structures have attracted vandals, muggers, and rapists in recent years. Courts have held owners, operators, and sometimes architects and engineers, responsible for security features or their absence. No amount of retrofitting can replace good original design practices.

Chapter 5 deals with structural design. Parking structures have unusual proportions, compared to most buildings. A typical cast-in-place structure might have one-way post-tensioned slabs spanning 18–24 ft, supported by post-tensioned beams spanning 54–62 ft. A precast structure typically has double tees spanning the 54–62-ft dimension, supported by spandrel beams spanning 18–30 ft. The floors in adjacent bays slope, so that the beams join the columns at staggered levels. The interior columns between sloped floors may be short and stiff because of the building proportions.

In plan, the structure is relatively large. Structures 200–300 ft long and 110–130 ft wide are common. Many structures are larger.

To this rigid framing system, add the combined effects of camber in the beams and floor elements due to prestressing, vertical and horizontal deflections due to car and people loads, the structure's own weight, wind, earthquake, and construction. To further complicate matters, next add the effects of structure volume changes due to shortening from the prestressing forces, shrinkage, and creep. Finally, add the effects of severe weather and climate fluctuations. Now we have a system with which many engineers are unfamiliar and others understand only partially. Even engineers relatively experienced in parking structure design may not always avoid some of the complexities inherent to a particular structure under certain combinations of conditions.

Because the parking structure is open, yearly temperature extremes can affect the floor elements, beams, columns, and walls of an open parking structure. These elements are accessible in varying degrees to rain, snow, and sun. In colder climates where pavements are salted, that salt will increase the number of freeze-thaw cycles and lead to corrosion of the steel reinforcement in floors and the lower parts of walls and columns.

Other than bridges, no other structure type has to resist such a variety of attack by corrosive environments and deteriorating forces. Unlike most buildings, parking structures have no protective envelope. Unlike most bridges, which rain can wash clean, only the roof of a parking structure is entirely open to rain. Designing a parking structure according to highway bridge codes will make it cost more than it should; however, if you design the structure according to building codes, without special attention to its unique exposure, use, and requirements for durability, it will not perform well.

In a parking structure, there are no carpets, ceilings, or wall finishes to conceal mistakes in forming or finishing. Extra care must be taken, then, to construct the building properly. There is also less leeway with respect to quality control of the concrete and reinforcement to achieve a durable structure. Finishing and curing require more care.

Parking structures require at least as much attention to maintenance as any other building, perhaps more. Though there may be only bare concrete to maintain, that concrete is exposed to severe weather fluctuations.

1.4 ORGANIZATION OF THE BOOK

We hope that the three major parts of this book will help you deal successfully with the problem areas described in Section 1.3.

No matter how well the structural framing is designed, no matter how durable, drivers must be able to enter and exit the structure, circulate, and park with safety and convenience. Patrons, whether driving or walking, should feel safe and secure. The first part's following three chapters address these matters by dealing with functional planning, parking space layout, parking efficiency, entrance and exit planning and control, security, and safety.

Having dealt with first things first, the book's second part, in four chapters, expands on the subjects of structural design, construction materials and durability, specifications, and construction.

To complete the subject matter, the third part's two chapters treat maintenance and repair. If you are not familiar with parking structure terminology, you will find the Glossary following Chapter 10.

Chapter 2
Functional Design

MARY S. SMITH

2.1 INTRODUCTION

Parking structures have many things in common with buildings, but also have some unique differences. A very elemental one is that there must be some circulation system that provides access from one floor to the next; cars cannot use the elevators and stairs that provide circulation for pedestrians. The circulation system can be quite complex and difficult for a lay person to understand when looking at drawings. Just because it is complex does not mean that it will be confusing to the parker; on the other hand, some systems are confusing to the unfamiliar user. It is important, therefore, that the owner have a basic understanding of the issues in order to intelligently review and approve designs. An owner is going to have to live with the functional system on a day-to-day basis, and will quickly find out if the functional design is not successful.

Many factors affect the determination of the best functional design for a particular parking facility:

type(s) of users
floor-to-floor height
dimensions of site
parking *geometrics*
peak-hour volumes
flow capacity

This chapter provides guidelines for the functional design. If these guidelines are followed, the most frequent pitfalls can be avoided. It should be noted, however, that the guidelines do not cover all the minute details that must be considered in the preparation of documents

6

for construction; only experience can teach all the little tricks that maximize user acceptance while minimizing cost.

2.2   THE LEVEL OF SERVICE APPROACH

Over the years parking designers have developed quite a number of "rules of thumb" for elements of functional design. These rules prescribe, for example, the maximum number of turns or spaces passed in the path of travel (Rich and Moukalian 1983). Professional judgment is still required in order to apply the rules to a specific situation; some rules are more important with some types of users than with others. For example, it is desirable to route unfamiliar users by as many spaces as possible in a small-to-moderate sized facility. However, if most users park in the facility every day, it is desirable to get them in and out as fast as possible, which usually means minimizing the number of spaces passed; thus, no one set of design standards is suitable for all situations.

Traffic engineers have similar problems in designing streets and intersections; the degree of congestion that is acceptable to users and the community varies substantially. To overcome this problem, traffic engineers developed a system of classifying conditions by *levels of service* (LOS). For traffic at signalized intersections, conditions of virtually free flow and no delays are LOS A, the highest level of service. As congestion increases, the level of service decreases. The lowest LOS, F, is popularly (or unpopularly to those caught in one) called "gridlock." LOS E is the maximum flow of cars that can be accommodated before conditions begin to jam. Traffic conditions are considered minimally acceptable for design purposes with LOS D; at this level delays occur but are acceptable to regular users. However, designers often choose to design for a higher LOS, such as C or B, especially in suburban or rural areas where delays are more objectionable.

Parking facilities are, of course, an extension of the transportation system in a community. Adapting the level of service approach for parking facility design permits qualitative measures of such collective factors as freedom to maneuver, delay, safety, driving comfort, and convenience to be assessed; the design can then be tailored to the needs of the ultimate users. If one wants a more comfortable design for marketing or other reasons, the designer can use the standards for LOS B or A. If economy is paramount and the marketing plan does not demand superior comfort, LOS D or C can be used. Please note that

TABLE 2-1 Level of Service Criteria

Design Consideration	Chief Factor	Acceptable Level of Service			
		D	C	B	A
Turning radii, ramp slopes, etc.	Freedom to maneuver	Employees			Visitors
Travel distance, number of turns, etc.	Travel time	Visitors			Employees
Geometrics	Freedom to maneuver	Employees			Visitors
Flow capacity	v/c Ratio	Employees			Visitors
Entry/exits	Average wait	Visitors			Employees

"marketing" concerns do not apply only to commercial, for-profit projects; hospitals, municipalities, and other parking facility owners who frequently subsidize parking from other funds must always be concerned with meeting the needs of the end user, the parker.

The acceptable LOS varies depending on the type of user. As seen in Table 2-1, user acceptance is related to the frequency of occurrence for the driver. When the chief factor being measured is speed of travel and/or delay, such as for entry/exits and travel distance, regular users (such as employees of the associated generators) will demand a better level of service than infrequent users (such as visitors). On the other hand, when freedom to maneuver and driving comfort are the chief factors measured by the LOS, employees will accept a lower LOS because they are familiar with the design. The turnover rate in a facility also plays a role; when arriving and departing vehicle activity is sustained at high levels throughout most of the day, a better level of service should be provided than if there is one rush period of a half hour in the morning and another short one in the evening. Finally, the more urban and congested the setting of the facility, the more tolerant users are of lower levels of service. LOS D should only be used in the core areas of the largest cities (New York, Los Angeles, Chicago, San Francisco), where land values and parking fees are at a premium level.

The level of service approach is applicable to a number of design considerations in parking facilities, including entry/exits, geometrics, flow capacity, travel distance and spaces passed, turning radii, and floor slopes. The old "rules of thumb" have thus been transformed into levels of service for these areas as will be discussed in this chapter on functional design and the next on access design.

2.3   CIRCULATION SYSTEMS

2.3.1   The Building Blocks

Four very basic building blocks are used in any parking facility design: "level" *parking bays*, "level" drive aisles without parking, "sloped" parking bays, and "sloped" drive aisles without parking. (See the Glossary for definitions of level and sloped floors in parking connotations.) Sloped drive aisles without parking are also called ramps. There are three ramp sub-types: *circular helixes*, *express ramps*, and *speed ramps* (Figure 2-1).

Almost all functional systems are composed of the four basic building blocks assembled in one of two forms of a *helix* for circulation through the facility (Figure 2-2).

The basic forms of the helix are the single-threaded helix which rises one *tier* (usually 10 ft) with every 360 degrees of revolution, and

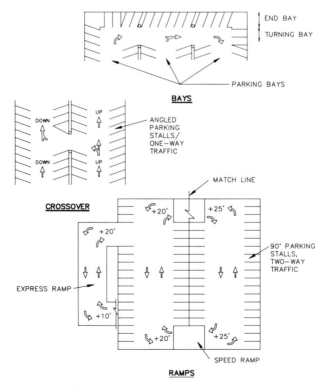

Figure 2-1.   Some basic parking terms.

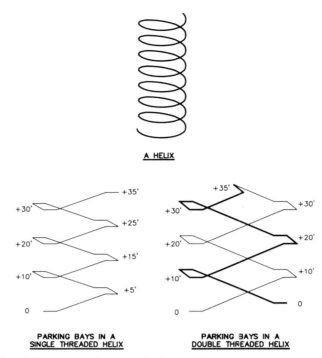

Figure 2-2.   Helixes are used to provide floor-to-floor circulation in parking facilities.

the double-threaded helix that rises two tiers with every 360 degrees of revolution. The reason the latter is called a double-threaded helix is because, by rising 20 ft per revolution, two "threads" may be intertwined on the same footprint (Figure 2-2). A double-threaded helix thus allows a vehicle to circulate from the bottom to the top (or the top to the bottom) of a facility with roughly half the number of turns. Express helixes can be either single or double threaded, just as parking bays can.

There is also a triple-threaded helix but it is seldom used. The triple-threaded helix rises 30 ft with each 360 degrees of revolution and has three threads intertwined.

2.3.2   Dimensions of Site

The floor-to-floor heights and slope of the elements are factors which impact the way in which our building blocks are assembled on a particular site. Floor-to-floor heights generally are dependent upon the

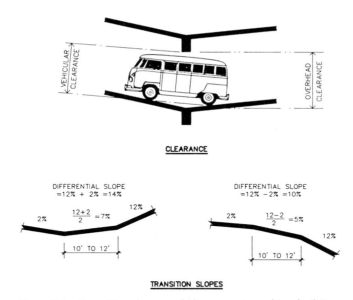

Figure 2-3.   Transition slopes and clearances in parking facilities.

depth of the structural system used (for beams and floor elements) and the desired clearance. It is generally advisable to limit vehicles entering a parking facility to 2 to 4 in. less than the overhead clearance because the vehicular clearance is impacted by sloping floors (Figure 2-3). In general most *long-span prestressed* parking facilities have a floor-to-floor height of 10'0" and overhead clearances of 7'2" to 7'4". Some *short-span* designs get by with floor-to-floor heights of 9'6" or less. With some structural systems, overhead clearances must also include allowances for light fixtures, signs, and piping. Lights and signs are tucked up between beams and/or stems in long-span, prestressed structural systems and thus do not affect clearance considerations.

It is a little known fact that model building codes originally specified a 7'0" **maximum** overhead clearance for parking facilities because this would keep out motor homes, heavier trucks, and other vehicles which weigh more than the design loads required elsewhere in the code. All standard production vans sold in America are 6'10" in height or less and can be accommodated in facilities with 7'0" vehicular clearance. However, vans modified with "pop tops," pickups with flashing lights, or sometimes just special antennas, may not be able to traverse a structure with a 7'0" vehicular clearance. Some communities and even some committees revising national codes did not realize the connection with design loads and have altered the requirement to a 7′0″ **minimum** to allow these vehicles access. Recently handicap-access proponents in some areas have been lobbying for minimum overhead clearances as great as 10′0″. The National Parking Association (NPA) Parking Consultants Council (PCC 1988) recommends that to provide access for vans modified for the personal use of handicapped individuals, an 8′2″ overhead clearance be provided. It is only necessary to provide this clearance for 25% of the handicap stalls. This standard will still exclude paratransit and camper vehicles, which are simply too heavy compared to the design loads employed for parking facilities.

It has only recently been acknowledged that the overhead clearance impacts the overall perception of a facility, by differences in visibility, lighting uniformity, and sense of space. Therefore LOS classifications for clearance have been developed. Once the clearance and floor-to-floor height are established, the slope of floor elements providing circulation from floor-to-floor will influence the design. LOS classifications for slopes (as well as for other design features) are shown in Table 2-2. To determine the minimum length of a site for a structure with parking bays of LOS D slope (6%), one must add the turning bay dimensions shown in Table 2-2 to the minimum runs necessary to achieve the desired rise and slope. Structural dimensions such as exterior walls must also be added to determine the length of a structure (Table 2-3 and Figure 2-4). Turning bays are usually kept level or slightly superelevated. When a site is tight however, crossovers can be sloped at the same rate as parking bays if the crossover is centered at the crossing point of the "X" formed by the two bays. Thus, in a single-threaded configuration which rises 5 ft along each parking bay, the length of the structure must be at least 142 ft for two-way traffic (120 ft for one-way traffic). A single-threaded system with a straight run that goes up a full 10 ft must have a site at least 204 ft long for one-way traffic (226 ft for two-way traffic). Double threads normally are not used with only a 5-ft rise in a straight run. For a 10-ft rise in a double-threaded helix, the structure length is also 204 if the crossover is sloped, but would be 220 ft if it is kept level. The triple-threaded helix previously mentioned requires a site of 287 ft.

If a sloping parking bay is desired but the structure length must be less than the dimensions needed to rise the desired run, speed ramps must be used at the turning bays to make up the difference. As long as the slope of the speed ramp does not exceed the slope used for parking bays, end-bay parking can be used. Therefore, if the speed ramp is 36 ft long and LOS D slopes are being used for parking areas,

TABLE 2-2 Recommended Design Parameters

Design Standard For:	LOS D	LOS C	LOS B	LOS A
Turning radii[1]	24'0"	30'0"	36'0"	42'0"
Driving lane width[2]				
Straight				
single lane[3]	11'0"	11'6"	12'0"	12'6"
multiple lane	10'0"	10'6"	11'0"	11'6"
24-ft radius				
single lane[4]	13'0"	13'6"	14'0"	14'6"
multiple lane[5]	12'6"	13'0"	13'6"	14'0"
Clearance from lane[6]	1'6"	2'0"	2'6"	3'0"
Turning bay[7]				
one-way	16'0"	17'6"	19'0"	20'6"
two-way, concentric[8]	27'0"	29'0"	31'0"	33'0"
two-way, non-concentric	30'0"	32'0"	34'0"	36'0"
PARC equipment lane width[9]	9'0"	9'4"	9'8"	10'0"
Circular helix				
Single-threaded[10]				
outside diameter[11]	60'0"	73'0"	86'0"	99'0"
inside diameter[12]	22'0"	32'0"	42'0"	52'0"
slope[13]	7.8%	6.1%	5.0%	4.2%
Double-threaded				
outside diameter[11]	73'0"	86'0"	99'0"	112'0"
inside diameter[12]	35'0"	45'0"	55'0"	65'0"
slope[13]	11.8%	9.7%	8.3%	7.2%
Parking ramp slope	6.0%	5.0%	4.0%	3.0%
Express ramp slope				
covered	14%	12%	10%	8%
uncovered	12%	10%	8%	6%
Speed ramp slope	16%	14%	12%	10%
Transition length	10'	11'	12'	13'
360° turns to top	7.0	5.5	4.0	2.5
Short circuit in long run	400'	350'	300'	250'
Travel "up" to crossover	750'	600'	450'	300'
Spaces passed				
angled	N.R.	1,100	750	400
perpendicular	N.R.	750	500	250
Flow capacity: ratio v/c[14]	N.R.	.8	.7	.6
Walking distance to elevator	300'	250'	200'	150'
Overhead clearance[15]	7'0"	8'0"	9'0"	10'0"

[1]Measured from center to edge of circle traced by outside front wheel of design vehicle.
[2]Based on AASHTO 1984.
[3]Add 5'0" to provide enough space to pass a vehicle which breaks down.
[4]At less than 10 mph, add 2'0" for higher speeds such as in circular helix; add 10'0" to allow vehicle to pass breakdown.
[5]At less than 10 mph; add 2'0" to each lane for higher speeds.
[6]To wall, column, or other obstruction per AASHTO 1984.
[7]Clear between face of columns; check clearance at parked vehicles with turning template.
[8]To be adequate for two-way traffic; if flow is predominately one-way, can reduce 3'0".
[9]For straight approach to lane; turning radii into lane must also be checked.
[10]For left-hand turns; add 5'0" to diameter for right-hand turns. (Not applicable to double-threaded designs.)
[11]Outside-to-outside outer bumper wall.
[12]Inside-to-inside inner bumper wall using recommended lane widths and clearances above and 6" walls. Reduce 10'0" to provide space to pass breakdown.
[13]At centerline of driving lane with 10'0" floor-to-floor height.
[14]Volume to capacity ratio; based on HRB.
[15]Straight vertical clearance from top of floor surface to underside of overhead obstruction.

TABLE 2-3   Structure Length Parameters

Traffic Flow/Rise	Required Site Length[1]
One-way[2,3]	
5-ft rise	120 ft
10-ft rise	204 ft
20-ft rise	370 ft
30-ft rise	537 ft
Two-way[4]	
5-ft rise	142 ft
10-ft rise	226 ft
20-ft rise	392 ft
30-ft rise	559 ft

[1] Using LOS D parking bay slopes and turning bay dimensions plus typical construction dimensions to achieve out-to-out structure length. Add 36–40 ft for end-bay parking at both ends.
[2] Add 18 ft if double-threaded and level crossover at center is preferred.
[3] Add 32 ft if end-to-end configuration is used.
[4] Deduct 6 ft if traffic flow is predominately one-way in peak hours.

a rise of more than 2 ft in the speed ramp will necessitate the elimination of end-bay parking (36′ * .06′ = 2.16′).

Any discussion of slope must include the *breakover effect*; when there is a difference in slope of 10% or more between two sections of floor slab, a *transition* slope is required to prevent the vehicle from "bottoming out" (Figure 2-3). In general, the transition area should have one half the slope of the differential in slope and should extend at least 10 ft to prevent the vehicle's wheels from straddling the transition area (Table 2-2).

Speed ramps are limited to rises of 5 ft, and even then they must be extended somewhat into the aisle because the overall ramp length, including transitions, must be 41.25 ft to achieve LOS D design (10′ * 8% + 21.25′ * 16% + 10′ * 8% = 5.0′).

In sum, the length of the site impacts the type of system used; speed ramps, however, can be used when the site is short.

The width of the site also impacts the selection of a system as it determines the number and *module* of parking bays which can be placed on the site. The out-to-out width of the structure includes not only the wall-to-wall dimension necessary to achieve the module, but also the structural dimensions of the walls.

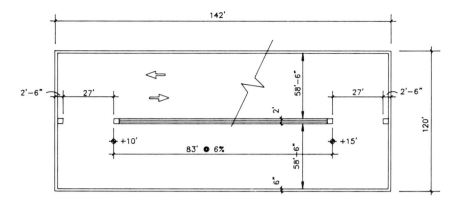

SINGLE THREADED HELIX

TWO WAY TRAFFIC, 90° PARKING
LOS D DIMENSIONS

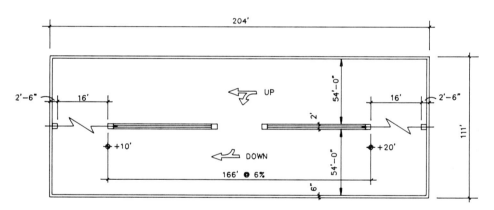

DOUBLE THREADED HELIX

ONE WAY TRAFFIC, 75° PARKING
LOS D DIMENSIONS

Figure 2-4. To determine structure dimensions, structural clearances, turning bay dimensions, module widths, and rise/run must all be included.

2.3.3   Traffic Flow

Most of the common circulation systems used in parking structure design have been given names that in general relate to the pattern of traffic flow and the number of parking bays.

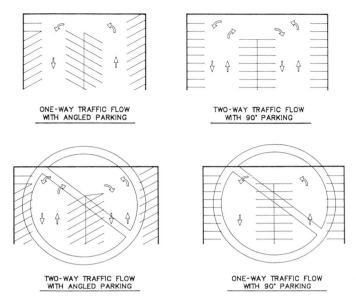

Figure 2-5.   One-way traffic flow systems should have angled parking. Two-way traffic flow systems should have 90-degree parking.

Traffic flow on single- and double-threaded helixes may be one-way or two-way. If one-way flow is desired, it is strongly recommended that angled parking stalls (not perpendicular to the driving aisle) be used (Figure 2-5). This allows the intended traffic flow to be self-enforcing. With perpendicular or 90-degree parking, one user who ignores the signs and proceeds the wrong way will cause problems for other drivers who are following the intended circulation pattern, especially if the turning bays are designed for one-way traffic. Therefore, if 90-degree parking is employed, design the system to accommodate two-way traffic. Conversely, if you want two-way traffic, do not use angled parking. The latter combination causes problems when a driver coming from one direction sees a space intended for the opposite approach and attempts several maneuvers to enter the stall. In addition to delaying traffic flow, this driver is very likely to park improperly, encroaching on an adjacent stall.

A two-bay single-threaded helix must have two-way traffic unless a circular helix or an express ramp provides a way down (Figure 2-6). Obviously, what goes up must come down. Single-threaded helixes may, however, be provided in combinations known as side-by-side helixes or end-to-end helixes to achieve one-way traffic flow (Figure

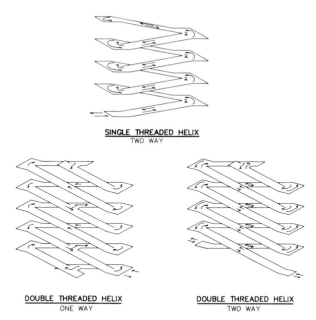

SINGLE THREADED HELIX
TWO WAY

DOUBLE THREADED HELIX
ONE WAY

DOUBLE THREADED HELIX
TWO WAY

Figure 2-6. When only two parking bays are provided, a single-threaded helix must have one way traffic, while a double-threaded helix can have either one-way or two-way flow.

2-7). Note that three-bay side-by-side helixes must have 90-degree parking and two-way traffic flow in the middle bay. Another type of single-threaded helix is the split level (Figure 2-8). In this system, the level parking bays are stepped at 5-ft intervals, with speed ramps making up the difference in elevation between the parking bays.

Single-threaded helixes are very repetitive and easy to understand for the user. In the split-level and the side-by-side with either three or four bays, much of the floor area is level, reducing the frequency of problems with users not finding the parked vehicle upon returning to the facility. Side-by-side single threads also tend to have the best visibility across the structure of any sloping parking bay design, thereby enhancing passive security (which is discussed in Chapter 4). Most architects prefer to work with level facade elements, although a creative architect can either hide or emphasize a sloping parking bay on the exterior.

A negative aspect of any single-threaded design is the number of revolutions required to go from bottom to top or top to bottom. Table 2-2 shows guidelines regarding the LOS for the number of turns. Double-threaded systems (shown in Figure 2-6), which by definition go

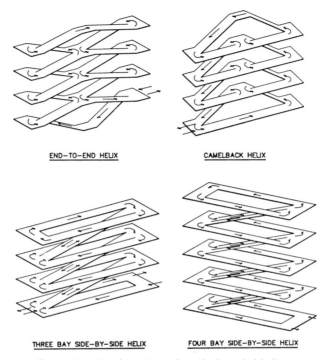

END—TO—END HELIX   CAMELBACK HELIX

THREE BAY SIDE—BY—SIDE HELIX   FOUR BAY SIDE—BY—SIDE HELIX

Figure 2-7.   Combinations of single-threaded helixes.

two tiers per revolution, thus become progressively more desirable as the number of floors increases.

One-way traffic flow can be provided in the two-bay configuration of a double-threaded helix. One thread goes up while the other thread goes down. To get from the inbound "up" thread to the outbound "down" thread (presuming the structure rises above street level), a crossover is provided. This is physically possible where the two sloping bays cross each other in the center of the structure. Depending on the type of user, crossovers are often provided only at every other tier. See the guidelines for travel distance "up" to a crossover in Table 2-2.

Two-way traffic is sometimes used on a double-threaded helix. In that case, the driver will travel back down the same thread, and crossover between threads is not required. Interconnection between the threads occurs only at the top and the bottom of the structure. Two-way traffic on a double-threaded helix results in two "up" threads and two "down" threads. This may or may not be advantageous, as will be discussed later.

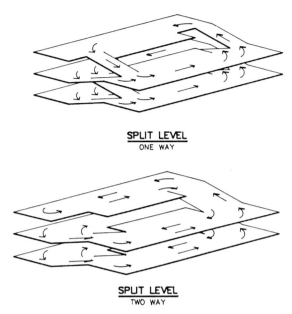

SPLIT LEVEL
ONE WAY

SPLIT LEVEL
TWO WAY

Figure 2-8.   Split levels may have either one-way or two-way traffic flows.

Like single-threaded helixes, double-threaded helixes may also be provided end-to-end or side-by-side when a structure is very large (generally over 1500 spaces). In such a case, the facility is usually treated as two separate structures, with crossover from one helix to the other allowed but not encouraged. With one-way traffic there will be two "up" and two "down" threads; with two-way traffic there would be four ups and four downs. This type of facility tends to be very confusing to the user, and "lost cars" are a frequent problem. Therefore, unless there is a capacity problem which requires more than one circulation route (to be discussed in Section 2.5), multiple double-threaded helixes are usually avoided.

Several common systems provide double-threaded helixes in two bays with additional level bays off the primary circulation patterns. In such a case the level bay may provide the crossover from the up circuit to the down circuit. Single-threaded helixes likewise may have "level" bays off the circulation route. In some cases the same sloping parking bay configuration can have different traffic flow, resulting in either a single- or double-threaded helix circulation pattern (Figure 2-9).

A structure with a very large footprint may also have just one or two bays sloping with the remainder level. In general the larger the

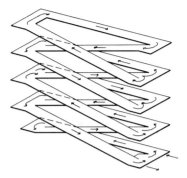

THREE BAY DOUBLE THREADED HELIX

THREE BAY INTERLOCKED HELIX
(SINGLE THREADED)

Figure 2-9. Sometimes traffic may be routed in either a double-threaded or single-thread pattern on the same configuration of parking bays.

number of level parking bays, the better the visibility for patron comfort and security. Express ramps are often employed when there is a goal to have all parking on level floors. Usually sloping parking bays or express ramps are combined with flat parking bays to achieve a single-threaded helix configuration (Figure 2-10). However, double-threaded systems can also be used when height and number of turns are considerations.

In general, concerns for patron comfort, visibility, and ease of orientation all encourage the use of single-threaded schemes for facilities that will serve large numbers of infrequent users. If the users are present every day, however, they will get to know the system and become frustrated by long searches and circuitous exit routes. Therefore, sloping parking bays in double-threaded patterns are generally preferred for office parking and other situations with predominately every-day users.

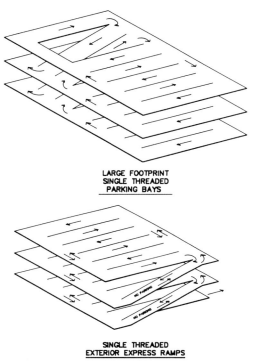

LARGE FOOTPRINT
SINGLE THREADED
PARKING BAYS

SINGLE THREADED
EXTERIOR EXPRESS RAMPS

Figure 2-10. Adding level bays off the primary circulation route results in more decisions for the driver.

2.4 PARKING GEOMETRICS

An important step in functional design is selection of the parking geometrics. The most critical dimensions are the stall width and the parking module. Parking designers consider the module dimension to be more important than the aisle dimension because the aisle is merely the space left when vehicles are parked opposite each other. The aisle is theoretical and varies in the field; the module is a real dimension.

The first major concern is the door opening dimension. For *long-term* parking (3 hr or more), studies (Parking Standards Associates, 1971) have shown that a door opening clearance of 20 in. between parked vehicles is acceptable. For high turnover parking, a door opening clearance of 24 in. provides a better level of convenience for the more frequent movements.

The second major concern is vehicle movement into the stall. As the angle of parking moves farther from 90 degrees (toward 45 degrees), the parking module may be reduced while providing similar maneuverability (i.e., one turning movement) into the stall. The module width is dependent to some extent upon the stall width. A narrower stall requires more module width for the same comfort in turning movement than a wider stall. Stall widths greater than the minimum provide higher levels of comfort for turning movement and door opening. Increasing module width is generally not as economical a method for increasing comfort as increasing stall width (Smith 1985).

2.4.1   The Impact of Downsizing

The trend to smaller cars began after the oil shortages of the early 1970s when the US Government fuel efficiency standards (Corporate Average Fuel Economy, CAFE) were adopted. At the same time smaller, more economical Japanese vehicles became popular. These two factors combined to force American automobile manufacturers to substantially reduce vehicle sizes (PCC 1985).

The first reaction to downsizing was to permit a certain percentage of the stalls in a facility to be designed for "small cars only." Because manufacturers use "compact" and "standard" to designate certain groups of vehicles of similar size, most parking consultants use the terms "small car" stalls and "large car" stalls. There is also some disagreement regarding the boundary between the two. A small car is defined herein as being either 5'9" or less in width or 14'11" or less in length, because a vehicle which exceeds either dimension will impact the use of adjacent small car stalls (Smith 1985). The Parking Consultants Council (PCC) uses a system based on the area of the vehicle in square meters (1985). The resulting classification of vehicles is largely the same.

During the late 1970s when smaller car sales were steadily increasing, many sources in the automobile and parking industries predicted that small car sales would continue to rise and that as many as 80% of all vehicles (on the road) would be small cars by 1990. This prediction, however, has not come true.

Small car/large car sales have been charted in each calendar year since 1970 (Smith, first published in 1985; updated annually through 1987). Small car sales rose slowly from 14–25% from 1973 to 1978, as shown in Figure 2-11. A steady rise in small car sales occurred through 1981, then stabilized with an average of 53% small cars sold each year from 1983 through 1987. Using data from the R. L. Polk Company on vehicle registrations, approximately 42% of the vehicles on the road as of January 1, 1988 are small cars. Presuming that the percentage of

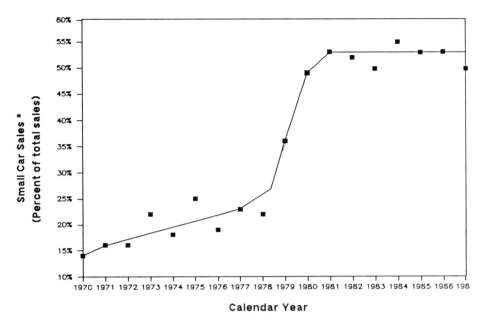

* Vehicles that are a maximum of 5'9" by 14'11".

Source: Compiled by WALKER Parking Consultants from "Automotive News", Market Data Issue.

Figure 2-11.   Small car sales since 1970.

small cars sold each year remains around 53%, the percentage of small cars on the road will continue to creep up at an estimated rate of about 2% per year, but can be expected to stabilize at just over 50%. There are, of course, definite regional variations. In many cases it is desirable to conduct a "vehicle mix survey" at the project site or at a similar land use in the same community to determine the expected small car/large car ratio for the project.

An even more important trend is that auto sales have begun to cluster around the small car/large car boundary in recent years. A recent study (PCC 1989) looked at automobile sales by class since 1980. Classes 5 through 7 are considered small cars; classes 8 through 11 are large cars (Figure 2-12).

The popular Ford Tempo and Taurus models are excellent examples of this clustering phenomenon. The Tempo is among the largest of the cars in the small car class; the Taurus is among the smallest of the large car class. In 1980 39% of all vehicles sold were in classes 7 and 8. This figure has been steadily rising so that in calendar year

ANNUAL AUTO SALES BY CLASS

Year	Small Car Classes			Small Car Subtotal		Large Car Classes			Large Car Subtotal	TOTAL
	5	6	7		8	9	10	11		
1980	33,201 0.4%	2,372,860 25.2%	2,556,835 27.2%	4,962,896 52.7%	1,015,406 10.8%	2,316,629 24.6%	1,086,576 11.5%	31,553 0.3%	4,450,164 47.3%	9,413,060
1981	56,174 0.6%	2,481,352 28.5%	2,105,792 24.2%	4,643,318 53.4%	1,144,123 13.1%	1,839,188 21.1%	986,817 11.3%	87,513 1.0%	4,057,641 46.6%	8,700,959
1982	56,879 0.7%	1,941,307 24.5%	1,942,424 24.5%	3,940,610 49.8%	1,383,807 17.5%	1,448,222 18.3%	1,036,090 13.1%	105,381 1.3%	3,973,500 50.2%	7,914,110
1983	27,756 0.3%	1,942,859 21.2%	2,522,645 27.5%	4,493,260 49.0%	1,617,857 17.7%	1,751,531 19.1%	1,240,607 13.5%	59,626 0.7%	4,669,621 51.0%	9,162,881
1984	58,631 0.6%	1,711,450 16.5%	3,447,243 33.3%	5,217,324 50.4%	1,864,253 18.0%	2,129,082 20.6%	1,057,527 10.2%	90,869 0.9%	5,141,731 49.6%	10,359,055
1985	88,023 0.8%	1,275,036 11.6%	4,728,474 42.9%	6,091,533 55.3%	2,138,597 19.4%	2,060,844 18.7%	613,521 5.6%	117,606 1.1%	4,930,568 44.7%	11,022,101
1986	151,012 1.3%	1,513,428 13.2%	4,744,133 41.5%	6,408,573 58.1%	2,495,661 21.8%	1,852,244 16.2%	546,513 4.8%	133,175 1.2%	5,027,593 44.0%	11,436,166
1987	149,161 1.5%	1,153,907 11.3%	4,020,479 39.3%	5,323,547 52.1%	3,060,943 29.9%	1,284,030 12.6%	411,860 4.0%	139,893 1.4%	4,896,726 47.9%	10,220,273

Figure 2-12. Automobile sales by class since 1980.

Source: Compiled by Walker Parking Consultants from Automotive News Annual Market Data.

24

1987, 69% of all vehicles sold were in classes 7 and 8. Therefore, the percentage of vehicles in classes 7 and 8 on the road will be slowly creeping up, as older cars which are more polarized in size are retired from service (PCC 1989).

In early 1987 and 1988, newspaper articles began to report that "large cars" are coming back, based on such things as:

> One Japanese manufacturer announced that it would build "large" cars for the American market. This vehicle however would be about 6' × 17', or in class 9.

> The "best-selling" models for some General Motors divisions are the largest cars in their line. However, these divisions, like Cadillac, Oldsmobile, and Buick, have almost abandoned the small car market to others to concentrate on bigger cars.

> Both GM and Ford have announced plans to increase the length of certain 1989 models, such as Buick Riveria, Cadillac Deville, and Thunderbird/Cougar. Other full-size and luxury models are expected to gain some length in the next few years.

The pendulum does seem to be swinging back toward larger cars; however, two things are important in this trend. The lengths are only going up to 18'0" to 18'6", not all the way back to 20' as existed in the early 1970s. Furthermore, the market share of these vehicles—whether they are 17', 18', or 20' in length—is still relatively small. The total sales of vehicles in classes 10 and 11 accounted for less than 6% of all vehicles sold in 1987. Even if the market share returns to that which existed prior to the downsizing of standard vehicles, only 12% of auto sales would be in classes 10 and 11. In an interview at the 1989 Chicago Auto Show (Mateja, 1989) GM Chairman Roger Smith stated that "there's no question" that standard and luxury models were downsized too much; however, he went on to say that the "dinosaurs of the past are gone forever."

Another trend is toward light trucks, vans and utility (LTVU) vehicles. Here again "downsizing" has had an impact; 43% of LTVU vehicles sold in calendar year 1987 are in the small car classes. When LTVU vehicles are added to the 1987 passenger auto sales discussed previously, the percentage of small vehicles reduces only by 3%. The mini-van (usually class 8) has largely replaced the old "family station wagon" that accounted for a substantial number of class 11 vehicles in the 70s (PCC 1989). Furthermore, the perception of trucks and vans as "big" is somewhat inaccurate; the largest light trucks and vans fall into class 10. The chief problem with LTVU vehicles in parking facilities is not a difficulty in aligning the LTVU in the parking stall, but rather, reduced visibility and comfort for the drivers of vehicles parked in adjacent spaces.

The impact of the downsizing of the automobile on parking dimensions has been pronounced; the change in vehicle length not only affects stall size but also requires less aisle for turning into the stall. Regulatory agencies such as local zoning boards, however, have been slow to respond. After experimentation with a small-car-only stall proved successful, many localities modified their ordinances to permit a certain percentage of small car spaces. Design standards for the large car stalls remained the same. Since then, however, nearly all intermediate and standard models have been reduced in size. (Exceptions are the Cadillac Fleetwood Brougham and the Lincoln Town Car, which retain their 1970s length of 18'5" and 18'3", respectively, and a few full-size station wagons.)

While many parking consultants promoted smaller, "one-size-fits-all" parking spaces, most zoning ordinances continued to use the old large car/small car standards. This has the impact of encouraging separate large car/small car stalls, which are difficult to enforce and frequently abused. Some small car owners will routinely choose to park in the more generous large car stalls, often leaving large car drivers no choice but to park in the small car stalls. One response has been to place the small car stalls in the "best" locations in the parking facility; intermediate and even large cars then try to use the small car stalls. In effect, the level of comfort for all users is reduced below what local officials desire (PCC 1989).

The current clustering of vehicles around the large car/small car boundary can only increase the confusion and/or abuse of small-car-only stalls. The PCC (1989) has therefore revamped its standards for parking design and strongly recommends that one-size-fits-all stalls be used. Other groups working on their own updated standards from the Institute of Transportation Engineers (ITE) and the Transportation Research Board (TRB), as well as the influential Eno Foundation, are also working on new standards. The time has come to work with local zoning boards to change their standards. The best approach is to bring together local developers, consultants, and officials to prepare a new ordinance. However, in the absence of an updated ordinance, owners should apply for variances. Two concepts have been developed to facilitate this process—the design vehicle (Smith 1985) and the level of service approach to parking dimensions (Smith 1987). Using these tools, many consultants have been successful in gaining ordinance changes and/or zoning variances.

2.4.2   The Design Vehicle

In 1983 Walker Parking Consultants performed an extensive study of parking dimensions. It was found helpful to select a theoretical vehicle size and then determine stall and module dimensions to accommodate the needs of this "design vehicle." To maintain the recommended level of comfort for users the design vehicle is selected as the 85th-percentile vehicle among the vehicles present.

It is highly unlikely that three 100th-percentile (i.e., absolutely largest) vehicles will be parked side-by-side with three 100th-percentile vehicles across the aisle. Use of the 85th percentile is still conservative with respect to the average condition (which would be the 50th percentile) while realistically representing the probable worst condition of parked vehicles. This approach parallels the standard design principle for traffic in which a roadway is designed for the 85th-percentile peak hour.

Parking dimensions were then developed to comfortably accommodate the design vehicle in both its parked position and its turning path for a range of mixes of small and large cars, from 20% small/80% large, 30% small/70% large and so on up to 80% small/20% large. Dimensions for designs with separate small and large car stalls were also provided, under the labels 100% small/0% large and 0% small/ 100% large. The recommendations were thus designed to allow parking dimensions to be tailored to the expected mix of small cars and large cars in any locality and to remain viable as the automobile population changes.

Several considerations were included in the determination of stall size, including the fact that vehicles often do not pull all the way to a wall, wheel stop, or other parking guide. It is also important to understand that the projections of the parked vehicles in the stalls determine the width of the aisle available in any parking bay rather than the rotation of a stall to the angle desired (Smith 1985). Many ordinances which call for rotation of the stall end up requiring a wider bay for a one-way aisle serving angled parking spaces than is required for a two-way aisle serving perpendicular parking spaces!

In 1985, the PCC also adopted a design vehicle approach based largely on the Walker research. Data on design vehicles has recently been updated (PCC 1989), as follows:

Design Vehicles		
	On the Road, 1983 (Smith 1985)	1987 Sales (PCC 1989)
Small Cars	5'7" × 14'8"	5'8" × 14'8"
Large Cars	6'7" × 18'4"	6'6" × 18'0"

The design vehicle for small cars sold since 1983 has remained quite stable, while the design vehicle for large cars has declined by only 4 in. in length. Therefore, the design vehicle length among those on the road may have reduced an inch or two. The expected increase in full-size vehicle length would, at most, shift the design vehicle back to that observed in 1983. Therefore, the design vehicles among those on the road today are, and will continue to be, quite similar to those in 1983. In turn, the dimensions developed in 1983 using the above design vehicles continue to be appropriate for parking design.

2.4.3   Levels of Service for Parking Geometrics

The level of service approach provides assistance in tailoring a design for the users of the specific project. In parking design there are virtually infinite combinations of stall width and module. LOS F designs result in extremely tight conditions where some parkers have to make several attempts to get into the stall. Encroachment into adjacent stalls may leave them unusable. The PCC (1989) dimensions are roughly LOS D, the minimum acceptable design. Since the design vehicle is the 85% vehicle in the mix, most users can turn into the stall in one movement. Larger vehicles may require a second movement but should be able to align in the stall properly. Regular users will become accustomed to and will accept the design. These dimensions would not, however, be acceptable in high turnover facilities or in areas where users are accustomed to very generous parking dimensions.

Table 2-4 shows the gradation of stall and module combinations from LOS D up to LOS A. In most facilities with predominately long-term parkers, LOS C would be appropriate. Projects of this type include office and manufacturing/industrial buildings, student and faculty parking facilities at universities, and other employee parking facilities. For a few projects, LOS D will be acceptable. A luxury office project might go to LOS B, but it would normally be unnecessary to go to LOS A.

For most mixed- or short-term parkers, such as retail, hospital, or municipal facilities, LOS B is the baseline. In a few cases LOS C might be acceptable; a few might merit LOS A.

TABLE 2-4   Recommended Stall and Module Widths

	LOS Angle	D Stall	D Module	C Stall	C Module	B Stall	B Module	A Stall	A Module
Large cars	45	8.25	49.58	8.50	50.58	8.75	51.58	9.00	52.58
only	60	8.25	54.67	8.50	55.67	8.75	56.67	9.00	57.67
	75	8.25	58.17	8.50	59.17	8.75	60.17	9.00	61.17
	90	8.25	61.00	8.50	62.00	8.75	63.00	9.00	64.00
30% compact	45	8.00	46.92	8.25	47.92	8.50	48.92	8.75	49.92
	60	8.00	51.83	8.25	52.83	8.50	53.83	8.75	54.83
	75	8.00	55.42	8.25	56.42	8.50	57.42	8.75	58.42
	90	8.00	59.83	8.25	60.83	8.50	61.83	8.75	62.83
40% compact	45	7.92	46.08	8.17	47.08	8.42	48.08	8.67	49.08
	60	7.92	50.75	8.17	51.75	8.42	52.75	8.67	53.75
	75	7.92	54.33	8.17	55.33	8.42	56.33	8.67	57.33
	90	7.92	58.50	8.17	59.50	8.42	60.50	8.67	61.50
50% compact	45	7.83	45.58	8.08	46.58	8.33	47.58	8.58	48.58
	60	7.83	50.17	8.08	51.17	8.33	52.17	8.58	53.17
	75	7.83	53.67	8.08	54.67	8.33	55.67	8.58	56.67
	90	7.83	57.67	8.08	58.67	8.33	59.67	8.58	60.67
Compact	45	7.25	41.25	7.50	42.25	7.75	43.25	8.00	44.25
only	60	7.25	45.17	7.50	46.17	7.75	47.17	8.00	48.17
	75	7.25	48.17	7.50	49.17	7.75	50.17	8.00	51.17
	90	7.25	50.00	7.50	51.00	7.75	52.00	8.00	53.00

Source: Smith 1987.

There are also regional and locality differences in the acceptable LOS. Totally aside from the vehicle mix, parkers in certain areas are already accustomed to downsized geometrics while others cling to their 9-ft and even 10-ft stalls. Thus a suburban office project in California might be quite acceptable at LOS D, while LOS C would be used in the Midwest. Likewise the density of development and activity can affect the LOS for a project. An urban area would accept LOS D for a low turnover facility, while a suburban location might require LOS C; a rural area might need LOS B.

Designers may reduce aisle (and consequently module) width 3 in. for each additional inch of stall width and maintain the same level of service. Increasing aisle and reducing stall width is not recommended. For example, in a 40% small car/60% large car mix, an 8'6" stall and 51-ft module for 60-degree parking would be LOS C.

For a more complete discussion of the details involved in laying out parking stalls the reader is referred to Smith (1985) and/or the PCC (1988).

2.4.4   Parking Efficiency

In addition to achieving the correct traffic flow, the selection of the angle of parking will depend on several factors; often the most critical is the *efficiency*. In most cases, efficiency has a direct impact on the construction cost per parking stall. Obviously if one design requires less floor area, it will cost less. Because parking bays with angled parking are narrower than those for 90-degree parking, one can sometimes put more bays of angled parking onto a site. For example, a site that is 110 ft wide is too narrow for two *double-loaded* bays of 90-degree parking but can comfortably accommodate two double-loaded bays of angled parking. More spaces can be accommodated on that site with angled parking than with 90-degree parking. It is something of a myth that 90-degree parking is *always* more efficient than angled parking. The case above, in which one bay in a 90-degree scheme would have to be *single-loaded*, is one example. Unfair comparison you say? Lay out two parking lots on a site 121 ft wide by 200 ft long. For simplicity's sake, ignore the need for parking equipment and assume that there are driveways at one end. As seen in Figure 2-13, one lot has 8′5″ stalls at 90 degrees on a module of 60′6″. This is classified as a level of service B for 40% small cars. For 75-degree parking, 8′5″ stalls require a module of 56′4″ for LOS B (40% small cars). Note that for 8′5″ stalls at 75 degrees, the dimension of the stall parallel to the aisle is about 8′9″. The turning bays at both ends are designed to provide similar comfort of turn in each layout. Both lots have the same number of cars, but the 75-degree design uses less floor area. The efficiency of the 75-degree layout is better than the efficiency of the 90-degree layout. In addition to the money saved on paving, the angled design provides a little more space on the site for landscaping, stormwater retention, and other design concerns.

While it is quite true that any 90-degree layout is more efficient in its use of the adjacent aisle, the larger turning bays required for two-way traffic in this particular layout counterbalance the efficiency advantage in the parking bays. Other design "tricks," such as parking along end bays, affect the efficiency of a design. Therefore, designers, always check several layouts before deciding which is the most efficient; owners, make sure your designers did it!

The actual efficiency figure depends on how it is calculated. The best approach is to use the *gross parking area* (GPA) (inside-to-inside of exterior walls) rather than the *gross floor area* (GFA) (out-to-out) because it most closely parallels the floor area dedicated to parking, which in turn determines the cost of building the parking space. The exterior panels are removed from the gross parking area because, in

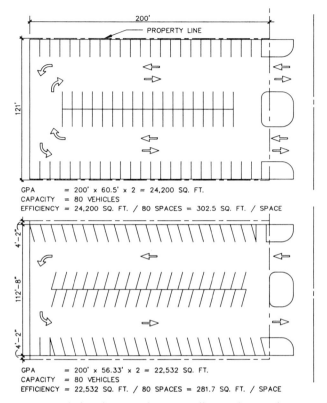

GPA   = 200' x 60.5' x 2 = 24,200 SQ. FT.
CAPACITY   = 80 VEHICLES
EFFICIENCY   = 24,200 SQ. FT. / 80 SPACES = 302.5 SQ. FT. / SPACE

GPA   = 200' x 56.33' x 2 = 22,532 SQ. FT.
CAPACITY   = 80 VEHICLES
EFFICIENCY   = 22,532 SQ. FT. / 80 SPACES = 281.7 SQ. FT. / SPACE

Figure 2-13. Angled parking *can* be more efficient than 90-degree parking.

many cases, the facade thickness is independent of any parking consideration. Although the panel in a precast structural system may be load bearing, it would not be fair to add it back in only for precast systems. Stair and elevator towers are removed from the gross parking area even if they are located *inboard;* the towers are present no matter what the efficiency of the parking area is, and the design of the stair/ elevator towers (be they spartan or palatial) should not impact the assessment of parking efficiency. Some designers make their efficiencies look better by removing interior columns, walls, and any openings between sloping bays to achieve a net floor area, but this is spurious at best. The gross parking area as defined herein is similar to the gross leasable area used for commercial buildings.

Using the stall and module dimensions recommended herein, efficiencies of 300 sq ft per car or less can often be achieved. Just using

these dimensions is no guarantee. A critical factor in the overall efficiency is the length of a row of parking stalls between turning bays or
crossovers. A structure 250 ft long will have better efficiency than one
150 feet long with the same geometrics. Therefore, it is generally more
efficient to lay the parking bays along the long dimensions of the site.

Structures that have fewer turning bays will also be more efficient
if other factors are equal. A single-threaded helix either alone or side-
by-side (two turning bays per level) will usually be more efficient than
a double-threaded helix (two turning bays plus a crossover). The latter
is usually more efficient than an end-to-end or a split level (with one-
way traffic), each of which has four turning bays. As noted previously,
however, it is always best to look at a couple of alternatives before
deciding which is most efficient.

2.5   FLOW CAPACITY CONSIDERATIONS

In the past, the relatively small number of parking structures with more
than 1500 parking spaces tended to be conservatively designed with
multiple circulation paths or high volume circular helixes. There were
no standards for the flow capacity of a circulation route except an old
"rule of thumb" that no more than 750 vehicles an hour should use a
circulation route that has parking alongside. It becomes apparent to
an experienced designer, however, that the type of traffic flow and the
design of the system does have an impact on the flow capacity. Several
other rules of thumb were developed, such as, it is better to use one-
way traffic flow than two-way traffic flow. However, how much better?
Where is the breaking point between acceptable and unacceptable?

Parking consultants generally agree that retail facilities with sloping parking bays should have one-way traffic flow because of the high
volume of vehicles arriving and departing at the same time. It is easier
to get in and out of one-way stalls; therefore, the delay to other users
is less. Also, departing vehicles have to wait for gaps in both departing
and arriving traffic streams, often from two directions. Conflicts between two vehicles, approaching an empty parking stall from different
directions, are eliminated. In most one-way systems, arriving and departing traffic are separated, and there is better visibility to watch for
approaching vehicles when the car is parked at an angle. The more
one direction of flow (either in or out) predominates, the more muddy
the waters become. Many consultants argue that 90-degree parking and
two-way traffic flow are perfectly acceptable for office parking, where
most vehicles arrive in the morning and depart in the evening. In fact,
a double-threaded helix with two-way traffic flow has twice the circulation routes (two up and two down) of a double-threaded helix with

one-way flow. Under the old rule of thumb of 750 vehicles per hour per route, the flow capacity has been doubled. The two-way design eliminates the need for a crossover in a two-bay configuration, improving efficiency. End of argument, right? No. Other consultants argue that one-way traffic flow is still better.

In 1986 Walker initiated a study to find a better way to assess the flow capacity of parking circulation systems. It turns out that the British equivalent of the TRB, the Transport and Road Research Laboratory (TRRL), did extensive research in 1969 and again in 1984 on this issue. Unfortunately, it was never widely published in the US. Also the test parameters employed were appropriate for British driving conditions, including a much higher ratio (virtually 100%) of small cars. Therefore, the TRRL equations are not directly applicable to American conditions.

The TRRL equations were therefore reviewed, and modifications made for American conditions (Smith 1989). It should be noted that the results have not been tested in the same way that the TRRL equations were. However, the results seem reasonable and provide a good basis for an "apples-to-apples" comparison of two circulation systems.

The intent of this book is to acquaint the reader with important information regarding the design of parking facilities, which in this case is that there is an analytical approach to determining the flow capacity of a parking facility. The equations and guidelines for assumptions are provided. The reader who wishes to question the methodology or the research which developed the procedure is referred to a more technical discussion (Smith) published in the Eno Foundation's *Transportation Quarterly* (1989).

2.5.1 Peak-Hour Volumes

The most critical variable in determining the adequacy of any circulation system is the volume, V, of vehicles expected to arrive and/or depart in the peak hour. Some references have recommended that a parking structure should be able to fill or discharge completely in 1 hr, with an even faster fill/discharge rate of 30 min for special events (Weant 1978). This standard is, however, substantially higher than that employed for designing streets. In standard traffic engineering practice, "trips" are generated for peak hours based on the square footage of the generating land use, such as office, retail, etc. (ITE 1984). When these figures are combined with the ratio of parking spaces required per 1000 sq ft, which have also been published (ITE 1987), a ratio of peak-hour volume as a percent of parking spaces can be determined for these land uses. Only in a very few cases do land uses generate peak-hour volumes equal to, or in excess of, the *static capacity* (de-

TABLE 2.5   Typical Peak Hour Volumes

| | Volume in one Hour[1,4] | | | |
| | Peak A.M. Hour | | Peak P.M. Hour | |
Land Use	In	Out	In	Out
Residential	5–10%	30–50%	30–50%	10–30%
Hotel/motel	30–50%	50–80%	30–60%	10–30%
Office	40–70%	5–15%	5–20%	40–70%
General retail/restaurant	20–50%	30–60%	30–60%	30–60%
Convenience retail/ banking	80–150%	80–150%	80–150%	80–150%
Central business district[2]	20–60%	10–60%	10–50%	20–60%
Medical office	40–60%	50–80%	60–80%	60–90%
Hospital				
Visitor spaces	30–40%	40–50%	40–60%	50–75%
Employee spaces	60–75%	5–10%	10–15%	60–75%
Airport				
Short term (0–3 hr)	50–75%	80–100%	90–100%	90–100%
Mid term (3–24 hr)	10–30%	5–10%	10–30%	10–30%
Long term (24+ hr)	5–10%	5–10%	5–10%	5–10%
Special event[3]	80–100%	85–200%		

[1]As a percentage of the static capacity of the parking facility.
[2]It is generally more accurate to determine what portion of the spaces are allocated to retail, office, and other uses.
[3]If 100% of the capacity leaves in 30 min, the equivalent volume in a full hour is 200% of capacity.
[4]As a general rule, the larger the facility and/or the more diverse the tenants of the generating land uses, the lower the peak-hour volume as a percentage of the static capacity.

noted N) required for that use (Table 2-5). One exception is the special event facility, which in most cases should be designed to fully "dump" in one hour or less. Convenience retail and consumer banking facilities also tend to turn over more than once an hour; however, the spaces associated with these uses are not normally a major component of demand for a multi-story parking facility.

It must be noted that there is a substantial variance from the ITE standards based on the specific characteristics of the land use served by the parking facility and/or the community in which it is located. In general, it is wise to estimate the peak-hour volume conservatively high. In addition to the peak hour of the generator as shown in the table, it is sometimes necessary to check volumes during the peak hour of street traffic, because the two may occur at different times. Also, the larger the facility, the more justification there is for doing a detailed study to more accurately determine the peak-hour volume for a particular case.

2.5.2   Flow Capacity

In most real conditions there are peaks and valleys in the flow during the course of the peak hour. To insure that the flow is not unacceptably constrained during shorter periods, the flow capacity (in vehicles per hour) of a system should be somewhat more than the expected volume in the hour. Traffic engineers use *peak-hour factors* to upwardly adjust the volume of traffic. The peak-hour factor (PHF) is usually determined by measuring volumes in 15-min intervals within the hour (V_{15}), selecting the highest 15-min volume, and converting to an equivalent hourly *rate of flow*, v. Therefore, $v = 4 * V_{15}$ and PHF $= V/v$. A peak-hour factor of 0.85 has been found to be reasonable for most traffic situations in the absence of complete data. However, if a special event facility is to "dump" in 30 min, the PHF would be quite different. One half of V would have to depart in 15 min, therefore $v = 4 * .5V = 2V$ and PHF $= V/2V = 0.5$. Also, a parking facility which serves a single employer with uniform starting and ending times, can have a peak-hour factor as low as 0.5.

The ratio of flow rate to the maximum flow capacity, v/c, provides a good measure for assessing the level of service for flow capacity considerations. Table 2-2 provides guidelines for v/c at each LOS. It is not recommended that an LOS D flow capacity be permitted in a parking facility and therefore no value is provided.

2.5.3   Non-Parking Circulation Components

Table 2-6 summarizes the capacities, c, of various non-parking circulation components as calculated using the TRRL equations (1969). The typical flow capacity of one-way aisles 10–12 ft in width, with no adjacent parking, ranges from 1850–1860 vph. TRRL does not provide equations for two-way aisles without parking alongside.

Capacity is decreased when vehicles flow through a turn or bend. In a parking facility, the most common bends are 90 degrees, 180 degrees, and 360 degrees (in a circular helix). In such cases, the TRRL capacity is expressed in relation to the radius of the path of the center of the front axle of the vehicle. The capacity is listed by the associated outside diameter as this is the common American terminology.

The flow capacity of a circular helix with an outside diameter of 60 ft is 1454 vph. It should be noted that a circular helix at the geometry for LOS D is acceptable for one turn and/or slow speeds, but for greater heights, or speeds, LOS C or better should be provided. Note also that the flow capacity LOS still must be determined by calculating v/c. c will be increased with increasing diameter; a helix with

TABLE 2-6   Capacity of Circulation Elements

Design Component	Flow Capacity, c (vehicles per hour)
Straight drive aisle or ramp[1]	
10-ft lane	1850
11-ft lane	1855
12-ft lane	1860
Circular helix[2]	
60-ft outside diameter	1278
73-ft outside diameter	1530
86-ft outside diameter	1661
99-ft outside diameter	1731
Turning bays[3]	
One-way	
50-ft parking bays	1473
55-ft parking bays	1552
Two-way	
60-ft parking bays[4]	
inside path	1435
outside path	1704

[1]Aisle with no parking alongside.
[2]Per Table 2-2.
[3]Turning bay adequate for one-way traffic; no parking on end bay; no merging traffic.
[4]Turning bay adequate for two-way traffic.
Source: TRRL 1969.

a diameter of 86 ft has 30% more capacity than one with a diameter of 60 ft. One note regarding circular helixes—they reduce travel distance and eliminate delays for the individual driver along the path of travel. Circular helixes therefore deliver the peak surges in activity to the exit area without the moderating effects of differing travel lengths. Systems with circular helixes thus require more careful design of the control lanes, stacking, and reservoir areas.

Also shown in Table 2-6 are c values for "U" turns in a parking facility based on TRRL (1969) equations. It should be noted that the TRRL equation does not include any component for width of lane; given the negligible differences in capacity for straight aisles of varying widths, it seems reasonable to neglect width of lane for curve capacity. The flow capacities at turning bays exceed those of adjacent parking bays during outbound flows, as will be presented in Section 2.6.4, and thus would not create bottlenecks. The turning bay capacities are, quite reasonably, much lower than the capacities of straight aisles or roadways without parking and thus would control the flow of vehicles in systems which combine straight non-parking ramps with 90- or 180-degree turns at the ends.

2.5.4   Capacity of Parking Bays

The TRRL procedures for determining the capacity of parking bays are substantially more complicated. As noted previously the TRRL equations and procedures are not reproduced; it is felt that a user must have the complete study in order to properly modify the TRRL equations for American situations. The procedures have been adapted for American conditions, and modified for use by personnel not trained in traffic engineering. To do this a number of assumptions have been made based upon recommendations for such variables as stall width and aisle width. The reader is cautioned that the values presented are thus predicated on certain specific assumptions that may not match those used in a particular project. Further, it is recommended that the LOS of flow capacity be kept at C or better. Therefore, the ratio of v/c should not exceed 0.8.

The 1969 TRRL equations for capacity of aisle/stall systems are based on field observations and measurements of situations where all the vehicles attempted to arrive or depart simultaneously. The term *tidal flow capacity* has been adopted for these conditions because, while waves of activity may occur, the volume is all inbound (or outbound) in the peak hour; c_{in} is the tidal flow capacity inbound and c_{out} is the tidal flow capacity outbound.

The TRRL research found a clear relationship between stall and aisle dimensions and flow capacity. Quite logically, vehicles can arrive and depart in a very comfortable stall/aisle system more quickly than in a system with tighter dimensions. A minor variation in capacity is due to the angle of parking; with approximately equal comfort of turning movement into the stall, the capacity decreases as the angle goes from 70 degrees to 90 degrees. The percentage of vehicles backed into stalls also affects capacity, reducing c_{in} and increasing c_{out}. TRRL also found in later (1984) research that short-span column designs reduce capacity compared to long-span designs. With a typical $30' \times 30'$ short-span grid system, inbound capacity is reduced by at least 30%, and outbound capacity by at least 15%. Standard statistical theory was used to develop equations reflecting these variables.

The TRRL found that the number of stalls in the system (or subsystem when there are multiple circulation routes) does not materially affect the flow capacity, except as it induces peak-hour volumes. That is, the flow capacity at a particular point in two facilities with the same

TABLE 2-7   Tidal-Flow Capacity of Parking Bays

| Geometric LOS | Tidal-Flow Capacity[1] (vehicles per hour) | | | | | | | |
| | D | | C | | B | | A | |
Angle	c_{in}	c_{out}	c_{in}	c_{out}	c_{in}	c_{out}	c_{in}	c_{out}
60° one-way	1349	1011	1500	1037	1500	1065	1500	1095
70° one-way	1189	994	1421	1018	1500	1043	1500	1071
80° one-way	866	941	979	961	1130	983	1343	1006
90° two-way	693	704	764	716	853	728	970	741

[1]One-hundred percent of static capacity arrives in 1 hr, or departs in 1 hr. See Table 2-4, compact only, for stall and aisle dimensions. Percent reversed: 0% for angle less than 90 degrees, 5% for 90 degrees. c_{max} = 1500 vph.
Source: TRRL 1984.

stall and aisle dimensions and similar traffic flow is the same when one facility has 100 spaces and the other 1000 spaces. Of course, the 1000 space facility is more likely to produce a volume of vehicles that exceeds the flow capacity.

Tidal flow capacity has been calculated, both inbound and outbound, for a variety of angles and the LOS of the geometrics provided (Table 2-7). These figures do not include a peak-hour factor. There is obviously an upper limit as to how much the degree of comfort of turn impacts the tidal flow capacity, but the TRRL apparently did not test dimensions generous enough to reach this upper limit. We have, therefore, arbitrarily set an upper limit on tidal flow capacity at 1500 vph.

The 1984 TRRL research was aimed at covering conditions during which spaces are turning over. The *turnover capacity*, c_t, was observed by dispatching vehicles at predetermined, pseudo-random intervals to enter the stall/aisle system and park; the driver was instructed to unpark and depart after one vehicle had passed. The TRRL then developed a procedure to calculate what maximum rate of turnover a facility with a certain number of spaces and a specific circulation system can handle in an hour. We have chosen a second approach to the capacity problem, that being whether the expected peak-hour volumes will approach the facility's flow capacity, thus allowing classification of flow conditions by LOS.

There is an obvious relationship between c_{in} and c_{out} under turnover conditions; the higher the inflow, the lower the capacity for outflow, and vice versa. The TRRL found that it is important to look at the *mean inhibiting period* of each vehicle passing the point at which capacity is critical. In a constant stream of vehicles coming as close together as they possibly can, there is an average *headway* or spacing that can be expressed in either length (such as feet) or time (such as seconds). In laymen's terms, each vehicle "uses" a certain amount of the available time. If c_{in} is 1000 vph, the average time used per vehicle at capacity flow is 60 min/1000 = 0.06 min = 3.6 sec. If the mean inhibiting period, t, is expressed in hours, it is the inverse of capacity, that is, t(hours) = $1/c$(vph); t is the average spacing (in time units) at capacity flow.

In a typical parking facility, some vehicles driving a primary circulation route may be searching for a stall; some vehicles may be unparking and exiting; others may be passing through to or from an area off the circulation route. The TRRL found that the t components will be different for each of these four types of movements.

To check flow capacity in a particular situation, one adds up the mean inhibiting period (expressed in hours) of each vehicle in the stream passing by the critical point in a peak hour. The equation for Σt by the vehicles passing through at capacity flow is as follows:

$$\Sigma t = p * t_p + u * t_u + s * t_s + e * t_e$$

where

p is the number of vehicles parking on the circulation route, and t_p is the mean inhibiting period of those vehicles, which is $1/c_{in}$,

u is the number of vehicles unparking and departing from the route, and t_u is the mean inhibiting period for those vehicles, or $1/c_{out}$,

s is the number of vehicles seeking a stall but parking off the circulation route being studied, with a mean inhibiting period of t_s, and

e is the number of vehicles that pass through from another area on the way to an exit, with a mean inhibiting period of t_e.

Note that the t components must be expressed in hours. Based on field observations of vehicle spacings at peak hours, a value of 1800 vehicles per hour is recommended for the flow rate of vehicles passing through on the way to an exit; therefore, t_e = 1/1800 or 0.00056 hr. Those searching for a stall but parking off the route are assumed to have t_s = 1/1500 or 0.00067 hr.

In a one-way system, there may be all or only one of these activities impacting the capacity at peak hours. In a two-way system, each of the two streams on the route will have its own Σt; however, the parking and unparking movements affect both streams. Another important consideration in applying this methodology is that the "subsystem" through which the vehicles pass may not be just one ramp or one leaf of a series in a typical sloping ramp parking facility. If there is basically continuous flow from parking bay to parking bay, the sub-system for the outbound flow would be the entire series of parking bays along the outbound route followed by the vehicle which must drive the farthest from parking space to exit. When bays act in parallel rather than in series, they would be separate subsystems. At most points at which traffic merges, the traffic from the minor leg is reflected in the e and s components. At points where traffic crosses, Σt for each stream are added to determine if the expected flows can be accommodated.

In the Appendix to this chapter there is an example problem to facilitate the use of this procedure.

2.5.5   Flow Capacity Level of Service

If Σt is less than one, there is theoretically some time available for more vehicles to join the stream. Remember, however, that at capacity, one has presumed conditions of absolutely constant flow rather than the peaks and valleys that occur in real conditions. Although a more complicated formula is used now, the 1965 *Highway Capacity Manual* employed the volume to capacity ratio, v/c, to determine the LOS. The t components represent V in time units; further, because we are looking at how much of the hour is used $V/c = \Sigma t/1 = \Sigma t$. The peak-hour factor and flow rate must also be considered before classifying the LOS. Since $v/c = V/(PHF * c)$, then $\Sigma t/PHF$ is equivalent to v/c. Σt for two of the most common PHF's is as follows:

LOS	v/c	Σt PHF = .85	Σt PHF = .5
C	0.8	0.68	0.40
B	0.7	0.60	0.35
A	0.6	0.51	0.30

2.5.6   Benefits of Flow Capacity Analysis

The type of circulation therefore does affect the flow capacity. In using this methodology, we have found it to be of substantial value in several ways:

Demonstrating that the LOS is quite good. Congestion for the user should be minimal and need not influence functional design.

Identifying "borderline" situations. This border cannot be treated as a "Berlin Wall" between acceptable and unacceptable conditions. Rather, as Σt approaches 0.7, further study should be made in an attempt to improve circulation.

Demonstrating that the system will clearly be overstressed and that additional circulation capacity is required.

Comparing two alternatives to determine which is the best from a circulation capacity standpoint. The analysis can provide an order of magnitude for the differences, answering the frequently asked question: "How much better is alternative X?" Using this analysis, one can say that alternative X will have 25, 50, or 100% more capacity, with a corresponding decrease in congestion, than alternative Y.

2.5.7   Functional System Capacities

When all components are included, virtually every different parking facility will have a different peak-hour flow capacity on its circulation routes. There is, however, some benefit to comparing circulation systems on an "apples-to-apples" basis; the general pattern of flow capacity can be observed, and factors which influence flow capacity can be determined. To do this, the static capacity, $N_{LOS\ C}$, which produces a v/c ratio at LOS C has been calculated for many common functional systems under four scenarios:

Special events where the volume of vehicles arriving before the event and the volume departing after the event are each equal to 85 percent of the static capacity. There is, further, no departing flow during peak arrival periods and vice versa (PHF = 0.5).

Retail usage with both arriving and departing volumes equal to 60 percent of the total number of parking spaces; these volumes occur simultaneously (PHF = 0.85).

Office usage where the volumes arriving in the morning or departing in the evening are each equal to 60 percent of the static capacity. The opposing flows (departing in the morning and arriving in the evening) are equal to five percent of the number of parking spaces (PHF = 0.85).

Airport parking conditions where both the volumes arriving and departing in a peak hour are equal to 30 percent of the total number of parking spaces; these volumes occur simultaneously (PHF = 0.85).

These percentages were selected to represent relatively high traffic volumes for those uses and thus would tend to be conservative. The results are shown in Table 2-8. Isometric views of most of the circu-

TABLE 2-8   Functional System Capacities

	N_{LOSC}[1]							
Use:	Sp. Event		Retail		Office		Airport	
PHF:	0.5		0.85		0.85		0.85	
Arrival/Departure Rate:	85%—0%		60%—60%		60%—5%		30%—30%	
Angle:	70	90	70	90	70	90	70	90
Two-Bay Systems:								
Single-threaded helix	N.A.	335	N.A.	420	N.A.	750	N.A.	840
Double-threaded helix	585	675	980	840	1360	1505	1960	1675
End-to-end helix	585	675	980	840	1360	1505	1960	1675
Split level	480	335	670	420	1090	750	1345	840
Three-Bay Systems:								
Interlocking helix	545	505	850	625	1255	1125	1695	1250
Double-threaded helix	635	830	1160	1135	1490	1880	2325	2275
Side-by-side helix[2]	480	410	710	560	1100	930	1425	1125
Four-Bay Systems:								
Side-by-side helix	585	675	980	840	1360	1505	1960	1675
S.T. helix	585	465	980	680	1360	1065	1960	1360
D.T. helix	655	930	1275	1360	1555	2125	2545	2720
Larger Systems:[3]								
5 Bays, S.T. helix	615	505	1080	775	1430	1160	2160	1555
5 Bays, D.T. helix	675	1010	1335	1550	1600	2320	2710	3110
6 Bays, S.T. helix	635	535	1160	865	1485	1270	2325	1730
6 Bays, D.T. helix	690	1070	1415	1730	1635	2540	2835	3460

N.A. = Not applicable.
[1]Static capacity which produces Σt/PHF = 0.8 or LOS C; geometrics also LOS C.
[2]70-degree values account for two-way, 90-degree ramps.
[3]Level bays except for floor-to-floor circulation as follows: S.T.: 90 degrees, one bay sloping and 70 degrees, two bays sloping. D.T.: two bays sloping.

lation systems have been previously presented. These table values work **only** for designs with flow patterns exactly as shown in the isometrics.

The tables all assume geometrics at LOS C, and adjustment for other conditions may be made by adding approximately 50 spaces in static capacity per step. That is, an LOS D design would have a capacity of the table value minus 50, and LOS A would be the value plus 100. Long-span conditions are also assumed; if a short-span design is used, capacity will be reduced substantially.

Designers can use this table as a guide to selecting a design that will provide LOS C or better in terms of flow capacity. Because of the range of the scenarios, the tables may be interpolated for different arrival and departure rates. For example, if a single-threaded helix is expected to have simultaneous peak hour arrival and departures equaling 40% of capacity, one can interpolate between values in the retail and airport columns ($N_{LOS\ C}$ = 420 + (840 − 420) * (0.4 − 0.3)/(0.6 − 0.3) = 560 spaces). It is important of course to check for the "weakest link in the chain," be it the parking aisles, a circular helix, or an express ramp.

When the size of the facility exceeds the static capacity which would produce a LOS C v/c ratio, alternatives or secondary routes should be developed. For example, if N is 800 compared with a table value which shows that a 700 space facility will be at LOS C, a different circulation system should be provided. The numbers are not so precise that exceeding the recommended static capacity by five to ten spaces will cause a major problem. The designer's judgment must always resolve "close calls."

2.5.8   General Implications

There are certain key considerations which maximize the ability to accommodate traffic. TRRL's research confirms in theory a number of rules for good design that many professional parking consultants have learned in practice. It is generally advantageous to provide separate, one-way, inbound and outbound routes even when there is little or no opposing flow. This is related directly to the fact that fewer spaces are located along the circulation route. For example, under office usage, $N_{LOS\ C}$ for 90-degree parking in a two-bay, single-threaded helix (in which the "in" route is retraced outbound) is 750 spaces compared to the 1360 space $N_{LOS\ C}$ of a 70-degree, one-way double-threaded helix which has separate in and out routes. The outbound vehicle encounters proportionately fewer delays from vehicles unparking along the outbound route in a double-threaded helix and thus the total static capacity can be greater.

Similarly, the flow capacity of a one-way, angled parking layout is greater than with a 90-degree, two-way design on the same system, if there are the same number of circulation routes. A one-way split level for example has 60% more capacity under retail use, and 45% more capacity under office use, than a two-way design. Therefore, although both designs may have adequate capacity, there will be less congestion and delay to users in the one-way system. When 90-degree parking and two-way circulation on the same ramping system increase the number of circulation routes (such as when two-way traffic is employed on a double-threaded helix), there is some increase in the capacity of the system when there is substantially one directional flow, and/or there are many spaces off the circulation routes. However, when there is opposing flow, one one-way system may have more capacity than two two-way systems. For example, when two-way flow with 90-degree parking is used instead of one-way 60-degree parking on a double-threaded helix, capacity under retail use reduces almost 15% ($N_{\text{LOS C}}$ is 840 versus 980). However, with office use, the two-way system increases flow capacity about 10% (1505 versus 1360). As spaces are added in bays off the circulation routes, the capacity benefit of one-way traffic flow is diminished. When a four-bay facility with a double-threaded helix plus two non-circulation bays is provided, the two-way system has 7% more flow capacity under retail use and 37% more capacity under office use, than the one-way system.

The rule of thumb which says "angled for retail, 90 degree for office" works well for some situations but not for others. In terms of flow capacity:

— Angled parking is "better" when there are equal numbers of circulation routes.
— Two two-way outbound paths are "better" than one one-way outbound path for office use.
— One one-way outbound path is "better" than two two-way outbound paths for retail use, if most spaces are on the paths of travel.
— Adding non-circulation bays increases flow capacity, especially under high turnover conditions such as retail.

In summary, the TRRL analysis method permits the capacity and congestion issues to be reviewed on an objective rather than subjective basis. Statements like "one-way traffic is better," or "two-way traffic providing more routes is better," can now be based on analysis rather than intuition. While the analysis procedure is still somewhat theoretical, it can be used to compare two circulation systems on an "apples-to-apples" basis.

2.6 PUTTING IT ALL TOGETHER

There are many subjective considerations in the selection of a circulation system, such as flow capacity, type of user, dimensions of the site, efficiency, spaces passed, and number of turns. For example, it may be desirable to route unfamiliar users past most of the spaces in a small-to-moderate sized facility, contrary to its effect on capacity. Side-by-side (with four bays), end-to-end, and double-threaded helixes with one-way traffic flow all route drivers past half the spaces on the way up and the other half on the way down. These systems are therefore good designs for retail with a capacity of up to almost 1000 spaces. The number of turns to the top, geometrics, and the size of the site would be other considerations in selecting among these circulation patterns. These systems are also excellent for airport and hospital uses. It should be noted, however, that while capacity restraints under "airport" peak-hour flow might permit these facilities to have as many as 2000 spaces, the search time for the available space is probably too long, especially for users who are in a hurry. While adding bays off the circulation route might increase capacity, the ability to find the available space is reduced.

In a larger facility serving unfamiliar users, it is desirable to break the system into smaller "compartments" with express (non-parking) ramping systems to speed users to the available space and return them to the street. The driver should then only have to search a limited area of perhaps 500 stalls for a vacant parking space. The compartments can have either angled or 90-degree parking as appropriate to the design. Two-way systems tend to have more accidents (it is harder to see all approaching vehicles), and have conflicts when two drivers coming from opposite directions want the same stall. One-way traffic patterns are thus "better" for turnover conditions. In general, if parkers can see across several bays to available spaces, two-way end bays should be provided whether or not the parking bays are two-way. Drivers then can get to the available spaces quickly and without frustration. If, however, parkers must be routed through an area a certain way, angled parking should be provided throughout the system. A second internal ramping system may be desirable to allow circulation between floors, and may be critical if the express system must be closed for maintenance. The secondary system may have parking along the path of travel and parking should be angled unless traffic flow on it will be minimal.

Some two-bay designs, such as a single-threaded helix or a split-level are very limited in application due to restrictive flow capacity. ($N_{LOS\,C}$ is about 420 spaces for retail use and 750 spaces for office use with 90-degree parking.) These systems are also considered less desirable by drivers than most other systems when the height exceeds three or four levels, due to the number of turns and the frustration of knowing that one is passing the same spaces on the way down as on the way up. Therefore, a design which has fewer turns, passes fewer spaces, and has less congestion (by virtue of having a greater flow capacity) will be "better" whether or not the capacity is exceeded.

For regular users, a double-threaded helix as the principal traffic route (whether one- or two-way) is almost always considered preferable with four or more tiers because the exit path is substantially shorter.

If the site permits, additional level bays of parking off the circulation routes are acceptable as they further reduce travel distance and congestion. Because most users are present every day, they will know where to find an available space at a particular time of day. It should be noted that a one-way system on a double-threaded helix is still preferable unless capacity problems require more than one "up" and one "down" route. Some additional capacity may be achieved by using two-way traffic on each of the routes in the double-threaded helix. Again, however, two one-way routes down will still be "better" than two two-way routes.

2.7 SUMMARY

The level of service approach to parking design provides a valuable tool for tailoring a design to the specific needs of the expected users. Guidelines using LOS have been provided for many considerations in parking facilities—turning radii, ramp slopes, travel distance, geometrics, and flow capacity. Using these guidelines, a comfortable, well-functioning internal circulation system can be developed for any parking facility. The next chapter "Access Design" deals with an equally important element in design—getting the vehicles you want into the facility, keeping out those you do not want, and collecting the established parking fee from each and every user.

2.8 REFERENCES

AASHTO, 1984. *A Policy on Geometric Design of Highways and Streets.* Washington, DC. American Association of State Highway and Transportation Officials.

Ellson, P. B., 1969. *Parking: Dynamic Capacities of Car Parks.* RRL Report LR221, Crowthorne, Berkshire, UK.: Road Research Laboratory.

Ellson, P. B., 1984. *Parking: Turnover Capacities of Car Parks.* TRRL Report 1126, Crowthorne, Berkshire, UK.: Transport and Road Research Laboratory.

Highway Research Board, 1965. *Highway Capacity Manual*, Special Report 87. Washington: Highway Research Board, National Research Council.

Institute of Transportation Engineers, 1987. *Trip Generation*, fourth edition, ITE No. 1R-016B, Washington: Institute of Transportation Engineers.

Institute of Transportation Engineers, 1984. *Parking Generation*, second edition, ITE No. 1R-034A, Washington: Institute of Transportation Engineers.

Klatt, R. T., Smith, M. S., and Hamouda, M. M., 1987. "Access and Circulation Guidelines for Parking Facilities," *Compendium of Technical Papers, 57th Annual Meeting*, vol. PP–012, Washington: Institute of Transportation Engineers.

Mateja, J., 1989. "A view from driver's seat at GM," *Chicago Tribune*, 19 February.

Parking Consultants Council, National Parking Association, 1985. *Parking Space Standards Report*, Washington: National Parking Association.

Parking Consultants Council, National Parking Association, 1988. *Recommended Standards for Designing Parking Facilities for Physically Handicapped People*, revised edition, Washington: National Parking Association.

Parking Consultants Council, National Parking Association, 1989. *Guidelines for Parking Geometrics,* Washington: National Parking Association.

Rich, R. C. and Moukalian, M., 1983. "Design of Structures," in *The Dimensions of Parking,* second edition, edited by the Parking Consultants Council, 61–76. Washington: The Urban Land Institute and the National Parking Association.

Smith, M. S., 1985. "Parking Standards," *Parking,* 24, no. 4 (July–August 1985).

Smith, M. S., 1987. "The Level of Service Approach to Parking," *Parking,* 26, no. 2 (March–April 1987).

Smith, M. S., 1988. "The Analytical Approach to Entry/Exit Design," *Parking,* 27, no. 2 (May–June): 47–56.

Smith, M. S., 1989. "The Analytical Approach to Parking Facility Capacity," *Transportation Quarterly,* July, Westport, Connecticut: Eno Foundation for Transportation, Inc.

Transportation Research Board, 1985. *Highway Capacity Manual,* Special Report 209. Washington: Transportation Research Board, National Research Council.

Weant, R. A., 1978. *Parking Garage Planning and Operation,* p. 71, Westport, CT: Eno Foundation for Transportation, Inc.

Appendix to Chapter 2

EXAMPLE 2-1

Suppose that a hospital parking facility is to have 650 parking spaces. A parking study was performed to determine the size of the facility; data from that study provide information for determining peak-hour volumes. Two thirds of the spaces will be used by employees who will be provided cards for access. Seventy-five percent of those spaces are for the use of the employees who work the day shift; the remaining are for employees who work the evening shift. Overlap between the shifts occurs at 3:00 P.M. From data on time cards it has been determined that 75% of the day shift works from 7:00 A.M. to 3:00 P.M.; most of the remaining day shift employees work from 8:00 A.M. to 5:00 P.M. The remaining spaces will be used by transients who will pay for parking based on the length of stay. Field studies indicate that the average length of stay for transients is 1.3 hr, and that the spaces allocated to transients will be fully utilized from about 10:00 A.M. until 4:30 P.M., when occupancy will begin to drop off. A negligible number of the transients will arrive before 7:00 A.M. Therefore:

The static capacity, N, consists of two components:

$$N = 650 \text{ spaces}$$

$$N_m = 650 * 0.67 = 436 \text{ spaces for monthly parkers}$$

$$N_t = 650 * 0.33 = 214 \text{ spaces for transients}$$

Peak morning arrivals are from 6:30–7:30 A.M. The volumes expected are:

$$V_m = 436 * 0.75 * 0.75 = 245 \text{ vehicles}$$

$$V_t = 214 * 0.05 = 11 \text{ vehicles}$$

$$V_{in} = 245 + 11 = 256 \text{ vehicles}$$

Peak afternoon departures are from 2:30–3:30 P.M. with these volumes:

$$V_m = 436 * 0.75 * 0.75 = 245 \text{ vehicles}$$

$$V_t = 214/1.3 = 165 \text{ vehicles}$$

$$V_{out} = 245 + 165 = 410 \text{ vehicles}$$

There will be some opposing flows—morning exiting, 6:30–7:30 A.M.:

$$V'_m = \text{say } 50 \text{ vehicles}$$

$$V'_t = \text{say } 10 \text{ vehicles}$$

$$V'_{out} = 50 + 10 = 60 \text{ vehicles}$$

The monthly arrivals in the morning are concentrated over 30 min. Transient arrivals and departures are at random throughout the day. Therefore, in the morning the total volume of vehicles in the peak 15 min is:

$$V_{15} = 245/2 + (11 + 60)/4 = 140 \text{ vehicles}$$

$$v = 140 \times 4 = 560 \text{ vph}$$

$$PHF_{A.M.} = (256 + 60)/560 = 0.54$$

The evening entering activity in the peak hour 2:30–3:30 P.M is:

$$V'_m = 436 * 0.15 = 65 \text{ vehicles}$$

$$V'_t = 214/1.3 = 165 \text{ vehicles}$$

$$V'_{in} = 65 + 165 = 230 \text{ vehicles}$$

The monthly exiting in the evening does not occur in the same 15 min as the arrivals of the evening shift. Therefore:

$$V_{15} = 245/2 + (165 + 165)/4 = 205 \text{ vehicles}$$

$$v = 205 \times 4 = 820 \text{ vph}$$

$$PHF_{P.M.} = (410 + 230)/820 = 0.78$$

Three circulation systems are possible: a single-threaded helix with 90-degree parking and two-way traffic; a double-threaded helix with 70-degree angled parking and one-way traffic flow, and the double-threaded system but with 90-degree parking and two-way traffic (Figure 2-14). In the latter case there are essentially two routes up and two routes down. In the other two cases

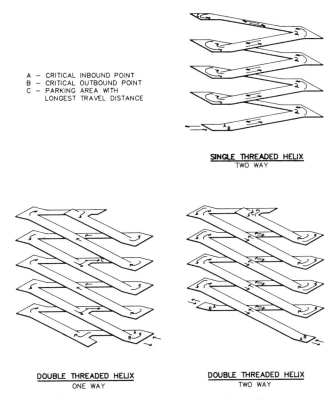

A — CRITICAL INBOUND POINT
B — CRITICAL OUTBOUND POINT
C — PARKING AREA WITH
        LONGEST TRAVEL DISTANCE

SINGLE THREADED HELIX
TWO WAY

DOUBLE THREADED HELIX
ONE WAY

DOUBLE THREADED HELIX
TWO WAY

Figure 2-14.   Isometrics for example problem.

there is one route up and one route down. In the double-threaded helix, however, the routes are separated, while in the single-thread the up route is retraced on the way down. The parking stall geometrics will be LOS B. Which is the "best" system from a circulation standpoint?

The afternoon peak hour has the highest total volume of vehicles in motion, with the following rate of arrivals (P_a) and departure (P_d).

$$P_a = V'_{in}/N = 230/650 = 0.35$$

$$P_d = V_{out}/N = 410/650 = 0.63$$

In the single-threaded helix, the critical inbound point is A, and the critical outbound point is B. There are no vehicles passing through this system to get to (e) or from (s) parking spaces off the system. e and s are therefore 0. While the vehicles in the A stream are not unparking, they will be delayed by the unparking vehicles and vice versa for stream B.

Therefore, the time used in each stream is:

$$c_{in} = 853 \text{ vph (from Table 2-7)}$$

$$c_{out} = 728 \text{ vph}$$

$$p * t_p = V'_{in} * (1/c_{in}) = 230/835 = 0.275$$

$$u * t_u = V_{out} * (1/c_{out}) = 410/728 = 0.563$$

$$\Sigma t = p * t_p + u * t_u = 0.275 + 0.563 = 0.838$$

$$\Sigma t/\text{PHF} = 0.838/0.78 = 1.07$$

$$\text{LOS} = \text{F}$$

As the LOS is well below D, the single-threaded helix should not be used. The double-threaded helix with two-way traffic is simply two separate single threads, each with half the total spaces. c_{in} and c_{out} are the same, but p and u are half of that above. Therefore $\Sigma t = 0.838/2 = 0.419$ and $\Sigma t/\text{PHF} = 0.419/0.78 = 0.54$ and the LOS is A$-$.

In the double-threaded helix with one-way flow, a car unparking at a location just after the last crossover point C has the farthest distance to travel from parking space to exit. This vehicle will pass about 425 spaces. Along this route:

$$c_{in} = 1500 \text{ vph}$$

$$c_{out} = 1043$$

$$p * t_p = P_a * 425 * (1/c_{in}) = 0.35 * 425/1500 = 0.099$$

$$u * t_u = P_d * 425 * (1/c_{out}) = 0.63 * 425/1043 = 0.256$$

s is very small at the critical point and is neglected.

$$e * t_e = (V_{out} - u) * (1/1500) = (410 - 0.63 * 425)/1500 = 0.095$$

$$\Sigma t = 0.099 + 0.256 + 0.095 = 0.45$$

$$\Sigma t/\text{PHF} = 0.45/0.78 = 0.58$$

$$\text{LOS} = \text{A} -$$

Both double-threaded helixes have roughly the same capacity which is adequate for the expected volumes. Note that providing a crossover at every floor will shorten the maximum travel distance and improve the v/c ratio, so that the level of service of the one-way double-threaded helix would then be better than the two-way double-threaded helix.

Chapter 3
Access Design

MARY S. SMITH

3.1 INTRODUCTION

The design of the entry/exit areas is critical to the ultimate acceptance and profitability of a parking facility. These areas provide the patron's first and last impressions of the facility. A positive or negative experience will be a very influential factor in decisions regarding future patronage.

To insure good design a number of things must be considered:

What type of parking access and revenue control (PARC) system, if any, is to be provided?

How many lanes are required to adequately handle peak and daily loads? Are there any special design requirements such as evening event parking?

What configuration of each lane is required to insure that the PARC system works as intended?

How much space is required to accommodate the lanes required, as compared to the space available?

What are the requirements for auxiliary spaces such as parking management offices?

This chapter will provide the reader with a basic understanding of the above considerations in the approximate sequence listed. Before we proceed, however, we must dispense with the first and most obvious decision—is a PARC system needed at all? A PARC system has one or both of two fundamental purposes—keeping unauthorized users out and keeping revenues in. If neither of these is an issue, a PARC system is not required. This situation sometimes occurs at a self-contained development that will provide free parking. The access points will be "free flow," at least to the extent that the surrounding street traffic system and the internal circulation design permit. If you know you are not going to have a PARC system, you can skip this chapter altogether. Otherwise we will proceed with determining what PARC system is right for the project at hand.

3.2   PARC SYSTEMS

The PARC industry is currently on the threshold of the computer age; a wide variety of equipment is available from the very simple to the extremely complex. Equipment similar to that employed in the 1960s is still available; it is largely mechanical/electrical with few, if any, electronic features. Such equipment requires a lot of human input and analysis for control. This primitive equipment can be considered the first generation of PARC systems (Donohue and Lathan 1983).

Second-generation systems use *solid state* electronics for parking applications. Some memory for data storage and automatic control of devices such as gates may be provided.

Third-generation equipment is controlled by microprocessors or other computer systems. The argument for computerizing parking equipment is the same as in any other industry—the information made available to management increases greatly while the time required of personnel at all levels is reduced dramatically. On the other hand, far more data are generated by a state-of-the-art, computerized system than many users want or need. To determine the right PARC system for each situation, it is critical to assess what is expected from the controls. Some or all of the following may be concerns and priorities of a client:

> controlling cash revenues
> detecting theft by employees
> detecting fraud by customers
> totaling and auditing cash revenues from several cashier stations
> maintaining an accurate count of spaces available
> providing activity counts for auditing purposes
> minimizing error
> controlling regular all-day parkers
> minimizing waiting time and/or delays
> providing passive or active security by cashier presence
> minimizing labor cost
> maximizing turnover, utilization, and revenues

As more of the above become priorities, the tighter the controls become. The need for tight controls generally increases as the fees and revenues increase; the incentive for patron cheating and/or fraud, of course, is directly proportional to the fees. Likewise, the more dollars that a cashier handles, the more he/she is tempted to try to divert funds for personal use. Employee theft is generally an even bigger problem than patron fraud and if uncontrolled can severely impact revenues. An additional, less predictable variable in the equation is the "computer hacker" who tries to beat the system merely for the challenge of doing it. The worst cases of theft often involve a few employees and/or a supervisor working with a hacker who modifies the programming to hide individual small thefts occurring regularly over a long period of time. It is critical, therefore, for the owner to determine the priorities of the PARC system. Once the needs and expectations are known the most cost-effective control system can be determined.

3.2.1   Non-Gated Systems

The parking *meter* was invented 50 yr ago to provide a means of keeping employees out of prime spaces intended for visitors and customers. The basic theory is that *short term* users are willing to pay a nominal amount for convenient parking; employees are theoretically not able to keep the meter current by leaving work every two hours to "feed" it. The hourly rate of the meter is intended to cover the cost of collecting the fees and maintaining the meter; in some cases a much lower rate is charged at spaces intended for long-term parkers. When used as intended, meters are quite effective, especially at widely scattered spaces on streets and in small lots.

In practice, however, the meter is frequently misused. Local governments may trim enforcement and maintenance expenditures to lower than acceptable levels, while diverting meter revenues to bolster the general fund. Hourly rates have generally not kept pace with inflation. Area employees find that they can get away to feed the meter a couple times a day and are willing to pay the meter fee and an occasional ticket. Cheating meters is a "folk crime": everyone does it if he or she thinks it will go undetected. If the municipality does not pursue enforcement vigorously, "scofflaws" ignore tickets and may eventually accumulate hundreds of dollars in unpaid ticket fines. Due to vandalism, poor maintenance, and the problem of time-consuming court appearances by enforcement personnel, a substantial number of tickets are thrown out of court. Thefts of the collected funds by both vandals and collection personnel are frequent problems. Under these conditions, the meter is an inefficient means of controlling parking.

Several variations on the meter have been developed (Figure 3-1). The second generation *electronic meters* are in appearance quite sim-

Figure 3-1. Parking meters are now available with electronic controls. Photo courtesy POM, Inc.

ilar to the old standard; however, the electronic workings require far less maintenance and provide audit information to detect and document theft. The cost of these meters is only 25% more than conventional meters which should pay for themselves in relatively short order through increased revenues and decreased maintenance. Enforcement requirements and problems, however, remain largely the same.

Another alternative to the conventional meter is the *meter box*. This usually consists of a box with numbered slots corresponding to each parking stall in the facility. The patron inserts the posted fee in the slot. Collection personnel then check that the appropriate fee is provided for each occupied space and issue tickets to those who have not paid the correct fee. Payment of these tickets is generally on an honor system. These boxes are most effective in perimeter all-day parking lots where a flat rate is charged and revenues can be checked and collected just once a day.

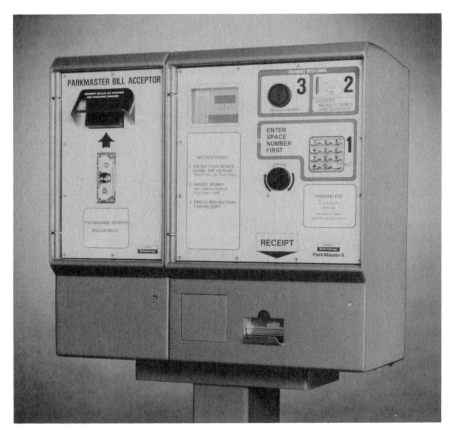

Figure 3-2. Electronic meter boxes can be used for both monthly and daily fee parking in large parking structures. Photo courtesy Paymatec-Schlumberger.

Electronic meter boxes have likewise been developed in recent years (Figure 3-2). Some are intended to replace small numbers of meters on a street or in a lot. In some cases, one unit has been used on each floor of a parking facility. The patron is required to park the vehicle, go to the meter, pay a variable fee for a certain amount of parking, and return to the vehicle to place the voucher on the dashboard. While somewhat less convenient for the patron than individual meters, these units have been successfully employed in Europe for many applications. A wider and more complex range of fee schedules is available, making electronic meter boxes applicable for short-term as well as long-term parking.

A third-generation *computerized meter box* has been developed for larger parking facilities. One pay station is centrally located, usually in the lobby. After parking, the patron approaches the unit, enters the parking space number, selects the desired amount of time, and pays the displayed fee. *Monthly parkers'* cards can be accepted by the machine with nearly all the features discussed later in this chapter for "on-line" card systems. Using special access to the unit, enforcement personnel receive a printout of currently paid spaces and circulate the facility issuing tickets to expired or unpaid users.

The unit also has audit information to provide accountability for the revenue collected. One primary benefit to such a unit is that one enforcement officer can check several large facilities on a frequent basis, eliminating full time cashiering at each facility. The ability to accommodate large volumes of vehicles during peak hours is also greatly improved by the lack of gates and cashiering operations. When combined with card capability, the computerized meter box finds its best application in large facilities that predominantly serve monthly parkers.

There are also some interesting possibilities in locating several pay units throughout a mall, downtown area, or airport with the units interconnected to each parking facility. The patron who has exceeded a time limit has only to go to the nearest station, enter the space and facility numbers, and purchase more time. One drawback is that, at least at the present time, a limited amount of space is provided in the units to store the cash received. If a facility has high turnover and high hourly rates, the cash boxes may need to be emptied several times a day.

One of the most recent ideas is a form of prepaid electronic meter. A parking patron purchases an amount of parking and is given an electronic device about the size of a credit card. This device is attached to the visor, turned on when the parker arrives at the parking stall, and turned off when leaving. The electronic device keeps track of the time parked and displays the paid time remaining for the enforcement officer to view. This is likely to be convenient and valuable for monthlies who can prepurchase several months of parking. However, it seems impractical for daily fee parkers. Unless the individual is a regular user of a facility or group of facilities, he/she must go to a cashier, pay for parking, and return to the vehicle to place the device on the visor.

The chief disadvantage to any form of meter is that all are essentially "honor" systems. Scofflaws may ignore tickets if enforcement is insufficient and private owners may lack any legal remedy for collection. It was chiefly for this reason that the parking gate was invented.

The gate keeps unauthorized users out and authorized users in until they have paid the appropriate fee. Parkers can be charged based on the actual length of stay, rather than an estimate made at the time of arrival. Patrons do not have to worry about whether or not a meter has expired.

In general, a gated system will yield more patron revenue than an ungated system, in most cases more than paying for the higher operational cost of the gated system. Most parking structures having a fee for parking are gated.

3.2.2 Gated Systems

The typical gated PARC system consists of a cashier system for daily fee parkers and a system for regular parkers who prepay on a monthly basis. In the most primitive systems, cash is kept in a "cigar box," with no *audit trail* whatsoever. The monthly parkers are issued permits in paper or decal format. Gates, if provided, are opened manually or by command of the cashier. These systems provide almost no revenue control and are really not "parking access and revenue control systems" at all.

In the first generation of true PARC systems, the gates are automatically opened by electrical signals sent by other devices in the lane, such as *ticket dispensers* and *card readers* (Figure 3-3). In most cases, the gate is closed by a signal from a *loop detector*. For parkers who pay a daily fee, tickets are issued at the entry lanes. At the exit, the cashier enters the fee in a standard commercial cash register and collects the fee due. Card readers are usually provided at both entry and exit for monthly parkers because the speed of the transaction is two to three times faster than if the cashier processes the monthly parker. Card systems therefore reduce the number of lanes and staffing requirements and are very cost effective. *Anti-passback* controls are provided by reversing a magnetic field in the card with each use at entry and exit. Once a card has been used at an entrance gate, it must be used at an exit gate before it will be accepted at an entrance again. If passback problems are not expected, the system may be designed as "card in, free out."

The negative aspect to first-generation gated systems is that control is not very sophisticated and substantial management time is required to achieve most of the usual goals of the PARC systems discussed previously. With a cash register, the only record of transactions is the *journal tape*. Substantial auditing time is required to find errors,

Figure 3-3. Gated systems generally have a ticket dispenser at the entry lanes for daily fee parkers. Photo courtesy Federal APD.

theft, and fraud. To lock out one card user who is no longer authorized to use the facility, all cards must be collected, recoded, and reissued. If the coding is not changed, cards can remain in circulation for months or even years after the card holder loses authorization to use the facility.

The next generation of PARC system eliminates all of these problems at relatively low additional cost. In fact, second generation systems are almost always cost-effective and should be used in most cases. Today's *fee computer* systems automatically print out summaries of activity each day, report transactions by type, reconcile cash that should be in the drawer, and raise red flags for *exception transactions* (Figure 3-4). Tracking exception transactions permits the manager to note, for example, that there are a lot of "lost tickets" when one particular cashier is on duty. The fee computer thus provides a good audit trail for auditing cash revenues, especially when the system software is specially designed for parking. The fee computer also allows the transaction to be completed more quickly since only "in" time and any validations are entered and the fee computer calculates the fee.

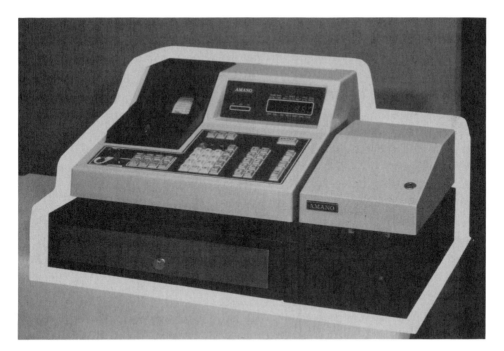

Figure 3-4. The fee computer provides an audit trail for control of cash revenues. Photo courtesy Amano America, Inc.

Errors and some types of cashier and patron fraud are minimized. The fee computer has a *fee indicator* specially designed for visibility by the exiting driver. Presumably, the driver will notify management if the fee quoted by the cashier is different than the fee displayed by the fee computer.

A common misconception is that fee computers eliminate the need to audit activity on a daily basis. For example, the tickets turned in at the end of each shift must be checked to be sure that a cashier is not entering a false time or falsely recording validations. This check is more easily done if each ticket is stamped with the transaction information processed by the fee computer. Daily, monthly, and annual reports for a facility, much less for a group of facilities, must also be prepared by manually totaling the paper reports from each lane.

The second generation of card systems allows owners to invalidate the cards of those who are no longer employed or have not paid, even if the card has not been recovered from the individual. These systems also prevent employees from ''losing'' a card while actually giving it to someone else. The card reader is "smart" in that it is microprocessor controlled. Periodically, management personnel can go to each card reader and, using a device similar to a hand-held calculator, program the reader not to accept specific cards (such as 113, 283, 139) and/or all cards in a certain block (such as 203–249). These card systems also have anti-passback capability.

Another important component of the second generation PARC system is a vehicle counting system. A *differential counter* maintains a count of the number of vehicles in the facility at all times. When occupancy reaches the preset "full" level, the unit automatically illuminates the *full sign* until occupancy drops off again. When card systems are used, it is generally desirable to set the "full" level a few spaces below actual capacity. The ticket dispenser is interconnected so that a ticket will not be issued until the occupancy drops below "full." However, the card readers continue to let card holders in, with the cushion of extra spaces between "full" and the actual capacity, insuring that the monthly parkers will find a space. The vehicle counting system also should have non-resettable counters, two for each lane, that automatically record gate uses and card uses. By comparing the total of card uses and cash uses (the latter as reported by the fee computer) with the total of gate uses for each cashier's shift, the manager can determine if cash transactions are being performed by hand, with the revenue going into the cashier's pocket.

3.2.3   Upgrades to the Basic Card System

In situations with numerous gates and especially when there are several lots and facilities (such as at a campus or hospital), a centralized computer system should be used for cards. All the readers are hardwired to a central microprocessing unit; therefore, the generic name is an on-line card system (Figure 3-5). The central computer may be either a standard microcomputer (such as an IBM PC), or a unit with a *CPU, operating system,* and *memory* designed specifically for this application. There are two variations of the on-line card system—the "dummy" reader and the "smart" reader. In the former case, whenever a card is used at any gate in the system, the number of the card is transmitted to the central unit which checks to see if it is valid. Authorization for every transaction is sent from the central unit. In the latter case, the wiring from reader to central computer is used to "download" changes in authorization, eliminating the need to go from reader to reader with the hand-held device. The authorization decision is made at the reader itself. The smart systems tend to be more expen-

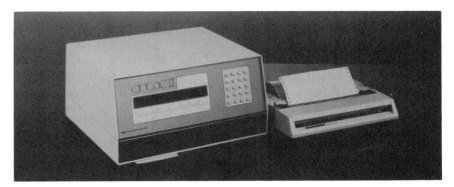

Figure 3-5. Card readers on-line to a central controller provide the best control of monthly parkers. Photo courtesy Cincinnati Time, Inc.

sive (because, of course, each reader is more expensive) and may require more maintenance. However, all readers are not shut down when the master unit goes down, as is the case with the dummy reader.

There is some disagreement in the industry over how tightly the anti-passback controls should be applied. "Misreads" of card numbers are a frequent cause of problems, often creating a chain reaction with other users. For example, if card number 301 is read at an entry lane as card number 311, the computer will not let card 301 out in the evening as it has not been considered "in." Meanwhile, card number 311 cannot get in because the computer thinks it already is "in." In some cases, the misread problem becomes progressively worse due to *degradation* of the cards.

Some manufacturers have reacted to this problem by designing the anti-passback software to be used in either a *soft* or *firm* mode, as selected in the field by the owner/operator of the facility. The soft control accepts a card that is properly paid but is out-of-sync with respect to the anti-passback mode, printing out an error message at the central controller for follow-up by the management. Follow-up and/or disciplinary action is generally taken only for repeated offenders. The argument for soft control is that it eliminates backups, delays, and complaints at lanes when a "good" card is rejected. The proponents of firm-only controls, which reject the out-of-sync card, argue that the correction to misreads is to eliminate the misread problem rather than to accommodate it. Furthermore, soft systems tend to encourage "lazy" users who pull a ticket on the way in when the card is not immediately at hand. The ticket is then discarded and the card is used to exit, avoiding the usually longer line at the cashier. The discarded ticket throws off the daily cash revenue audit. When challenged about the anti-passback violation, the lazy user plays dumb and blames "the computer" or shrugs his/her shoulders, and confesses to being unable to find the card when entering. A similar problem occurs when gates are lifted in off-hours and the card holder does not stop to card out. The next morning the computer thinks the user is still "in."

The firm mode proponents say that the peer pressure exerted at the lane when a card is rejected and other card holders are delayed is enough to insure that the lazy and/or fraudulent user only tries it once. After a certain acclimation period the system works with very few rejections, virtually all of which are intentional misuse. The firm-only system manufacturers tend to provide more reliable but not coincidentally more expensive systems, striving to minimize error and system problems. We believe firm anti-passback is thus better on a philosophical basis. In practice, however, an owner/operator may have conflicting goals. Some owners tend to pamper their users even at the expense of control of the system; these owners prefer the soft anti-passback feature. The need for firm anti-passback increases as the incentive to cheat the system—which is usually the price of parking—increases. If an owner is not yet sure of how tight he/she wants the controls to be, buying a system with both soft and firm anti-passback provides flexibility to determine what is the "best" system through experience.

There are of course times when many vehicles legitimately depart without carding out, such as when the computer is down or a special situation occurs. The redial feature was invented to correct this problem. When redial is activated, all cards are given one authorization in or out before firm anti-passback is restored. This feature can be misused; if done too frequently, the anti-passback feature is essentially voided. Both soft anti-passback and redial features can also be used for theft or kickbacks by the on-site personnel.

A good on-line card system has the capability to allow any individual card access at certain points and at specific times while denying access at others. Take, for example, a hospital with a number of different parking facilities. An employee card would not work at a certain lot until after 2:00 P.M., reserving those spaces for the evening shift. Doctors' cards might work anywhere or at only one location. When the status of a particular card is to be changed (for example, Jane Smith has been promoted to a position that allows her to park in a different area), the information is entered at the central console. Ms. Smith never has to turn in her card to be reassigned to the new lot.

Some of the systems on the market do far more. For example, in a commercial facility, the ledger for monthly payments is part of the system. If someone has not paid on time, the computer can automatically lock out that person at either the entrance or exit point. The cashier has no control but can certainly accept the individual's payment. If the individual does not pay the monthly bill, he can be charged for one day of parking at the transient rate.

Another possible feature is the so-called *credit card* option. One major tenant in a building leases a block of spaces for its employees and argues that many of its employees are outside salespersons who will not be present most of the time. Therefore, 500 cards are to be issued at the fee for 400. The computer keeps track of occupancy by these users and when the 401st patron enters, it begins charging hourly rates for that and all subsequent patrons from this group until the number drops below 400. At the end of the month a statement is issued to the tenant for the over-charges. This option can also be used for those who regularly come to a facility but stay for shorter periods, such as doctors and part-time employees. The CPU keeps track of the usage and bills are issued for parking charges accumulated over the previous month.

There is also a *nesting* feature available. If a user pays to park in a certain area on a monthly basis, he/she must pass a second card reader to the area within a set period of time after entering. If the user fails to park in the assigned area, he/she can be refused exit that day, or refused entrance to the facility the next day.

3.2.4   Upgrades to Fee Computer Systems

Upgrades to the cash control system include *machine read* tickets and/ or *pay-on-foot* systems. Machine read systems substantially reduce keying error and the potential for theft by employees. Audit requirements are reduced as tickets do not need to be checked on a daily basis unless damaged or mutilated. Random auditing and tracking of exception transactions are still recommended. The speed of the exit transaction is also somewhat faster, sometimes allowing a reduction in personnel and/or equipment needs.

There are two types of pay-on-foot systems: *manned* and *automated*. In the manned system, the patron merely pays "on-foot" to a strategically located cashier rather than from the car (Figure 3-6). He/ she receives a *token* or *exit ticket* and inserts it into a device at the exit lane to open the gate. In most systems, the token or exit ticket is programmed to be valid only for a set period after it is issued, such as 15 minutes. This allows the patron time to retrieve the car and return to the exit, but not enough to go to a meeting, shopping, or dining and use the exit token much later.

In an automated system the patron approaches a machine, inserts

Figure 3-6. Pay-on-foot machines are expected to become more popular if manufacturers can solve the dollar bill problem. Photo courtesy Federal APD.

the ticket, is informed of the fee by digital readout, and tenders payment to the machine which makes change. The exit ticket is then issued. These machines are widely used in Europe but have not been particularly successful in the United States for one simple reason—the dollar bill. Most European countries have widely circulating coins with relatively high value, so that substantial parking fees can be reasonably paid, and change made, in coins. The common large American coin is the quarter; it takes a lot of quarters to pay a $1.50 fee, and even more for a fee of $6 or $7. Likewise, Americans are somewhat disconcerted by receiving a couple of Susan B. Anthony dollars or eight quarters as change. The European machines have been unable to effectively deal with American currency to date. At best, they accept most common bills and make change in coins.

At least one American manufacturer is working to incorporate the same dollar bill receiver/changer used in Automated Teller Machines (ATM). However, modifications must be made since most ATMs only deal with one or two types of bills, such as 5s and 20s. Other manufacturers think the solution is designing the machines to accept major credit cards. Presumably very few transactions with cash or change-making are required. The machine must be "on-line" to a bank or other computer which checks the validity of the card. The on-line feature has an additional benefit in that transactions are reported automatically and quickly to the bank and deposits are made directly to the owner/operator's bank account. Unfortunately, this development is somewhat dependent on the banking and credit card industries getting together and developing a nationwide system of on-line credit card approval for all cards. As of now, the machines only accept the cards accepted by the bank which handles the processing.

A primary advantage to pay-on-foot systems is that both the processing of the token at the exit lane and the on-foot cashier transaction are generally faster than a cashier-at-exit transaction, reducing the number of exit lanes and the cashier staffing required. Because of the likelihood that errors, theft, and fraud will be drastically reduced, and substantial savings will be realized in personnel expenses, an automated machine may pay off the additional capital cost in less than a year. We certainly have no doubt that automated pay-on-foot systems will be widely adopted in the near future and, as has happened with ATMs, most people will become familiar and comfortable with their use. Unfortunately, the technology is available but the adaptation of it for parking use is not 100% complete or field tested yet.

One other interesting feature is available when tickets are read directly by either the fee computer or a pay-on-foot station. *Declinating* systems involve the prepurchase of a ticket for a certain sum, say $20 worth of parking. With each use, the fee is credited on the coded strip on the ticket until the prepaid amount is used up. It usually is desirable to have a reader process tickets at both entry and exit, bypassing the cashier. A light on the reader warns when the fee remaining is low. Cashiering needs can thus be reduced. This option is quite similar to the latest technology employed for commuter rail systems like the Washington, DC Metro and can replace the monthly card for regular users. Declinating systems are valuable for visitors who will have ''in'' and ''out'' activity over several days such as at a hospital, hotel, or seminar. Prepaid parking tickets can also be issued months in advance for use on a specific day and time such as at a special event. Advance ticketing substantially reduces the number of lanes and cashiers required for major events.

3.2.5 Other Options/Upgrades

Another desirable option is the *controller/monitor* (CM). This unit, located in a central location such as the parking management office, monitors activity in each lane, and signals the fact that the gate remains up too long or that the ticket dispenser is running out of tickets. The CM is especially beneficial for remote lanes where the cashier cannot see problems. The oldest systems are electro/mechanical, but now electronic and on-line computerized systems (called *computerized count controllers* (CCC)) are available.

Intercoms to remote lanes can be installed directly into, or attached to, certain pieces of equipment. These devices greatly reduce frustration by the patron and eliminate some breakage of gate arms.

CCTV monitoring of entry/exit lanes is also valuable in systems with high revenues. Such a system would not have to be continuously monitored. Rather, alarms triggered by exception transactions would summon management, turn on the appropriate camera, and start up a videotape recorder.

License plate inventory (LPI) systems were developed for airports to thwart the parker who has "lost" the ticket and claims to have come in "just an hour ago." In facilities where parkers don't stay for more than one day, the patron with a lost ticket is charged the maximum daily rate. At an airport, however, cars can be parked for days or even weeks. If the parking rate is $10 per day or more, there obviously is substantial incentive to pull the lost ticket trick. Cashiers also may try to charge the patron the correct fee for the full stay, but enter the transaction into the fee computer as a lost ticket. With an LPI, the license plate is entered into the fee computer which checks to see if the vehicle has been present for the number of days and the elapsed time calculated either from the "in" time and date as entered by the cashier or from a machine read of the ticket.

Some systems use the LPI only for exception transactions, such as a lost ticket; others require that the license plate be entered for every transaction. The entry of the license plate number obviously slows the transaction, especially in states without a front license plate. In that case, the cashier must wait until the vehicle is pulled up to a gate for the plate to be read via CCTV. Compared to a transaction without an LPI, the processing time for an LPI transaction may be only 75% as fast with a front plate and 50% with only a rear plate. There is then a corresponding increase in the number of lanes, both in terms of equipment and staffing on a day-to-day basis. Therefore, the number of lanes/cashiers may increase 33% (1–1/0.75) with LPI/front plates and 100% (1–1/0.50) with LPI/rear plate only. In some cases, the revenue loss due to fraud and/or theft on "normal" transactions (that is, with a ticket given to the cashier) may not merit the additional capital and operating costs of the compulsory (every transaction) LPI system. However, there are a number of ways patrons and cashiers can "pull" tickets, substituting a ticket of much shorter duration. Most airports with an LPI on every transaction are convinced that the compulsory system is cost-effective.

3.2.6   What Technology is Right for You?

As systems for automatic card and ticket reading have been developed, different manufacturers have used different technologies. As with other control considerations, the choice of technology for a specific project depends largely upon how "tight" a system is desired.

First generation, off-line card systems usually have metallic slugs buried in the card in a certain pattern for reading by a magnetic device. All cards are permanently coded with a single code for each facility. The dominant technology for individually coding cards is the magnetic stripe developed by IBM and often used on credit cards. The major problem with "mag stripes" is that the information can be changed, copied, or recoded. While it takes a sophisticated user to purposely recode a card, a card can be easily copied by a simple electronic skimmer. Firm anti-passback, of course, minimizes the benefit of a copied card, since two users with the same ID number can not be in the facility at the same time. More critically, the information on a mag stripe can be scrambled by rubbing against a number of magnetic devices, including, on occasion, a card with magnetic spots. The latter cards are formed by a center core of a magnetic material such as barium ferrite, sandwiched between layers of plastic. The increasing use of barium ferrite cards could cause substantial problems for all mag stripe card systems. Wiegand-effect cards employ a magnetic reaction to read a unique code created by the placement of individual wires in each card. No power is needed at the card reader to read the code, but power is required to check the validity of the number. Also, cards can be copied by a sophisticated individual with the right tools. Infrared cards were developed by Citibank to minimize fraud and theft with their credit and ATM cards. Infrared systems tend to be more reliable and more difficult to copy or tamper with (but more expensive), because cards and reading devices can only be purchased through licensees of Citibank.

All of the preceding systems require the insertion of the card into the reader. Proximity systems read a card from a distance, in some cases without the driver removing the card from a wallet or other carrier. The distance at which the card can be read varies between systems, again based on the technology employed. Some systems require the card to be held up within a few inches of the reader; the technology of this type usually involves scanning by very low power radio frequency signals. Other systems can be read at a distance of 10, 20, or even 30 ft. Most of the longer distance systems can read while a vehicle is in motion, at speeds of 10 mph or more, and were developed for regular users of toll highways and bridges. At least one system uses a decal with a bar code placed on the lower left front windshield that is read by laser. Bar codes are familiar to most people, even if not known by that name, because of their use on virtually every product sold by grocery stores. Another system recently introduced by 3M reads the license plate number of the vehicle, using the reflectivity of the plates (in most states). This system can read plates on vehicles up to 75 ft away, and/or traveling up to 60 mph. The speed of transaction for proximity readers is faster, especially with the longer distance systems where the vehicle does not need to come to a full stop, which can reduce the number of lanes required. Proximity readers also eliminate many weather problems that can occur with insertion readers.

Most machine read systems for tickets currently use one of three technologies—hole punch, mag stripe, or bar codes. Hole punch systems are substantially less expensive, but the coding, once placed, is permanent. Data for features such as declinating tickets cannot be added with each use. Mag stripes, of course, can accommodate additional information; in many pay-on-foot systems, the original ticket issued at entry is recoded with the grace period for use as the exit ticket. In another system, a bar code, usually in the form of a series of random numbers, is preprinted on each ticket. The ticket dispenser "reads" the bar code and tells a central computer when that ticket was issued. When the bar code is read again at the exit, the computer searches its memory for the data on that ticket. The relevant information is all kept in the computer, not on the ticket. Thus, the chief concern is insuring that there is enough memory to store all the necessary data on each currently valid ticket.

The choice between these reading systems again comes down to what kind of problems and what level of cheating are expected. If the only information necessary to calculate a parking fee is the "in" time on a single trip, the less expensive hole punch system should be perfectly acceptable. Mag stripe technology works well for tickets which require some rewriting and in most cases can be combined with other control techniques to provide an effective system. If the situation requires the tightest possible controls and/or it will be on-line to a *facility management system* (FMS) anyway, the bar code approach may be more desirable. Likewise, Wiegand and other coding technologies may be good for most situations, but a supervisor who has access to the computer could skim a lot of revenue by copying and selling counterfeit cards. The infrared system will provide the best defense against most types of theft and fraud involving cards.

The license plate reading systems hold substantial promise for future systems. Paper tickets could be eliminated entirely if the license plate of every vehicle is read at the entry gate and then read at the exit. The computer would check its memory for the entry time and calculate the fee or confirm authorization as a monthly.

3.2.7   Facility Management Systems

The "ultimate" in gated systems now available is a system fully on-line to a central computer. The primary reason for going totally on-line is to allow management of the parking system, be it one facility or a dozen, from a central station. As more complex logic and sophistication is added to any system, it is capable of greater control and management with less human input. The information available increases greatly while personnel time dramatically decreases. One of the generic names for these systems is thus a facility management system (FMS). Using *data management* software, the FMS can generate just about any type of report imaginable. While this information has always been available to parking managers, the amount of time required to track trends in utilization and revenues was cost-prohibitive. Now, computers can do the searching and compiling, allowing management to improve performance (Figure 3-7). Some specific management functions which can be performed are (Smith and Surna 1988):

Figure 3-7.   Facility management systems allow managers to maximize revenues, monitor and project performance, and plan equipment maintenance and replacement. Photo courtesy Federal APD.

Revenue Maximization. This term refers to a step-by-step refinement of management procedures with the goal of maximizing revenues through improvements in facility performance and elimination of fraud and theft. An integral component of revenue control is the feedback provided to local facility supervisors and employees from a series of timely reports. Information in such reports is derived from the transaction data received from "intelligent" *peripherals,* which include the card reader controller, the fee computers, and the gates.

Facility Utilization. Analysis of peripheral transaction data can also reveal patterns of usage which are vital in the preparation of *overbooking* plans for the facility. Such information is also valuable for setting empirically-based rate structures and in formulating expansion plans.

Equipment Maintenance Control. By tracking malfunction data returned from peripherals, objective judgments can be made regarding which devices are failing, the nature of the failure, and environmental factors related to the failure. This information is useful for scheduling preventive maintenance and for deciding which pieces of equipment are due for replacement.

Revenue Forecasting. Information obtained from statistical analysis of peripheral transaction data can be extrapolated for revenue forecasting and management planning. By monitoring specific data, "trends" may be identified early to optimize management response. In addition, hypothetical situations can be analyzed and should provide management insight for business planning.

Alarm Reporting. Communication lines can provide status information, exception transactions, or failure conditions for the various peripherals (gate arm stuck).

One significant feature of these on-line systems is that they are designed to control a number of different lanes and/or different facilities from one central computer. Several parking structures can operate independently even with different commercial operators, but all transaction data is "off-loaded" to a central computer for analysis and management action by the owner. The owner can program the system to poll each parking facility overnight, tabulate and summarize the activity, and print out reports before management arrives in the morning. Substantial clerical time can thus be saved in tabulating activity at several facilities, while management can spot a new trend in minutes. Fee schedules and other programming changes can be *down-loaded* to any individual facility or to all facilities in the system. The on-line system is thus most effective for an owner who has multiple parking structures (such as a parking authority or an operator) or structure(s) with many lanes and high revenues (such as an airport).

3.2.7.1 Configuration of an FMS. The number of facilities connected to an FMS has a bearing on the design of the system. When multiple facilities are connected to a single FMS, each individual facility may have a *local facility computer* (LFC) which collects and tabulates the data for that facility before sending it to the *central facility computer* (CFC). The LFC provides the facility's on-site manager with the data necessary for day-to-day operations while allowing central management to track data and monitor the full system. In other cases one CFC is connected to peripherals at several facilities. The following terminology has been developed to describe the configuration of different types of systems (Smith and Surna 1988):

> *Simple System.* Peripherals connected to a CFC serving only that facility; used when a facility is likely to always stand alone and never require connection to a remote location.
> *Compound System.* Peripherals at several facilities connected to a single CFC; used when a group of several facilities are managed jointly so that all data need be reported to only one location.
> *Complex System.* Peripherals connected to an LFC at each facility; used when a group of several facilities are managed locally/individually on a day-to-day basis, but from which system wide performance data is desired.

Figure 3-8 shows the basic configuration of each of these system types. In each of these cases the FMS may be purchased initially in a comprehensive package or as a planned add-on at a later date. The latter is termed "phased" for each of the above configurations. The "black box" converts the data from the peripheral to the *protocol* required by the FMS. It may only be required in the phased situations since the initial software provider can design the devices to communicate directly with one another.

With a CCC, a facility may be divided into zones and occupancy within each zone can be monitored. These functions may be provided in the LFC if one is provided and/or there is enough memory available on the system. In the compound system without an LFC at every facility, it is desirable to provide a CCC unit at each location.

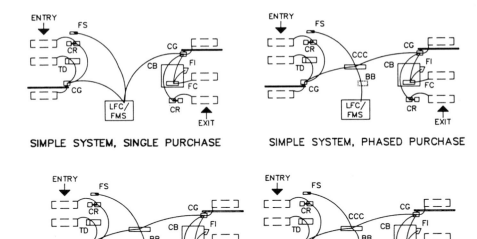

SIMPLE SYSTEM, SINGLE PURCHASE   SIMPLE SYSTEM, PHASED PURCHASE

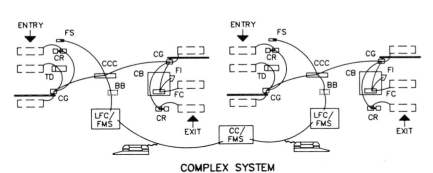

COMPOUND SYSTEM

COMPLEX SYSTEM

LEGEND

CR	CARD READER		TELECOMMUNICATION	BB	BLACK BOX
TD	TICKET DISPENSER	CB	CASHIER BOOTH	CC	CENTRAL COMPUTER
CG	CONTROL GATE	FI	FEE INDICATOR	FMS	FACILITY MANAGEMENT SYSTEM
FS	FULL SIGN	FC	FEE COMPUTER		DETECTOR LOOP
CCC	CONTROLLER/MONITOR	LFC	LOCAL FACILITY COMPUTER		

Figure 3-8. The configuration of an FMS depends on how many facilities will be connected, and whether or not purchases will be phased.

The size of the system and the features desired also have a bearing on the computer hardware required. When each peripheral is "smart," the FMS may only need to handle one task at a time, such as preparing a certain report or downloading new programming to peripherals. In those cases a personal computer with a *buffer* added would be adequate; however, other situations require *multitasking*. For example, an LPI and the CCC may need to be working at the same time. Most personal computers (such as IBM-AT and -XT compatible models) are simply not big enough or fast enough to perform these tasks "simultaneously." (Of course, no multitasking computer actually does tasks simultaneously, but when each task takes a fraction of a second, the human user does not perceive any delay.) A larger multitasking computer may then be required; at least two parking FMS vendors use a *minicomputer.* An alternative is to use a network of several microcomputers, each performing certain non-simultaneous tasks. The network concept should not be discarded out of hand; if one already has an online card system, a microcomputer CCC, and microcomputer fee computers, a PC based FMS might be perfectly adequate for many users. There are also "in between" cases where a PC model with the faster 386 CPU can handle the project requirements.

Computer experts usually advise "green" buyers to pick software first and then buy the hardware that runs that software. Because system needs such as multitasking determine the size of the computer required, the same advice holds true for the FMS purchaser.

3.2.7.2 Coordinating Peripherals with an FMS. Owners often want to have the flexibility to choose from many vendors for their PARC peripherals. When an FMS system is designed to run on a standard computer, one is not, in theory, married to a particular brand of peripherals. If the peripherals are microprocessor controlled and have the capability to communicate to a central computer, one can always design a black box to handle protocol. In practice, however, every FMS system now on the market is tied to specific PARC devices due to the intricacy of communication between devices. Customizing the FMS package for a different set of peripherals or developing a black box can be very time consuming and thus would not be cost competitive with the peripherals that already work with the FMS. Only a few parking vendors make all the necessary peripherals and FMS systems; that software package may not meet the requirements of a specific project even with substantial customizing. Furthermore, if one chooses peripherals now from a vendor in anticipation of a future purchase of that firm's FMS, one might not be happy with the local service or another FMS may become available later that is more attractive. All of these factors make it difficult to specify and competitively bid PARC/FMS systems, especially when the FMS will be added later.

When components come from more than one vendor, someone must pay for the coordination of interfaces, protocol, and if necessary, black boxes. This work is called *integrating* the system, and may be included in one of the following contractual modes:

In a single combined package
In the software design contract (whether custom or off-the-shelf)
In the peripheral specifications

No matter who does the work, most of the cost of integrating a system will usually be passed along to the owner of the project combining those specific components. However, the more integrating work in the parking field the firm does, the lower the cost to the purchaser.

Some in the parking industry have proposed that a recognized group endorse "standards" for computerized parking devices. Standards, when in place, cover such things as the CPU, interfaces, and protocol. If there were standards, integration would be much simpler and less expensive. The Parking Consultants Council of the National Parking Association recently explored the standards issue and declined to adopt any standards. There are too many different needs for one type of hardware, such as an IBM PC-compatible computer, to be designated "standard." The PCC also felt that the market place will determine if standards are appropriate and which approaches will become standard. This is essentially what happened with the IBM PC; no professional group has actually adopted the IBM PC as the standard. Instead, both hardware and software manufacturers recognized that hitching their wagon to the IBM caravan was the best market position.

In the absence of standards, one must either purchase a package that has already been developed or pay for the integration of components that individually have the features desired. The more explicit the specifications, the smaller the pool of qualified bidders becomes. Smith and Surna (1988) provide a number of guidelines for selecting the right components for differing circumstances:

If only one vendor provides an FMS with the features desired, or customizing is required, it is best to negotiate a contract with the vendor rather than attempt to bid the project.

If the FMS is preselected but the peripherals are to be bid, specify that either the peripherals must meet the required protocol for communication with the FMS or the cost of the modifications (including any black boxes) must be included in the peripherals bid.

When an FMS may be added later, plan to use a black box in the future rather than writing a specification around a certain FMS now available. Of course the peripherals must be microprocessor controlled and be able to send data and receive instructions from a computer, otherwise, new peripherals would have to be purchased later.

Smith and Surna (1988) also provide guidelines for writing specifications for bidding systems.

3.2.8   Summary of Considerations in PARC System Selection

In general, a good fee computer system, and/or a programmable or on-line card system is appropriate for most parking facilities. Owners who are particularly interested in monitoring activity closely and will use the voluminous reports that can be generated, may be interested in having an FMS in an individual facility. The FMS, however, will be most cost effective to those who own or operate a number of parking facilities. The purchaser of PARC system equipment must also address how tight a system must be to meet the facility's needs and select manufacturers and technology appropriate to those needs. With an FMS it is desirable to select the software package or packages that best meet the project needs and purchase equipment appropriate for running that FMS. In some cases, project needs may be best served by negotiating with one or two vendors rather than open bidding. If an FMS is to be added later, plan on using a black box rather than writing a specification around a particular FMS now available.

How much to spend on a PARC system is another consideration. In facilities with relatively low revenues, the PARC system is usually designed to keep unauthorized users out more than to retain revenues. At the time of this writing, a basic second-generation system will usually run $30,000 to $50,000 (installed) depending on the number of lanes. Upgrades to the system can easily push the cost to $100,000 or more. When a facility or group of facilities has annual revenues exceeding $1,000,000, an investment in the PARC system equal to 10% of the annual revenues has been found to be appropriate.

3.3   DETERMINING LANE REQUIREMENTS

The traditional method for determining the required number of lanes of PARC equipment involves estimating the number of vehicles expected in a certain peak period and dividing that by the "capacity" of the equipment in the same period. In recent years, however, the average size of parking facilities has dramatically increased. Consultants with extensive experience with larger facilities have found that this methodology can be very inaccurate, resulting in a very overdesigned system in one case and a very underdesigned system in another. While overdesigned systems merely result in wasted capital resources, underdesigned systems can result in user frustration sufficient to cause patrons to choose another facility. Crommelin (1972) first adapted standard traffic engineering theory for queueing at traffic signals for use in PARC lane design. Smith (1988) has further developed this approach, updating and expanding the procedures for conditions common today.

3.3.1   How Many Lanes?

The number of lanes needed is estimated by dividing the *volume* of vehicles expected in the *peak hour*, V, by a *peak-hour factor*, PHF, times the *service rate*, μ, of one lane as follows:

$$n = V/(PHF * \mu)$$

When the peak-hour volumes and peak-hour factors (as discussed in Chapter 2) are estimated conservatively, fewer lanes can be equipped initially to accommodate a more realistic estimated volume. Then, if worse comes to worse, equipment can be added later for additional lanes. It is extremely difficult and expensive, however, to add lanes later when no consideration of additional lanes has been made in the initial design.

The queueing model discussed later will provide a better picture of the peak and average activity in the hour, but the PHF allows for the number of lanes to be quickly estimated. In general, the higher the volume and the greater the number of lanes required, the higher the PHF that can be used. This is because the peaks and valleys in activity tend to be moderated as overall activity increases, and because the bursts in traffic can be distributed over several lanes.

If one lane (either in or out) is provided at a certain location, the peak-hour factor should be no higher than 75%. As additional lanes in the same direction are added, the PHF can be increased, to about 85% for two lanes, 90% for three lanes, and 92% for four or more. When these peak-hour factors are used, any fraction should be increased to the next highest number, i.e., if 1.2 lanes are calculated, two should be provided.

The service rate is determined by using the inverse of the average time per transaction $\bar{s}$ and converting to hours. Thus, the service rate,

TABLE 3-1   Parking Control Service Rates

	Service Rate μ (vph)	
	Easy Approach	Sharp Turn
Entrance and/or Exit		
Clear aisle, no control	800	379
Coded-card reader	400	257
Proximity card reader (2–6 in. distant)	511	300
Coin/token	140	116
Fixed fee to cashier	270	164
Fixed fee—no gate	424	270
Entrance		
Ticket spitter—automatic	522	303
Ticket spitter—push button	480	257
Ticket spitter—machine read	400	232
Exit		
Variable fee to cashier	144	120
Validated ticket	300	212
Machine read ticket	180	144
Machine read with license plate check		
Front plate—manual	110	NA
Rear plate—camera	80	NA
Pay-on-Foot		
Cashier	200	NA
Machine	212	NA
Exit token	400	257

NA = Not applicable.
Source: Klatt, Smith, and Hamouda 1987.

$\mu = 1/\bar{s}$. If a cashier can process two vehicles per minute, $\bar{s} = 30$ seconds and $\mu = 120$ vph.

Sharp turns in the approach to equipment lanes have a significant impact on μ (Klatt, Smith, and Hamouda 1987) (Table 3-1). It should be noted that the service rates of equipment also vary from one manufacturer to another depending on the mechanical and/or electrical technology employed. Certainly if the manufacturer is known at the time of the design, the actual service rates should be obtained and used. However, when several manufacturers are possible bidders, it is neither practical nor advisable to calculate the required number of lanes for each manufacturer. If service rate is that critical, it would be more desirable to specify that the equipment must achieve the desired service rate. In any event, the determination of an accurate design-hour volume will be far more critical and valuable to the analysis than fine tuning the service rates according to each manufacturer.

3.3.2   Queueing Analysis

The proper design of access points requires additional information. For example, vehicles may back into the street if there is not enough space between an entrance gate and the street, even though there are enough lanes to comfortably process the peak flows in an hour. Problems may also occur if the queue of vehicles waiting for one lane blocks vehicles trying to get to another lane; in such a case the second lane is not effective. Designing sufficient lanes only to meet peak hour factors may result in an unacceptable level of service in the field, especially in larger facilities. Therefore, additional traffic engineering theory must be employed to provide a good design. Traffic engineers have developed queueing theory using standard statistical procedures to model flow patterns over the course of an hour.

Queueing equations are available for two types of conditions: "single-channel" and "multi-channel." Single-channel equations are intended for use where one lane is provided at the access point. The multi-channel equations are used when the driver has a choice of two or more similarly equipped lanes at an exit or entry area.

A simple graphical approach avoids the need to use the actual equations. Figure 3-9 shows single-channel situations. Note that the vehicle at the equipment is in the *service position* and is not counted in the queue. Curves are shown for various probabilities; for example,

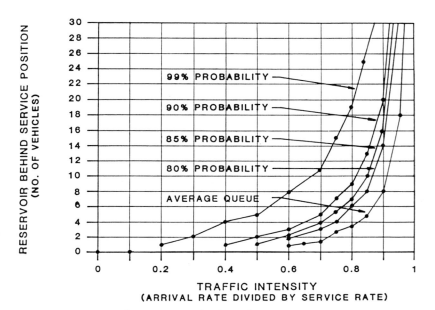

Figure 3-9.   Single-channel queue curves.

if the 90% curve is used, there is a 90% probability that q_{90} will be the maximum queue. Ninety percent probability does not imply that this queue will be exceeded by 10% of the minutes in the peak hour. A better translation is as follows: If one went out and observed many different lanes, each with this flow intensity for a full hour, and recorded the queue several times each minute, 90% of the recordings would be less than q_{90}.

In traffic engineering, it is generally acceptable to design for an 80–90% probability, and as seen in the curves, there is relatively small variation in reservoir size within this zone. Substantially larger reservoir space and/or more lanes would be required to be adequate for essentially all conditions, as depicted by the 99% curve. Therefore, most systems should be designed for q_{90}. Even if the queue does exceed that indicated by the 90% probability curve, it will most likely be quite rare and short-lived.

Also shown for information purposes is the average queue curve, $\bar{q}$, which can be used to determine the average wait, $\bar{w} = \bar{q} * \bar{s}$. Because the average time per transaction, $\bar{s} = 1/\mu$, $\bar{w} = \bar{q}/\mu$ can also be used. If the service rate, μ is in vph, $\bar{w}$ will be of the order of magnitude of 10^{-3} or smaller; conversion to minutes or seconds will make $\bar{w}$ more readily understood. Use of $\bar{q}$ for determining the levels of service will be discussed later.

The traffic intensity, λ, is V/μ. Thus, when a $V = 300$ vehicles is expected to arrive at a card reader, with a service rate $\mu = 400$ vph, $\lambda = 0.75$. When a mixture of users, such as one volume of monthly parkers (V_m), and one volume of transient parkers (V_t) is anticipated at a lane, the service rate must be a weighted average of μ_m and μ_t as follows:

$$\mu_{wa} = \frac{V_m * \mu_m + V_t * \mu_t}{V_m + V_t}$$

The designer thus calculates λ, moves vertically up to the acceptable probability line (e.g., 90%), and then traces horizontally across to determine the queue, q. Because the queueing equation models the approach of vehicles to the lane, a peak-hour factor is not used in the queueing analysis.

For the multiple channel equations, the queues for various combinations of service rates and number of channels have been calculated and plotted with the design queue, q_{90}, in Figure 3-10, and the average queue, $\bar{q}$, in Figure 3-11. Both q and λ in the graphs are per lane. For example if $V = 600$ cars is expected at two lanes, each with a $\mu = 400$ vph, $\lambda = 600/(400 * 2) = 0.75$. q_{90} at each lane is just under four vehicles and $\bar{q}$ is one vehicle per lane.

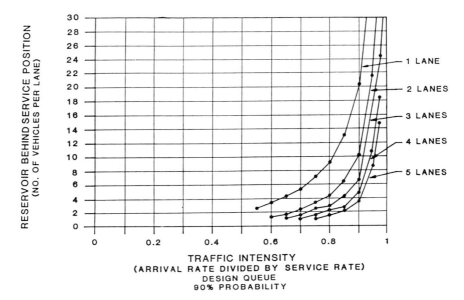

Figure 3-10.   Multi-channel design queue curves.

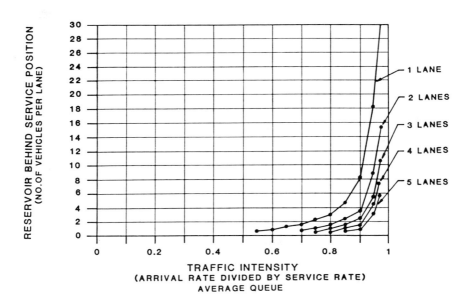

Figure 3-11.   Multi-channel average queue curves.

It can be seen that using the single-channel equation instead of the multi-channel equation becomes more conservative (that is, over-estimating the queue) as the number of lanes (n) increases and also as intensity increases. For example, at $\lambda = 0.6$ and $n = 2$, q_{90} with the multi-channel equation is two vehicles less (3 versus 1) than that with the single-channel equation. At $\lambda = 0.9$ and $n = 2$, the multi-channel design queue is more than ten vehicles less than the single-channel design queue.

The multi-channel equation is, of course, only applicable when multiple lanes are located side-by-side. If two exit lanes are provided with one at each end of a facility, each lane should be designed using the single-channel equations.

3.3.3 Level of Service Classification

A question often raised is what queue is acceptable to patrons. The level of service (LOS) approach is a useful concept in this case. At a traffic signal, the LOS is related to the average delay encountered in the design or peak hour. This concept is easily applied to the delay at a parking gate. The acceptable average delay of each LOS at exit/entry lanes was determined to be slightly longer than that at traffic signals (Smith 1988). This modification is based on the fact that delay at entry/ exit lanes is a single occurrence in each trip to or from the facility. The same delay at each of a series of traffic signals would be more frustrating and less acceptable.

Table 3-2 displays the definition of acceptable delay for each LOS and the associated *average* queue. The design queue (maximum ex-pected with 90% probability) would be substantially longer. For ex-ample, an exit area with two card-controlled gates, each having aver-age queues of 6.66 vehicles (LOS D), would have design queues of 16 vehicles each. Two cashier-controlled exit gates with LOS D would each have average queues of 2.4 vehicles and design queues of ap-proximately six vehicles.

This approach, therefore, takes into account the fact that some transactions are considerably slower than others, such as a variable fee paid to a cashier versus a card reader. While the length of the line does have some psychological impact, the critical factor to user ac-ceptability is the delay time.

As with other traffic and parking conditions, the acceptable LOS at a facility's entrance/exit depends on the type of user. However, it should be noted that one user type does not require the same LOS at every point in the facility. As discussed in Chapter 2, short-term visitor

TABLE 3-2   Entrance-/Exit-Level of Service

	Average Queue Length, $\bar{q}$ (vehicles)			
Level of Service	D	C	B	A
Average Delay, $\bar{w}$ (minutes)	1	0.5	0.25	0.08
Coded card				
easy approach	6.66	3.33	1.66	0.55
sharp turn	4.28	2.14	1.07	0.35
Coin/token				
easy approach	2.33	1.16	0.58	0.19
sharp turn	1.93	0.96	0.48	0.16
Ticket dispenser—auto-spit				
easy approach	8.7	4.35	2.17	0.72
sharp turn	5.0	2.5	1.25	0.41
Ticket dispenser—push button				
easy approach	8.0	4.0	2.0	0.66
sharp turn	3.53	1.76	0.88	0.29
Variable fee with license plate				
easy approach	1.83	0.91	0.45	0.15
sharp turn	1.33	0.66	0.33	0.11
Pay-on-foot				
exit	6.66	3.33	1.66	0.55

parking stalls should have a higher LOS (typically B) than in an employee area (typically C), because of the frequency of turnover and the lack of patron familiarity with the design. At entry/exits, regular monthly parkers demand a higher LOS than less frequent users. Users who encounter the same delay day after day are more likely to complain or to choose a facility with a better LOS. Employees or monthly card holders generally want LOS A, but will accept B. More irregular users, such as shoppers, will accept LOS C or B. There are relatively few cases where LOS D is acceptable, unless it occurs on a very infrequent basis, such as at a parking facility under special event conditions. Even then, the season ticket holder will find and regularly use a facility with a faster exit time, while the single-game patron will accept a longer delay. The urban environment also plays a role. If the facility is located in a congested, downtown area, longer queues will be tolerated than in a rural setting where the driver waiting in line can see an almost empty street just outside the facility.

3.3.4 Entry/Exit Layout

When the type of PARC equipment and the number of lanes are known, entry/exit layouts are fairly routine. Typical entry and exit lanes are shown in Figure 3-12. The configuration of "card only" lanes is the

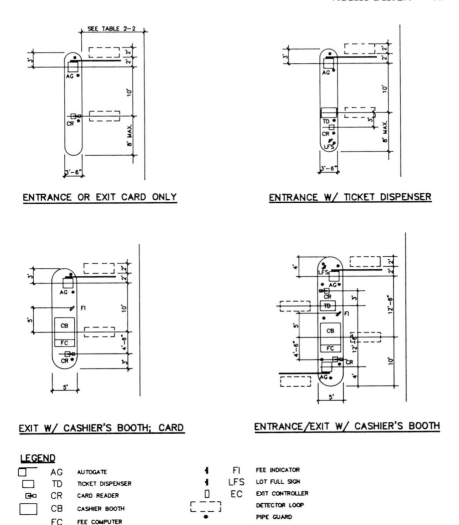

Figure 3-12. Typical entry/exit layout.

same at both entry and exit. At entry lanes ticket dispensers may be placed just before card readers. In this configuration, the patron must press a button on the machine to dispense a ticket. When space is available to separate the dispenser and the card reader, a detector loop in the floor can sense a vehicle approaching the dispenser and "spit" the ticket before the vehicle even comes to a full stop. "Auto-spit" ticket dispensers therefore are faster, as well as less confusing to unfamiliar patrons. The card reader must be placed at least 10 ft in front

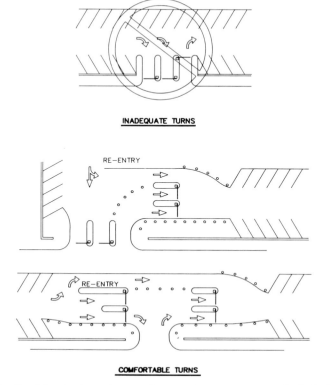

INADEQUATE TURNS

COMFORTABLE TURNS

Figure 3-13.   Providing inadequate space for turns is a common error in designing access points. Pulling the control equipment inside the facility will provide a much more comfortable arrangement.

of the ticket dispenser so that the vehicle stopping at the card reader does not activate the auto-spit function of the ticket dispenser. If the vehicle has used the card reader before reaching the ticket dispenser, the auto-spit function is bypassed. A similar bypass control is used when cash and card customers use the same exit lane, but separation of equipment is not required.

"Reversible" lanes which serve as entry lanes in the morning and exit lanes in the evening can be very space efficient when peak-hour volumes are predominantly one way. Reversible lanes are less confusing if they are monthly only, but with proper signage can be cashier-equipped as well.

A common error in entry/exit layout is providing inadequate space for the driver to turn into the lane and get aligned with the ticket

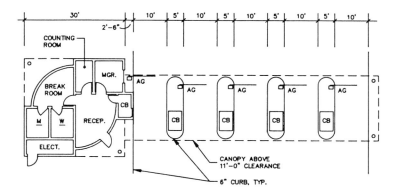

MULTI-LANE DESIGN

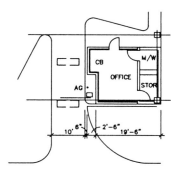

SINGLE LANE DESIGN

Figure 3-14.   Management office layout.

dispenser or card reader. See the discussion of turning radius in Chapter 2. Overhang beyond the wheel track must also be considered. The sharper the turn, the slower the processing of vehicles as documented previously. Figure 3-13 shows a poor design of an entry/exit point along with recommended layouts.

3.3.5   Auxiliary Spaces

In many parking facilities it is desirable to provide an office for management purposes. In smaller facilities, an enlarged prefabricated booth that combines a cashier station and a counter and/or wall space for various panels (such as the facility intercom, the vehicle counting system, etc.), can meet project requirements. However, the design requirements may also include restrooms (especially if handicap accessibility is required by state or local law), security stations, storage, coat/locker facilities, and management workspace. A custom-built office may then be desirable. Figure 3-14 shows custom designs at two ends of the spectrum: a relatively simple combined cashier/management office and a complex with multiple offices, employee lunchroom, lockers, etc., such as might be required at an airport.

3.4   REFERENCES

Crommelin, R. W., 1972. "Entrance-Exit Design and Control for Major Parking Facilities." Paper presented to Los Angeles Parking Association (October), in Los Angeles.

Donohue, L. and Latham, R., 1983. "Revenue Control Procedures and Equipment," In *The Dimensions of Parking*, second edition, edited by the Parking Consultants Council, National Parking Association, 119–125, Washington: The Urban Land Institute and National Parking Association.

Institute of Transportation Engineers, 1987. *Trip Generation*, fourth edition, ITE No. 1R-016B, Washington: Institute of Transportation Engineers.

Institute of Transportation Engineers, 1984. *Parking Generation*, second edition, ITE No. 1R-034A, Washington: Institute of Transportation Engineers.

Klatt, R. T., Smith, M. S., and Hamouda, M. M., 1987. "Access and Circulation Guidelines for Parking Facilities," *Compendium of Technical Papers, 57th Annual Meeting*, vol. PP-012, Washington, DC, Institute of Transportation Engineers.

Smith, M. S., 1988. "The Analytical Approach to Entry/Exit Design," *Parking* 27, no. 2 (May–June): 47–56.

Smith, M. S. and Surna, W. L., 1988. "The Hi Tech Approach to Parking Access and Revenue Control," *Parking* 25, no. 5 (September–October): 35–49.

Weant, R. A., 1978. *Parking Garage Planning and Operation*, p. 71, Westport, CT: Eno Foundation for Transportation, Inc.

Appendix to Chapter 3

EXAMPLE 3-1

Consider the hospital parking structure example from Chapter 2. Employees will have cards and transients will pay based on length of stay. There will be one entrance/exit area which is located very close to the street, requiring a sharp turn of vehicles from the street to the entrance lanes. All entry lanes will be fitted for both monthlies and transients. There will not be room to separate the ticket spitter from the card reader, necessitating a push-button mechanism on the spitter. The approach to the exit lanes will be straight and easy. How many lanes of equipment are required for this facility? The capacity (N) is:

$$N = 650 \text{ spaces}$$

$$N_m = 436 \text{ spaces for monthly parkers}$$

$$N_t = 214 \text{ spaces for transients}$$

Peak morning arrivals are from 6:30–7:30 A.M. The volumes (V) previously established are:

$$V_m = 245 \text{ vehicles}$$

$$V_t = 11 \text{ vehicles}$$

$$V_{in} = 256 \text{ vehicles}$$

From Table 3-1, the service rates for the proposed equipment are:

$$\mu_m = 257 \text{ vph}$$

$$\mu_t = 300 \text{ vph}$$

$$\mu_{wa} = (245 * 257 + 11 * 300)/256 = 259 \text{ vph}$$

The approximation of the number of entry lanes, assuming a peak-hour factor of 0.85, is:

$$n = V_{in}/(PHF * \mu) = 256/(0.85 * 259) = 1.16 \text{ lanes}$$

Therefore, two lanes are needed for peak arrivals. The traffic intensity is then:

$$\lambda = V_{in}/(n * \mu_{wa}) = 256/(2 * 259) = 0.49$$

Referring to Figure 3-10 for the 90%-probability queue, and to Figure 3-11 for the average queue, under multi-channel conditions:

$$q_{90} = 1 \text{ vehicle per lane}$$

$$\bar{q} = 0.25 \text{ vehicle per lane}$$

The average wait $\bar{w}$, can then be calculated:

$$\bar{w} = \bar{q}/\mu_{wa} = 0.25/259 = 0.00097 \text{ hr} = 0.06 \text{ min}$$

From Table 3-2, $\bar{w} = 0.06$ minutes indicate that the level of service

$$LOS = A+$$

Peak afternoon departures are from 2:30–3:30 P.M.

$$V_m = 245 \text{ vehicles}$$

$$V_t = 165 \text{ vehicles}$$

$$V_{out} = 410 \text{ vehicles}$$

$$\mu_m = 400 \text{ vph}$$

$$\mu_t = 144 \text{ vph}$$

$$\mu_{wa} = (245 * 400 + 165 * 144)/410 = 297 \text{ vph}$$

Approximation of number of exit lanes, if dedicated to one group:

$$n_m = V_m/(PHF * \mu_m) = 245/(0.85 * 400) = 0.7 \text{ lanes}$$

$$n_t = V_t/(PHF * \mu_t) = 165/(0.85 * 144) = 1.4 \text{ lanes}$$

Approximation of number of exit lanes, if shared:

$$n = V_{out}/(PHF * \mu_{wa}) = 410/(0.85 * 297) = 1.6 \text{ lanes}$$

Therefore, use two lanes fitted for both:

$$\lambda = V_{out}/(n * \mu_{wa}) = 410/(2 * 297) = 0.69$$

The 90% probability and average queues, for multi-channel conditions are:

$$q_{90} = 2 \text{ vehicles per lane}$$

$$\bar{q} = 0.75 \text{ vehicles per lane}$$

$$\bar{w} = \bar{q}/\mu_{wa} = 0.75/297 = 0.0025 \text{ hr} = 0.15 \text{ min}$$

$$LOS = A-$$

Check to see if one lane can be reversed (inbound A.M., outbound P.M.). Morning exiting, 6:30–7:30 A.M.:

$$V'_m = 50 \text{ vehicles}$$

$$V'_t = 10 \text{ vehicles}$$

$$V'_{out} = 50 + 10 = 60 \text{ vehicles}$$

$$\mu_m = 400 \text{ vph}$$

$$\mu_t = 144 \text{ vph}$$

$$\mu_{wa} = (50 * 400 + 10 * 144)/60 = 357 \text{ vph}$$

$$\lambda = V'_{out}/(n * \mu_{wa}) = 60/357 = 0.17$$

Queues are negligible. One lane okay in A.M. Evening entering, 2:30–3:30 P.M. (does not have to be the same hour):

$$V'_m = 65 \text{ vehicles}$$

$$V'_t = 165 \text{ vehicles}$$

$$V'_{in} = 230 \text{ vehicles}$$

$$\mu_m = 257 \text{ vph}$$

$$\mu_t = 300 \text{ vph}$$

$$\mu_{wa} = (65 * 257 + 165 * 300)/230 = 288 \text{ vph}$$

$$\lambda = V'_{out}/(n * \mu_{wa}) = 230/288 = 0.80$$

90% probability, single channel:

$$q_{90} = 9 \text{ vehicles}$$

$$\overline{q} = 3.5 \text{ vehicles}$$

$$\overline{w} = \overline{q}/\mu_{wa} = 3.5/288 = 0.012 \text{ hr} = 0.73 \text{ min}$$

$$\text{LOS} = \text{C}-$$

LOS C is not acceptable for this use; provide two lanes in and two lanes out in afternoon peak hour.

Chapter 4
Security and Safety

MARY S. SMITH

4.1 INTRODUCTION

Security design in parking facilities deals with minimizing the risk of incidents which threaten the safety of parking patrons and parking attendants. Additional concerns include the protection of cars, personal property, cash receipts, and the facility itself. Psychology plays a big role in security design; a good design uses perception to influence people. Obviously, the more secure a facility appears, the more likely parkers will be to accept and use the facility. A potential wrongdoer will normally analyze the situation before committing a crime to determine the odds of being seen, and if seen, of being recognized and apprehended. He/she is less likely to commit the crime in a facility where security features are apparent.

Furthermore, courts often hold owners and operators liable for injuries suffered in criminal attacks when the defendant did not take adequate steps to reduce foreseeable risks (Lubben 1987). Of course no security system guarantees safety or protection of property. There are also no hard and fast rules about what systems should be provided in specific situations. Negligence rather than omission is a key to liability. Courts will generally not find an owner or operator liable when security risks were thoughtfully assessed and appropriate measures were taken, even if the expert witnesses disagree about what the "best" system would include.

An additional but parallel issue is that of design hazards that directly threaten the safety of the patrons of a facility without the involvement of a third party (the criminal). In some cases an element can be a hazard to all users, while in others it may be a hazard to those of underdeveloped reasoning faculties, such as young children. Building and life safety codes are the general minimum standard, and if the design of an element conforms to the code in force at the time of construction the owner may be relieved of liability. However, the codes do not cover all the potential hazards in a facility nor do they cover reasonably foreseeable special circumstances. For example, if a parking facility is to serve a hospital with psychiatric services, suicide prevention may need to be addressed.

The following sections discuss a variety of security and safety measures for parking facilities addressing security issues first, followed by safety concerns.

4.2   SECURITY DESIGN ISSUES

4.2.1   The Security Audit

The selection of the appropriate security features depends on the history of incidents in the area of the facility, and the likelihood of different incident types occurring in various locations within the facility. The neighborhood in which the facility is located will usually have the greatest impact on the degree of potential risk.

The higher the general level of crime in a neighborhood, the higher the risk for incidents in a facility. The process of assessing the risk is called the *security audit*. The first step in a security audit is an analysis of the risk of different incident types. If there is an on-site security staff, obtain the annual incident reports for the last five years and data on any personal injury incidents prior to that. Develop an incident history and profile for the neighborhood by contacting the local police and the operators of nearby facilities. Using this information, classify the facility as one of the following:

> Low risk facilities are those in which only minor vandalism and juvenile theft problems but no personal injury incidents and no professional theft activity are expected.
>
> Moderate risk facilities are those where there may be an occasional suspicious person or vehicle theft in off-hours, but there is no reason to anticipate personal injury attacks.
>
> High risk facilities are those where incidents of personal injury or a pattern of thefts that might escalate to personal injury have occurred previously at the site or in the neighborhood.

The second step of the security audit is an evaluation of the design features and constraints of the facility that impact security, either positively or negatively. In an existing facility one can walk the facility to identify problem areas in the security program. In a new facility, however, visualizing a "walk thru" of the facility is necessary to assess the strengths and weaknesses in the design.

Two types of security measures, *passive security* and *active security*, are employed to maximize security in a parking facility. Passive security measures are a physical part of the facility, such as lighting and glass-walled elevators and stairtowers. The common thread among all passive features is visibility—the ability to see and be seen while in a parking facility.

Active security measures invoke an active response by the management and/or employees of the facility. Examples of these measures include active security patrols and monitored Closed Circuit Television (CCTV) systems. Active systems are often needed to solve problems created by constraints on the passive security features. For example, some building codes require enclosing exit stairs with little or no glass. This requirement creates a closed space with little or no pedestrian activity; in short there is no visibility. Active systems, such as intercoms and/or CCTV may then be needed. Eliminating the enclosure is really the better solution, because the threat to life safety by attack is far greater than by fire in a parking facility. For this reason some codes have been changed to reduce the enclosure requirements for open parking facilities. Local officials may also be receptive to modifications and/or variances when security risks are obviously greater than fire risks.

General guidelines for correlating risk levels with the need for passive and active systems are enumerated in Table 4-1. As the risk level increases, the priority of passive features in the overall design should increase. Passive features are still "good" design features in low risk facilities, if only to add to patron comfort. Furthermore, retrofitting passive features is often expensive and sometimes impossible. Labor-intensive active systems may then be necessary. Therefore, many passive security features can and should be provided in parking facilities of all risk levels.

Active systems are generally not necessary in low risk facilities, but may be provided for patron perception and comfort rather than

TABLE 4-1   Guidelines for Relating Design Features to Risk Levels

Risk Level	Passive Features	Active Features
Low	As many as possible	For patron perception not prevention
Moderate	High priority in overall design	To correct defects in passive systems
High	Highest priority in overall design	Comprehensive program including CCTV, patrols, etc.

prevention of incidents. If not provided initially, plan for later installation of additional security systems in case circumstances change, and the facility moves to a higher risk level. Providing conduit in initial construction for future CCTV, for example, can be relatively economical.

The security audit will highlight the most likely locations for problems in moderate and high risk situations and will guide the selection of active systems. In a moderate-risk facility, there may be higher risk locations, such as an enclosed stair. Active systems are generally only provided in these specific locations. In high risk facilities a comprehensive security program is necessary to achieve a reasonable level of security.

The owner and/or operator of the parking facility must be integrally involved in the design of security systems. Owners and operators are in the best position to determine how many dollars can reasonably be spent on security systems. They should be very concerned with the cost-effectiveness of the expenditures, both from capital and life-cycle perspectives. In a life-cycle analysis, do not forget to include the impact of good security design on liability insurance premiums. In the end, the owner/operator will carry the lion's share of the liability for an attack. Therefore the owner/operator, rather than the architect, engineer, or parking consultant, must make the final decisions on security features and resolve conflicts between security planning and architectural, structural, or other design considerations.

4.2.2 Structural Design

During the design process the structural system of the facility must be evaluated from the security aspect as well as the engineering aspect. *Long-span* construction and high ceilings create an effect of openness and aid in lighting the facility. *Shear walls* should be avoided, especially near *turning bays* and pedestrian travel paths. Large holes in shear walls can help to improve visibility. When vision obstructions are unavoidable, strategically placed mirrors will allow patrons to see around corners where potential attackers may be hiding. Mirrors are a last resort, however, because they can be broken or stolen and not be in place when needed.

4.2.3 Lighting

Lighting is universally considered to be the most important security feature in a parking facility. Good lighting deters crime and presents a more secure atmosphere to parkers.

TABLE 4-2   Recommended Illumination Levels For
Parking Facilities

	Horizontal Illumination (Footcandles)	
	NPA[1]	IES[2]
Vehicle entrance	40	50[3]
Vehicle exit	20	–
Stairwells, exit lobbies	20	10/15/20[4]
Parking areas		
general parking areas	6	5
minimum at bumper walls	2	–
ramps and corners	–	10[3]
Roof and surface	2	3.6/2.4/.8[5]

[1]Minimum 30 in. above floor.
[2]Average on pavement; uniformity ratio (average to minimum) 4:1.
[3]Daytime only; 5 footcandles at night.
[4]See IES/ANSI 1983.
[5]Average footcandles for high/medium/low activity areas.

The Parking Consultants Council of the National Parking Association (1987) recommends the minimum levels of light shown in Table 4-2 for open parking structures. The Illuminating Engineering Society Subcommittee on Off-Roadway Facilities' (1985) recommendations are also shown. These recommendations should be carefully evaluated for each individual facility design. Closed areas or closed garages should generally have higher light levels, and special attention should be paid to lighting corners and perimeter walls to improve the patron's feeling of security. CCTV may also require higher light levels than shown in the table for picture clarity.

Many factors affect lighting levels and must be considered in the design process. Uniformity of light is extremely important and can be critical for CCTV. Non-uniform lighting creates shadows and hiding places. Sunlight at the perimeter of the structure or near light wells can cause other areas to appear dark by comparison. Rather than turning all the lights off, use photoelectric cells to cut off lights in areas with natural light in the daytime, such as along the perimeter of an open structure.

Another important consideration is glare which can affect vision by reducing the contrast of objects against the background (Chism 1986). The result is a lack of depth perception which is a potential hazard for all drivers. Glare is particularly dangerous for senior citizens and others whose vision may be weak or impaired. Glare can be minimized by careful selection and positioning of the fixture. Position lights over the parked vehicles rather than in the center of the drive aisle. Also, with one-way traffic, lights can be positioned near beams, using the latter as a shield for the approach angles which create the most glare.

The lighting fixtures selected for a parking facility must do more than provide ample, glare-free lighting. As a key component of the security system, fixtures must be reliable, be able to withstand the elements, be protected from vandalism, and be easy to maintain.

Staining concrete has proved to be a cost-effective method of increasing the general brightness and creating a sense of well-being (Hundt and Arons 1985). White stain on ceilings and beam soffits reflects the light and increases its uniformity. A good-quality concrete stain will last at least ten years in these locations. Paint creates the same brightness but requires much more maintenance. White stain on walls further improves brightness; however, it seems to encourage graffiti which tends to hurt the perception of security more than help it. Instead, anti-graffiti coatings may be used on walls, if desired, to aid quick and easy cleaning.

4.2.4   Stair Towers and Elevators

Rule number one in security is to design stair towers and elevator lobbies as open as the code permits. The "ideal" solution is a stair and/or elevator waiting area totally open to the exterior and/or the parking areas (Figure 4-1). If a stair must be enclosed for code or weather protection purposes, glass walls will deter the incidence of both personal injury attacks and various types of vandalism. Elevator cabs should have "glass backs" whenever possible (Figure 4-2). Elevator lobbies should be well lit and visible to the public using the facility and/or street. Try to get the local code personnel to approve an automatic fire door, or for a larger opening, a rolling fire shutter with an access door, so that the area is wide open during normal use. The door of the shutter will be closed by a smoke detector—a better alternative to a fire-rated door that remains closed all the time. Also, eliminate nooks and crannies and seal potential hiding places below stairs.

4.2.5   Restrooms

Parking owners, operators, and consultants all agree that public restrooms in a parking facility are nothing but nuisances. A restroom may also be a security trouble spot because use is infrequent and places of

Figure 4-1.   Open stairs provide the best security and can be an architectural feature as well. Photo courtesy of Walker Parking Consultants/Restoration Engineers.

concealment abound. Public restrooms should therefore be provided at the destination itself (office building, shopping center, etc.), where there will be more use and activity. If provided in a parking facility, design restrooms with ''maze'' entrances instead of outer/inner door arrangements that could trap a victim (Engineering Professional Development 1987).

4.2.6   Perimeter Security

Locate any attended booth or office in such a way that activity at pedestrian and vehicular entry points to the facility can be monitored. Likewise, locate a security station, if provided, where it is visible to the public. Provide security screening or fencing at points of low activity to discourage anyone from entering the facility on foot. In high risk cases, design a system of fencing, grilles, doors, etc., to completely close down the entire facility in unattended hours and to limit entry points during attended hours (Figure 4-3). Any ground level pedestrian exits that open into unsafe areas should be emergency exits only and

Figure 4-2. Glassback elevators and glass walls in stair towers provide visibility when these spaces must be enclosed. Photo courtesy of Walker Parking Consultants/Restoration Engineers.

fitted with panic bar hardware for exiting movement. Consider installing alarms that activate when a ground level door is opened. It is very desirable to consider the future implementation of perimeter security controls in the initial design stage, in case the facility's risk level should change.

4.2.7   Landscaping/Maintenance

Landscaping should be done judiciously so as not to provide hiding places. It is desirable to hold plantings away from the facility to permit the observation of intruders. The pruning and trimming of shrubbery are equally important. General maintenance and upkeep are of utmost importance in the overall security program. Trash, beer cans, graffiti, etc., may leave the impression that the facility is not policed or managed well.

4.2.8   Signs and Graphics

Careful design and placement of the general *signs* and *graphics* can eliminate confusion and delays for the patron. Help the patron get to

Figure 4-3. Security screening can be unobtrusive but provide control of the facility perimeter. Photo courtesy of Walker Parking Consultants/Restoration Engineers.

his/her destination quickly and efficiently, thereby minimizing the time for an incident. Color-coding and/or unique memory aids can help patrons locate the parked vehicle quickly upon returning to the facility. Signs and graphics can also assure the user that his/her safety is being monitored. Likewise, a perpetrator may be deterred by a notice that he/she is under surveillance. A disclaimer of liability for valuables and property left in vehicles should be located at or near the entrance.

It is critical for an owner/operator to back up any claims of security on signs with the services promised. If a sign says that conversations may be monitored for security, a person must be able to monitor the system, at least during higher risk hours. The latter usually occur at night when activity is lower but the facility is still open.

4.2.9   Cash Security

A number of important security features help to protect the cash receipts of the facility and relieve the attendant of the accompanying responsibility and theft hazards (Boldon and Sharitz 1983). A *drop safe* is most important since it makes the cash unavailable to the potential robber. Second, in moderate to high risk facilities, post a sign at the cashier booth(s) stating that all cash is deposited in a safe, and that the cashier has minimal change on hand.

In high risk situations, install foot-operated *duress alarms* that sound at the police station and/or the security office. Dollar bill alarm activators that are triggered by removal of all bills in a compartment of the cash drawer can also be useful in this application. Cash receipts should be removed on a regular basis, preventing the accumulation of large amounts of cash.

For liability control purposes, cash security should not be emphasized more than patron security.

4.2.10   Security Personnel

The visible presence of uniformed security officers is one of the best preventions of crime and should be considered in high risk facilities. Keep patrols unscheduled and vary the routes taken throughout the shift. In very high risk situations, check-in stations should be provided at key locations to monitor and record the frequency of patrols. Medical certification training for security personnel is also highly recommended, particularly CPR and Advanced First Aid Training.

All personnel charged with any security responsibility must be trained to monitor, operate, and respond to all security equipment provided in the facility, no matter what the risk level.

4.2.11   Emergency Communication

Alarm systems of this type come in many forms: panic buttons, emergency telephones, two-way intercoms, and two-way radio. Panic buttons are often located in elevators, lobbies, stairs, and occasionally in parking areas. However, their use is dependent on the victim of the attack reaching the button and sounding the alarm. Panic buttons also seem to be irresistible to pranksters. A "cry wolf" syndrome can develop among those monitoring the system. Emergency telephones make it even more difficult to sound an alarm plus they are more expensive to install and maintain in working order. Emergency communications are therefore not a complete solution in high risk facilities.

On the other hand, intercoms used together with panic buttons, motion and/or sound surveillance, or CCTV can be very practical security features. Two-way intercoms make it possible to zero in on an incident and communicate to the victim that help is on the way, possibly deterring the criminal. Voice-activated intercoms with panic buttons should be installed in all elevator cabs and partially or fully enclosed stairwells. In high risk facilities, intercoms with panic buttons and lighted "emergency aid" signs may be installed as frequently as every 100–150 ft in parking areas. Standard voice-activated systems are generally not practical in parking areas due to background noise (vehicles driving by, honking horns, etc.). Intercoms should also be installed in all cashier booths and at remote entrance/exit lanes. Connect all intercom stations to a master at the nearest point of observation with a provision to switch to a manned security office or police station during unstaffed hours.

4.2.12   Closed Circuit Television Systems

Closed Circuit Television (CCTV) can provide any level of surveillance an owner wishes to provide. However, it is important to recognize the inherent strengths and weaknesses of CCTV systems in order for CCTV to be an effective component of the overall security plan. While CCTV will not be able to replace all security personnel, it will frequently permit a reduction in, and provide invaluable support to, the security force.

CCTV monitoring can be very effective both to deter and detect incidents in progress in the enclosed areas (such as stair towers), which are historically at the highest risk for incidents (Figure 4-4). Parking areas may also be monitored by CCTV in high risk facilities. However, there is great difficulty in positioning cameras to fully cover all areas— lighting shades and shadows, external light sources, vehicles, and sloping floors—all restrict the ability to monitor activity in parking areas by a CCTV system. Even with a "state-of-the-art" system, only a small proportion of the incidents in a high risk facility may be first detected on the CCTV system.

Therefore, relying on CCTV to eliminate all patrols when there remain inadequately covered areas will invite greater liability problems than not providing CCTV at all. In fact, CCTV is best used in combination with other systems to support security personnel. When a report of a suspicious person, incident, or a door alarm is received, the person monitoring the CCTV screen searches for the location of the individual and directs the responding officer to the scene. If the CCTV operator sees an incident involving a personal injury attack in progress, the intercom system is used to scare off the offender and let the victim know that help is on the way. The officer continues to monitor the moves of the suspect to assist security or police in apprehension.

Figure 4-4.   CCTV camera with pan and tilt. Photo courtesy of Vicon Industries Inc.

CCTV can also be useful in apprehension and conviction following an incident by providing an accurate description of a vehicle or suspect. Using a video tape of an incident for apprehension and conviction will in turn convince habitual criminals to choose another facility for their unlawful activities.

Often several well-equipped cameras strategically located will enhance security system efficiency. Therefore emphasis should be placed on the location of cameras and the acquisition of good capabilities (i.e., telephoto lens, *pan-and-tilt*, *zoom*, etc.).

Advances in solid-state electronics are revolutionizing CCTV technology just as they are changing every other aspect of modern life. Solid state cameras are called charge-coupled devices (CCD) in the CCTV industry. Some major advantages of CCDs over those with conventional tube technology are longer operational life-spans between maintenance and repairs (tube life is ten years versus two years) and resistance to image burn and picture distortion. Installation of CCDs is less expensive than conventional tube cameras; no adjustment or fine-tuning is needed in the field. The primary negative aspect to CCD cameras is that the technology is relatively new and some problems such as *resolution* are still being worked on. All in all, CCD cameras are rapidly becoming the camera of choice for parking facilities.

A variety of lenses can be used to achieve the coverage needed. Wide-angle and super-wide-angle lenses cover small enclosed areas such as an elevator cab or room. Standard, semi-telephoto, and super-telephoto lenses view scenes at progressively greater distances.

The environment in which the CCTV system is placed must be considered (Kapinos 1987). Two basic types of housings are available. The traditional rectangular-shaped housing covers only the camera and lens. Domed, circular units cover and protect the entire assembly. Both have models for indoor and outdoor use and can be used with a pan-and-tilt mechanism. One advantage of the domed unit is that all camera parts, including the pan-and-tilt mechanism, are enclosed and protected from the elements and vandalism. The domed units are discrete and the patron may not realize that a camera is present (Figure 4-5). Conversely, the rectangular housings covering only the camera and lens "hold no secrets."

Remote positioning devices or controls allow the attendant to adjust a pan-and-tilt or a zoom lens on a camera from the monitoring station.

A multi-camera system may require a video *switcher* which allows the operator to select the scene from any one camera and display it on the monitor (Kapinos 1987). The switcher may also change from camera to camera at predetermined intervals, saving money since it allows one monitor to service multiple cameras. The switcher also makes the system more manageable for the observer.

Figure 4-5. CCTV camera housings may be discrete or clearly in view. Photo courtesy of Vicon Industries Inc.

Motion detectors and alarm-activated devices may likewise be cost-effective, as areas without activity are neither displayed nor recorded on tape. A panel design might provide several smaller monitors which automatically switch from camera to camera and one large monitor that the operator can switch to a particular camera. A developing situation can then be observed in greater detail.

The video recorder is an important part of the system as it allows the attendant to pinpoint when an incident occurred. The VCR generally records whatever is displayed on the monitors. For example, following a car theft, the tapes may show that the car was stolen from the parking space between 11:30 and 11:32 P.M. By reviewing the activity at that camera or other cameras at that time, a description of a suspect can be provided to the police. Following an arrest, the tapes can help get a conviction.

The central CCTV monitoring station, including the operator watching the monitors, should be visible to the parking patrons. It may not be necessary to station an operator at the monitors during hours of lower risk, such as the typical daytime activity hours. However, dummy or completely unmonitored systems should never be used. Also, CCTV systems require constant maintenance and upkeep to maintain picture quality. A decision to install a CCTV should include an ongoing budgetary commitment to maintain the system and replace parts as they wear.

4.2.13   Security Management

Planning for security in the design and operation of the facility is not enough. If active systems are provided they *must* be monitored by trained personnel. Policy standards to handle all situations must be established in writing and must be followed.

Although booth attendants are not usually security personnel, proper training of these individuals can significantly enhance the security program. Emphasis should be placed on being another set of eyes for security and reporting suspicious activity immediately. Booth attendants should concentrate on developing a good physical description of suspicious persons or an attacker if the crime is observed. Also, all booth attendants should be instructed on what to do should a theft or other crime occur.

A checklist for all security equipment and practices should be developed and regularly completed and filed. Records of incidents should be catalogued by type (vandalism, juvenile theft, rape, etc.) and an annual report should be prepared. While these issues are be-

yond the scope of the design of a new parking facility, owner(s)/op-
erator(s) must be aware that good professional security management
and documentation are some of the best defenses against liability
claims. With a high risk facility, if the owner/operator does not have
a professional security staff, a security management consultant should
be retained to develop policy and training manuals.

4.3   SAFETY CONSIDERATIONS

As noted in the introduction to this chapter, design hazards can create
liability problems as great as (if not greater than) attacks by a third
party. The increasing tendency to sue for damages over what used to
be accepted as an accident has started to change some design phi-
losophies. In many cases the patron's own actions, such as drinking
alcohol, substantially contribute to an accident in a design that meets
all codes and standard practices in the industry. Even so, a jury may
hold a facility owner liable because of the perception of the insurance
company's "deep pockets." Insurance companies, however, pass these
costs on to policy holders in the form of increased premiums.

At the same time, it is important to weigh the possibility of an
accident occurring with other important life-safety considerations, such
as durability, structural integrity, and fire safety. Some features that
enhance security also enhance safety, such as lighting, visibility, and
openness. Also, good maintenance is critical—it will yield substantial
dividends on the investment in good design. The following paragraphs
discuss some of the most common safety design errors in parking facilities.

4.3.1   Tripping and Slipping

Ice (in snow-belt areas) is one of the most frequent causes of falls in a
parking facility. Good drainage design is the first line of defense. Some
areas of icing may not be preventable. The most common one occurs
at covered/uncovered ramp junctures. First, the sun melts the ice on
the uncovered sections. The water then runs down onto the covered
section and refreezes. Floor drains can help, but the water tends to run
in a sheet across the floor, so that it freezes before it gets to the drain.
The owner must be vigilant to monitor and sand all icing spots as they
occur.

A slick floor is another potential hazard, especially in a sloping
ramp facility. The skilled concrete finisher may take pride in creating
a perfectly smooth floor, but it belongs in an industrial plant, not a

parking facility. A broom or swirl finish (recommended in Chapter 6) provides both good traction and a durable floor surface.

A roughened surface should be carried into the stair/elevator tower because snow and rain are often tracked in and may cause slippery spots, especially in unheated towers. Rubber stud flooring can be applied to lobbies to correct this problem. Abrasive nosings are also desirable on stairtreads. Wherever possible the cast-in-type should be used rather than pressure-applied strips, which are less durable.

Expansion joints must be carefully designed, and installation coordinated with temperature conditions, if possible, to minimize bubbling, buckling, and other tripping hazards. Good maintenance of expansion joints and replacing broken or missing drain grates are equally important.

Of late, liability concerns have made it important to eliminate curbs and wheel stops in areas where pedestrians are likely to be present. When adjacent bays are ''level'' (sloped for drainage, of course), pedestrians are likely to cut across the structure between cars. The cars necessarily create shadows and a curb or wheel stop then becomes a potential tripping hazard.

Curbs may still be appropriate in certain situations. Pedestrians very rarely walk to the perimeter of a facility merely to look at the surrounding area. Conversely, the durability and structural integrity of the structure can be greatly enhanced by employing a curb to cover connections between exterior panels and floor slabs. Curbs are also desirable at parking equipment islands. Tripping hazards in these areas are generally reduced to a minimum level by the high level of lighting otherwise necessary. Painting the faces and edges of curbs will further reduce the hazard.

4.3.2 Head Knockers and Other Projectiles

Most codes prescribe minimum overhead clearances for pedestrians and vehicles. These standards should be adhered to, even in isolated areas. Whenever substandard (with respect to the code) clearance exists, the international ''hazard'' symbol (alternating diagonal bars of yellow and black) and a notice of a low clearance should be affixed to the obstruction, even if it is not intended that pedestrians or vehicles pass underneath. Watch especially for reduced clearance at curbs.

Another ''head-knocker'' problem occurs all too frequently when patrons walk down parking control equipment lanes and are struck by descending gate arms. Sidewalks should always be provided with groups of entry/exit lanes, and should be well marked as available.

"No Pedestrians" messages, perhaps using international symbols, may also have to be added to entry/exit lane signs.

Clearance bars at all entrances stating the minimum vehicular clearance have become a standard in the industry. It is not recommended however, that a fixed, heavy obstruction be employed. In one recently reported case, a cashier was knocked unconscious by a falling clearance bar while making a tour of the facility at closing time. He was found, still unconscious, more than 30 min later. Luckily the person who found him was honest and summoned help rather than leaving him there and absconding with the day's receipts. In other cases, a main beam has been deliberately designed at the posted clearance height to keep oversized vehicles out. This tactic not only raises patrons' tempers, but also does not eliminate liability if inadequate advanced warning is shown to exist. Most parking consultants now use a long, large diameter (10 in.) PVC tube hung from chains for a clearance bar at each and every entrance lane. This tube provides a certain stiffness and creates a sufficient racket when hit while minimizing damage to vehicles or pedestrians.

Other devices may become projectiles. Sand-filled oil barrels are commonly used as inexpensive traffic control devices. However, if knocked over, these drums will roll down a sloped parking ramp with substantial speed and momentum. Some manufacturers make flat-sided, plastic barrels striped with reflective sheeting for high visibility. Ballast in the form of sand bags can be added as required to keep the barrel in the desired location.

An error commonly made in parking facilities is designing stair towers with doors which swing into driving aisles. One solution, of course, is to eliminate the door, which also enhances security. When doors are required, a careful design can achieve both vision to the aisle and a protected area to open the tower door (Figure 4-6). A similar problem occasionally occurs when elevator waiting areas are located too near driving aisles.

4.3.3   Vehicular and Pedestrian Barriers

In recent years, many building codes have begun to address the issue of preventing out-of-control vehicles from breaking through exterior and interior railings at areas of grade separation. There is however no uniformity between standards. The NPA Parking Consultants Council (1987) recommends the following:

> Vehicle restraints should be placed at the perimeter of the structure and where there is a difference in floor elevation of greater

Figure 4-6.   Both vehicular and pedestrian barrier requirements must be addressed. This detail provides both safety and visibility for security.

than one foot. Vehicle restraint systems should not be less than two feet in height and should be designed for a single horizontal ultimate load of 10,000 lb applied at a height of 18 in. above the floor at any point along the structure.

Openings in railings or spacing of components should conform to other sections of the local governing code. If vehicle restraints and handrails are used, no other barriers such as wheel stops or curbs should be necessary.

We recommend that the NPA standard be followed except when the locally adopted code has a higher standard. Unfortunately some codes have well-intended but misguided standards. For example, they specify that the barrier must stop a vehicle moving at a specific speed. To calculate the force applied to the barrier, which is necessary for design, one must use an energy equation which requires assumptions that 99% of all designers are not qualified to make. Some negotiation with local code officials regarding the standard may be required.

Another need for vehicle restraint occurs at entry/exit locations. Inadequate design for turning movements threatens not only the parking control equipment but also a cashier in a booth. The most frequent problem is not providing enough space for turning into the lane *and* getting aligned properly before reaching the ticket dispenser or card reader.

In addition, it is considered good practice to provide a concrete-filled steel post, solidly anchored in the curb, at each piece of parking equipment. Casting a pipe sleeve in the curb facilitates replacing the post should it be damaged. One word of warning: check for all possible angles of approach. For example, vehicles backing out of nearby stalls can hit the gate from the back side of an island.

In recent years life safety and/or building codes have been substantially tightened to require handrails at a spacing no greater than 6 in. In a few cases the standard is 5 in. This standard is designed not only to prevent a toddler from falling through the rails, but also from getting his/her head stuck. A facility designed under the immediately prior standard (generally, 9-in. spacing) will usually not have any liability to upgrade to the current code. However, an unsafe condition which is clearly apparent should be corrected.

Codes are often unclear regarding what degree of grade separation requires a handrail. The grey area tends to be that between normal curb height and a differential in a grade of 18 in. Good professional judgment should be exercised in this area. A handrail should always be provided if there is any possibility of a severe, life-threatening injury.

Codes also prescribe a minimum height of a railing. Courts have tended to hold owners to literal compliance with the code; that is, if the handrail is even a $\frac{1}{2}$ in. too low, the owner is liable for an accident. Therefore, handrail heights should be very carefully designed for some of the conditions common only to parking facilities, such as sloping, cambered, and warped floor areas, etc. Attention to minor details such as these will minimize liability for the owner.

4.3.4 Vehicular/Pedestrian Conflicts

Vehicular/pedestrian conflicts are inherent to parking facilities. Thoughtful design can, however, minimize owner/operator/designer exposure to liability. Pedestrians have a tendency to take the shortest possible route rather than a designated pedestrian walkway, especially when a nondesignated route is encountered first. For example, some people will always walk down the middle of a gated entrance/exit lane instead of crossing to the far side of the lane grouping and using the sidewalk provided, as previously recommended. If a sidewalk is provided and clearly visible and/or marked for the user, a court will consider the patron to have used the driving lane at his/her own risk. Beware, however, of designs that expect a pedestrian to take an unnecessarily long route.

4.4   SUMMARY

The key to good security is visibility. Passive security features should be a high priority in virtually all parking facility designs because:

> good passive design maximizes visibility at the lowest possible cost
> circumstances and risk levels change
> retrofitting is very expensive if not impossible
> labor- and equipment-intensive active systems are generally needed because of shortcomings in passive security features

Progressive reaction to incidents or changes in the risk level, such as adding access control, CCTV, or active patrols will greatly reduce the potential for crime and the liability should a criminal act occur.

As evidenced by the interest in and attendance at security sessions at the conventions of various parking groups, security is one of the biggest problems in the industry today. Security is an ongoing process that good design alone will not achieve. Training and management of security forces by a professional are equally important. A comprehensive security program will provide the owner and parking patron with a safe parking facility in all but the highest risk situations.

Governing codes do not provide a complete guide for avoiding safety hazards because of the unique characteristics of a parking facility. Personal injuries due to tripping, ''head knockers,'' and lack of consideration of other hazards in the pedestrian's path of travel, generate surprisingly large awards in suits against parking facility owners and operators. Experience in parking facility design and attention to details will minimize the risk of safety hazards, reducing liability exposure to a minimum level.

4.5   REFERENCES

Boldon, C. M. and Sharitz, C. J., 1983. ''Security,'' in *The Dimensions of Parking*, second edition, edited by the Parking Consultants Council, National Parking Association, 105–108, Washington: The Urban Land Institute and the National Parking Association.

Chism, R. W., 1986. ''Lighting—First Line of Defense in Parking Structure Security,'' *Parking* 25, no. 5 (September–October): 77–79.

Engineering Professional Development, The College of Engineering, University of Wisconsin, Madison, 1987. ''Inhibiting Crime Through Design?,'' *Building Design and Construction Newsletter* 3, no. 4 (Fall): 1–3.

Hundt, R. M. and Arons, W. C., 1985. ''Planning Parking Protection,'' revised from article in *Security Management* (February): 44–47.

IES/ANSI, 1983. *American National Standard Practice for Roadway Lighting*, IES/ANSI RP-8-1983, New York: Illuminating Engineering Society of North America.

IES Subcommittee on Off-Roadway Facilities, Roadway Lighting Committee, 1985. "Lighting for Parking Facilities," *Journal of IES* (April): 616–623.

Kapinos, T. S., ed., 1987. "CCTV . . . A Primer for the Security Administrator," *Security Systems* 16, no. 8 (August): 27–29.

———, 1987. "CCTV Switchers," *Security Systems* 16, no. 8 (August): 30.

———, 1987. "Housings," *Security Systems* 16, no. 10 (October): 13–14.

Lubben, C. H., 1987. "How Safe is Your Parking Lot?," *Business Digest* (November): 63–65.

Parking Consultants Council, National Parking Association, 1987. *Recommended Building Code Provisions for Open Parking Facilities* (revised), Washington: National Parking Association.

Chapter 5
Structure

ANTHONY P. CHREST

5.1 INTRODUCTION

Section 5.2 will discuss the various factors influencing the structural design for any parking facility, or any building, for that matter. Among the items covered will be cost, schedule, and building codes.

Section 5.3 is a guide to help you narrow down the selection of structural systems. Some systems are better than others for parking structures, while some should be avoided.

Section 5.4 helps you identify and design for volume-change effects in parking structures. Their large plan areas and exposed structure make them more susceptible to these effects than many other building types.

Certain design items are somewhat unique to parking structures. Discussed in Section 5.5 are beam-column joints, variable height columns, torsion, and the relationship of stair and elevator structures to the main structure.

Finally, in Section 5.6 specific loads on parking structures are examined, followed by a chapter summary in Section 5.7.

Please note that this chapter addresses only structural design as it pertains to parking structures.

5.2 DESIGN

5.2.1 General

Structural design should satisfy requirements for strength, flexibility, durability, ease of maintenance, and repair. Equally important are function, cost, appearance, and user comfort.

111

5.2.2   Factors Affecting Design

5.2.2.1   Cost. Factors which will influence structural system cost are:

5.2.2.1.1   Local Construction Technology. For example, say your design calls for a column-concrete strength of 14,000 psi, and no cast-in-place building in that locale has ever been built with concrete strength exceeding 10,000 psi. Bids for your cast-in-place structure will be higher than you expect; you may not get any bids. However, assume for this example that precasters in this same locale have had experience with concrete strengths as high as 14,000 psi. Unless you are willing to increase the cross-sectional dimensions of your columns to permit the use of 10,000-psi concrete, you should select a structural system which uses precast concrete columns, not cast-in-place ones. Another alternative would be for you to embark on an education program with the local ready-mix supplier, which has already been done as seen below.

A similar example relates to silica-fume concrete, also called micro-silica concrete. The material is stronger and less permeable than conventional concrete, but if it has never been used in the project city, the bids will probably again exceed expectations or not even materialize.

A final example concerns pretopped double tees. If no precaster in the project area has ever built a structure with pretopped tees, it is likely that none will be interested in bidding on your pretopped structure. Or, bids will be too high to cover the precaster's and erector's learning curves.

All of the above should not lead you to believe that there is no innovation in the construction industry. It is true that the engineer may have to spend more time educating builders and/or precasters in a given area, but once understanding is reached, these people are almost always ready to move into new technology or practices. As one example, with a high-strength concrete consultant, we helped a ready-mix supplier and builder team consistently produce 14,000 psi concrete for columns in a 15 story multi-use parking structure. As another example, we have helped a number of precasters fabricate and erect their first pretopped double tee parking structures. Having done their first, all of these precasters wanted to do more.

5.2.2.1.2   Who Will Bid? If there is a general contractor around who likes to pour concrete and is short on work, he is likely to give you a very good bid on a cast-in-place structure. Conversely, if no precast projects have been built recently and there are hungry precasters around, a precast job will attract a lot of interest and low bids.

On the other hand, if there are no precasters in an area or all available precasters are committed to other work, it would be unwise to put a precast structure out for bids.

In some areas there may be a shortage of construction tradesmen. A structural system which is less site-labor intensive, such as pre-topped precast concrete or structural steel, might be a good choice in such circumstances.

To achieve an economical structure, poor soil conditions will require the lightest structural system possible.

5.2.2.1.3 *Quality of Construction.* A concern similar to that above is that even in certain metropolitan areas of North America, there is a lack of builders capable of producing quality cast-in-place concrete construction. A related problem is that trade union rules in at least one metropolitan area require all rebar bending to be done on site. These two circumstances, alone or combined, have nearly eliminated cast-in-place concrete parking structure construction in those areas.

5.2.2.1.4 *Budget.* The project budget may eliminate some systems from consideration. As an example, though long-span construction is best for parking structures because it permits easier parking and flexibility in striping and in resizing parking spaces, short-span construction in which cars park between columns is less expensive. If the building code requires parking as part of the total project and the project will not be built unless the parking structure cost is rock bottom, then long-span structural systems will probably have to be discarded from further investigation.

5.2.2.1.5 *Building Codes.* Local building codes and ordinances, especially those that have not recognized downsizing trends in automobile manufacture, may mandate parking module wall-to-wall dimensions of 65 ft. If the owner's and engineer's desires are to use clear-span construction, then some systems will not work. For instance, if 24-in. precast concrete double tees are the deepest available within economical shipping range, a double-tee system will be eliminated because the available members do not have the capacity to carry typical parking structure loads for the required 65-ft-clear span.

5.2.2.1.6 *Durability Costs.* Durability requirements will raise the cost of certain structural systems more than others because some systems are inherently more durable than others. For additional discussion, see Section 5.3.

5.2.2.1.7 Schedule Costs. Speed always costs more, no matter what the structure is. If a compressed schedule is necessary and the project budget cannot be increased, the increased costs may dictate a compensatingly less expensive structure, which leads back to the above discussion on budget.

The second aspect of schedule is that in cold climates, winter construction is significantly more expensive than construction in moderate temperatures. Summer construction is more expensive in hot climates. There is not always the freedom to choose when to start construction, but if there is a choice, scheduling with the seasons in mind can effect significant savings or permit extra features to be added to the facility while keeping the project cost within the original budget.

5.2.2.1.8 Appearance Costs. The facade may be of limitless variety, and corresponding costs can range from zero dollars over the cost of the ''bare-bones'' structure to tens of percent of the total project cost. As with schedule, there may be freedom in selecting the facade treatment, or there may not. As an example of the latter situation, it is not unusual to find a hospital campus surrounded by an urban or suburban residential neighborhood. In planning a parking structure to serve such a hospital, the planners often encounter objections to the structure from neighborhood groups. The objections must often be overcome, either because the hospital wants to be a good neighbor or in order to get a building permit, or both. Overcoming the objections usually requires that the parking structure facade hides the cars inside from the view of the surrounding streets, that its architectural appearance blends in with the hospital campus and/or the neighborhood, that the structure not be so high as to tower over surrounding residences or block sunlight, and that lighting does not spill outside the structure. Satisfying each of these objections will add to the cost of the project to some degree.

If the structure must park enough cars and the plan area is limited, several stories of parking will be required. But if the structure height must be limited, the only solution to both requirements will be to go underground. On a per-car basis, all else being equal, underground parking construction costs are roughly double those for an aboveground parking structure.

5.2.2.2 Schedule. The project schedule should encompass the entire project, with durations assigned for all design phases; whatever reviews are required by parties such as the client, local citizens' groups, the building department, zoning, other public agencies, the financing body and others; bidding or negotiating, and construction. Take long lead-time items, such as elevators, into account. Remember that winter weather in cold climates will affect the schedule for an open parking structure more than it would for a conventional building. A conventional building will be closed in at some point, permitting completion of finish items which require moderate temperatures. Examples of these finish items are sealers, membranes, sealants, paints, curing compounds, and defect patching.

5.2.2.3   Building Code. Applicable building code requirements vary in their details across North America. They affect:

5.2.2.3.1   Loading Criteria. Magnitudes of live loads on floors and of live load reductions to floor members and columns will vary. (Car-bumper-impact load magnitudes and height above the floor to barrier walls vary more.) These loads in turn will influence the structural design.

5.2.2.3.2   Barrier Requirements. Handrail heights, live load, and railing-spacing requirements all may vary and all affect design. Openings in barriers to resist car bumper impact are limited in size by codes, but these limits also vary.

5.2.2.3.3   Fire Ratings. Fire ratings vary considerably from one jurisdiction to the next. Protection requirements have the effect of dictating structural element thickness and, therefore, weight, which in turn affects the structural design of all superstructure members as well as the foundations.

5.2.2.3.4   Standpipes and Sprinklers. Almost any parking structure of more than two or three stories will be required to have dry standpipes. Some codes will require all standpipes, dry or wet, to be interconnected so that they may be filled or drained from any hose bibb. Hose-bibb threads must meet local fire department requirements. Underground garages and some multi-use parking structures, whether open or not, will be required to have an automatic sprinkler system. An example of the latter might be a multilevel parking structure taller than a mandated height, meeting the requirements to be considered *open* but having office space on its top floor or floors, even though the "roof" of the parking structure (bottom floor of the office space) meets the separation requirements for mixed-use occupancy.

5.2.2.3.5   Ventilation. As stated in Chapter 1, requirements for open space around the exterior walls of a parking structure vary. If the structure meets or exceeds the particular openness requirements, no mechanical ventilation of the parking and driving areas will be needed. There may be separate code requirements for ventilation of ancillary facilities, such as restrooms, maintenance areas, stairs and elevators, attendants' booths, offices, isolated or dead-end areas, or a low point where carbon monoxide could collect and endanger patrons or staff.

If the structure does need mechanical ventilation in large areas or throughout because it does not meet code openness requirements, the project cost will of course increase.

5.2.2.3.6   Storage Room. A parking structure will often contain a storage room for maintenance equipment and supplies, parking and revenue control equipment parts and supplies, and odds and ends. As a corollary to the sections on fire protection and ventilation requirements above, if the storage room is truly for storage only, and is not used for maintenance or anything other than storage, it is important to label that room as *Dead Storage* on the contract documents, otherwise you may find yourself having to provide heat, mechanical ventilation, and sprinklers for that room.

5.2.2.3.7   Height and Area Limits. These too will vary across the continent and are often tied to fire rating requirements. As in the example of the hospital campus located in a residential neighborhood, the height limitation may not be a previously set limit, but may be subject entirely to the opinions of a special interest group.

5.2.2.4   Appearance. Appearance is not within our scope; however, the nature of parking structures often requires a structural element to also serve an architectural function. An example is the typical exterior beam: it carries live, dead, and bumper loads, and may be part of a rigid frame which carries wind or earthquake forces, yet it must also contribute to the appearance of the structure. A second example is the often required resolution of the conflict between the functional design, which may require sloping floors, and the desire for horizontal elements on the building facade. As a third example, on a recent project, the architect wanted to reflect the 20-ft module of an adjacent existing building in the structure of the new parking structure project. A 20-ft column spacing is economical for a cast-in-place post-tensioned one-way slab and beam system, but not for an 8- or 10-ft-wide precast prestressed double tee system. The architectural facade in this case decided the structural system. It is also important to resolve the possible conflict between the requirement for a stiff structure to support a rigid facade, such as masonry, and the flexibility of the main structure.

5.2.3 Associated Design Elements

5.2.3.1 Snow Removal. In cold climates a snow removal system may be required. This provision may involve a design for special vehicle loading in addition to the snow loads. Consider provisions for snow melting equipment and chutes or containment structures.

5.2.3.2 Drainage. Proper drainage is *essential* to structure durability. If water is allowed to stand (*pond*) on the parking structure floors for long periods, deterioration will accelerate in the concrete beneath the ponds. If the water which collects is salt-laden, chances for accelerated deterioration are much greater. If the structure is to be durable then, it is important that the drainage system rapidly carry away all water, whatever its source, and not allow ponding anywhere. Drainage design requires attention to three areas: proper slopes, proper catchment area sizes, and proper drains.

First, no parking structure floor should ever be flat, even if no rain can ever fall directly on it. Windblown rain will come into floors below the roof. Cars will carry water into lower floors also. Heavy rains may overload top floor drains. The overflow will run down ramped floors until the lower floor drains can carry it away.

For drainage, the absolute minimum slope should be $\frac{1}{8}$ in. per ft, or about 1%. Preferred slope is $\frac{1}{4}$ in. per ft, or about 2%. Note that cement finishers will have a very difficult time consistently achieving a slope of less than $\frac{1}{8}$ in. per ft, if they can achieve it at all.

When setting slopes on design drawings, be sure to take expected prestressed member camber and deflections of all members into account. Either could reduce or eliminate design slopes if not recognized and if not compensated for. Pay particular attention to cantilever spans so that water drains off them instead of collecting at the free ends.

Prevent ponding anywhere by controlling the deflection of slabs and beams. In addition to sloping top surfaces, deflection control may be achieved by prestressing and by setting slopes or cambers into forms.

At the same time, do not forget that pedestrians will be walking across your slopes and swales. Keep the steeper slopes away from pedestrian paths; do not make walking uncomfortable.

Drain catchment areas should not exceed about 4500 sq ft on floors which are nominally flat, that is, sloped one or two percent only for drainage. On floors which have more than minimum slopes, drains may be spaced to drain more than the 4500 sq ft maximum. However, drains should be located so that runoff does not have to cross an isolation (expansion) joint seal, or turn a corner to reach the drain.

Drains must be adequately sized or slightly oversized for the design storm. Locate them in gutter lines or other low points. Recess the drain 1 in. below the adjacent floor surface.

The drain basin should be shaped to generate a vortex in the water which will speed the flow of runoff out of the drain and into the downspout.

The top grate should have sufficient openings to admit the design runoff, but individual opening size cannot be so large as to present a hazard like catching a shoe heel. The grate should be permanently attached to the drain body with a hinge on one side and a tamperproof screw on the other. With a hinged grate, lifting off the grate to clean out the drain will not run the risk of losing the grate. With a tamperproof screw to hold it, vandals will not be able to open the grate.

The drain should also have a sediment bucket which can be removed and emptied during regular drain maintenance. Drains at lower floors should include backwater valves to prevent flow from the downspouts backing up and overflowing into the lower levels.

Circular and square drains are usually of a size which is small enough so as not to interfere with structure. For example, in a post-tensioned floor, the slab tendons can easily be curved around the drains. Sometimes, though, smaller drains are inadequate, particularly at a top floor (roof) or at the bottom of a ramp or ramped floor leading to a roof. Segmented or continuous trench drains may be necessary to accommodate runoff volume in such cases.

Segmented trench drains are simply short sections of trench drain, say 12–24 in. long, separated by a foot or two of floor slab structure. Slope the floors between the drain segments into them. Below the floor slab, the trench drain segments are continuously connected to a downspout. The segmented trench drain permits less disruption of the floor structure. A continuous trench drain running completely across a bay will effectively create a structural isolation joint. The resulting separation must be treated like any other isolation joint.

5.2.3.3   Electrical Conduit. If there is a choice, do not cast electrical conduit into the structure. While exposed conduit is initially more expensive than embedded, maintenance will be far easier. Rewiring, if ever needed, will be less expensive if the conduit is initially installed exposed.

Concrete durability will be improved. When conduit is cast into the structure and moisture gets into it through leaks or condensation, deterioration of the concrete around the conduit may be accelerated. If the moisture freezes and the ice is not free to expand within the conduit, the conduit may split and the surrounding concrete may spall. Moisture inside or around the conduit may cause the conduit to rust. The rust in turn will exert pressure on the surrounding concrete, spalling it. Some designers may object to exposed conduit on esthetic grounds, but if properly installed, the average patron will never notice it.

If conduit is to be exposed, provide formed holes in floor beams and/or tee stems to permit straight and economical conduit runs.

5.2.3.4 Expansion Joint Seals. Even the best isolation (expansion) joint seals for parking structure floors will present problems sooner or later. If floor isolation joints can prudently be avoided, do so, and you will save yourself and the owner trouble later.

Expansion joint spacing is discussed in a later section. Seal types, uses, and pros and cons are outlined in Table 5-1 and discussed in the remainder of this section.

The trouble with expansion joint seals is that the requirements for the perfect seal are contradictory. The perfect seal should:

be leak-free
not be under tension or compression most of the time
not be a tripping hazard
not collect water, ice, dirt, or debris
stand up to normal abrasion from car tires
resist damage from snow plows
not deteriorate under normal use

One of the more common seals is the Tee joint, named for the shape of the concrete around it. Most Tee joints look like Figure 5-1. The Tee joint satisfies most of the list above, except that it is almost always in tension or compression, as opposed to bending. It also is not particularly resistant to damage from snow plows. It is better, when possible, to install the Tee-joint seal in summer, when the expansion joint gap is at its narrowest. Otherwise, it will bulge in hot weather and become a tripping hazard. If necessary, the seal may be temporarily installed during cold weather and then permanently installed when summer comes.

TABLE 5-1 Common Types of Isolation-(Expansion-) Joint Seals

Type	Description		Pro or Con
Tee joint	Figures 5-1, 5-2, 5-3	Economical, good track record when used for the right applications.	Exposed to tire or snow plow damage, joint movement puts seal in tension, maintenance more difficult.
Strip	Figures 5-5, 5-6	Seal protected from tire or snow plow damage. Joint movement puts seal in bending, not tension. Easy seal replacement for maintenance.	More expensive.
Compression	Figure 5-7	Usually protected, relatively economical. One newer type is performing well.	Older types have a poor track record.
Foam compression	Figure 5-8	May be protected, economical.	Poor track record, not recommended.

A variation of the Tee joint is shown in Figure 5-2. It is used to seal the gap between a floor slab and a vertical surface, such as a stair tower wall. Where the detail of Figure 5-1 meets that of Figure 5-2, ignoring strain compatibility will lead to seal failure. Figure 5-3 is a second variation of Figure 5-1. It may be used where there are no possibilities of tripping hazards, traffic, or snow plow damage. Its ad-

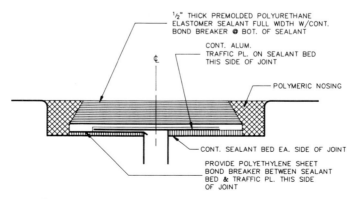

Figure 5-1. Tee-joint expansion-joint seal—floor to floor.

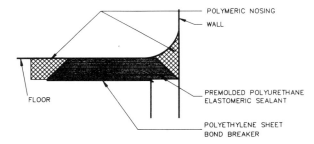

Figure 5-2. Tee-joint expansion-joint seal—floor to wall or column.

vantage is that it does not need a blockout in the concrete. The detail of Figure 5-3 may be combined with that of Figure 5-1 to seal the condition shown in Figure 5-4 where the expansion joint has to go around a column.

Figure 5-5 shows an armored strip-seal. There are many variations but the basic concept is the same. It satisfies most of the requirements in our list. During cold weather, when the expansion joint opens to its widest, the seal may be a tripping hazard, however. Figure 5-6 shows the variation of Figure 5-5 similar to that of Figure 5-2—the slab-to-wall seal.

Figures 5-7 and 5-8 illustrate two variations of the compression seal. There are many variations of Figure 5-7 available and a few of Figure 5-8.

Whichever seal you select, you can improve its chances of performing well following the advice below, if at all possible.

Locate expansion joints at high points in the floor drainage pattern; keep expansion joints out of the path of runoff; locate expansion joints so as to avoid columns and other elements which will complicate installation and impede performance. If there are conditions which

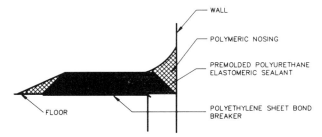

Figure 5-3. Tee-joint expansion-joint seal—floor to wall or column. Mounted on top of slab.

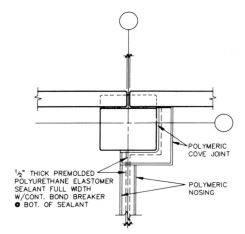

POLYMERIC
COVE JOINT

½" THICK PREMOLDED
POLYURETHANE ELASTOMER
SEALANT FULL WIDTH
W/CONT. BOND BREAKER
⊙ BOT. OF SEALANT

POLYMERIC
NOSING

Figure 5-4. Plan of expansion-joint seal around column.

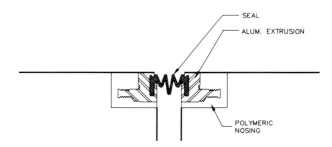

SEAL

ALUM. EXTRUSION

POLYMERIC
NOSING

Figure 5-5. Armored strip seal for expansion joint—floor to floor.

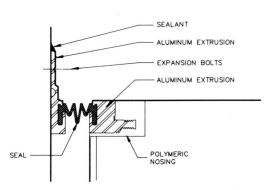

SEALANT

ALUMINUM EXTRUSION

EXPANSION BOLTS

ALUMINUM EXTRUSION

SEAL

POLYMERIC
NOSING

Figure 5-6. Armored strip seal for expansion joint—floor to wall.

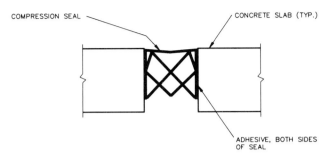

Figure 5-7. Compression seal—floor to floor.

require that the seal go around a column or up a curb or wall, provide the details and views necessary to show what is to be done.

Again, if possible, eliminate the need for the expansion joint so that you can prudently eliminate the joint. (See the following section on Structure Vibrations.)

5.2.3.5   Structure Vibrations. Structures producing an efficient parking layout often have column spacing with the inherently limber long-span beams in one direction and a multiple of the parking spacing in the other. Alternately, if the floor is precast, the dimension should be a multiple of the precast module, usually 24 or 30 ft, or if cast-in-place, an economical span in the 18- to 24-ft range. These long-span floors must have the necessary stiffness and mass to reduce vibration and noise to acceptable levels. Perceptible vibrations are a normal consequence of the span-depth ratios found in modern parking structures. As a rule, such vibrations are not detrimental to the use of the structure; what is acceptable is somewhat subjective; no code requirements exist. The best resolution might be to visit several existing parking structures with the client, reaching an agreement as to what is an

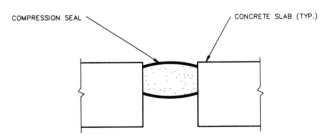

Figure 5-8. Foam compression seal—floor to floor.

acceptable level of vibration. Remember that facility pedestrians will feel vibration more than motorists will.

A specific consideration is the effect of vibrations on an isolation (expansion) joint seal. Normally concern is necessary only when the seal bridges two cantilevered elements. For instance, an expansion joint may be located between two cantilevered slab ends, which are in turn located midway between two beams or double-tee stems. If differential vibration/deflection of the cantilever ends will expose the seal to premature wear from car tires or snow plows, then reduce or eliminate the differential. Provide a shear connection, which will permit free expansion and contraction of the joint in the horizontal plane, while preventing relative movement in the vertical plane.

5.2.3.6 Tensile Stress Control. Areas of tensile stress-induced cracking are among the first to yield to weathering deterioration. The rate at which concrete deteriorates and steel corrosion begins will be proportional to the amount of concrete cracking.

Minimize bending stresses in exposed reinforced concrete design by attention to structural depth and reinforcement clear cover, quantity of reinforcement, and provision of closely spaced small diameter bars at the tension face. Limit factor z (ACI 318, Section 10.6) to 55 or less under dead load.

Reduce or eliminate tensile bending stresses and, in turn, tensile cracks, by the judicious use of prestressing. Use only as much prestressing as is required to reduce stresses to noncracking levels under service load conditions. Stiff walls and columns and wide beams can reduce prestressing forces applied at a building perimeter. Account for this reduction by increasing the applied forces, or by using temporary hinges to reduce member stiffness. Higher levels of prestressing will only increase problems due to elastic shortening and creep. See the section on Design Measures under Volume-Change Effects.

5.2.3.7 Future Construction. Scope-of-work discussions with the owner should always cover whether or not the design will allow for future horizontal or vertical additions. If future additions are planned, provide adequate structural capacity. Define the nature and extent of the future addition in the contract documents.

If the original structure is to be precast concrete, consider whether or not a precast addition will be feasible. There must be room for erection cranes. Will a crane be able to reach far enough to place the precast pieces? There have been cases where a cast-in-place addition

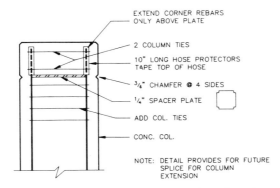

EXTEND CORNER REBARS
ONLY ABOVE PLATE

2 COLUMN TIES

10" LONG HOSE PROTECTORS
TAPE TOP OF HOSE

³/₄" CHAMFER @ 4 SIDES

¹/₄" SPACER PLATE

ADD COL. TIES

CONC. COL.

NOTE: DETAIL PROVIDES FOR FUTURE
SPLICE FOR COLUMN
EXTENSION

Figure 5-9.   Detail to permit easy extension of a column.

was built atop a precast original structure because a precast addition was not feasible.

Include details to permit easy future additions in the contract documents for the original structure. Figure 5-9 shows a detail which will accept a future column extension. Figure 5-10 shows a detail which permits future extension of a supported post-tensioned slab.

5.3   STRUCTURAL SYSTEM SELECTION

5.3.1   General

Selection of the structural system for a parking structure will be influenced by the factors cited in Section 5.2. The designer provides the client with design options which he evolves into a final concept to satisfy the client. Though the engineer must certainly recommend a structural system alternative, together with associated costs, the client ultimately decides which system will satisfy his/her requirements.

Important considerations in selecting a structural system are availability, cost, expected quality of construction, expected life, function, and appearance. The first five factors are equally important; the sixth may be less so. However, if the owner is a developer who intends to sell the project soon after it is completed, he may only be concerned about first cost. He may not appreciate the values of expected life and function in establishing an asking price. Educating this type of owner may be difficult. In the extreme case, you may be better off not taking the project.

Concern about quality has been much discussed recently. If a

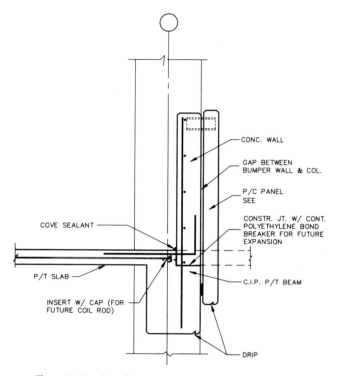

Figure 5-10.   Detail to permit easy extension of a floor.

structure meets the requirements set for it in the scope-of-work statement, which must be part of the design-services agreement, then it is
a quality structure. Obviously, stating the project requirements clearly
in the design-services agreement and the scope-of-work statement is
of prime importance. Failure to achieve agreement on project requirements should cause you to drop the project.

5.3.2   Foundations

Keep in mind that parking structures tend to move more than other
building types. Here is one common error as an example.

   A designer considers the unbraced length of the first-floor columns as the dimension from top of column footing to bottom of beam
at first-supported level. He uses that dimension in his frame analysis
and column design. The foundation details, however, show a foundation wall between footings which rises 3′6″ above grade. These de-

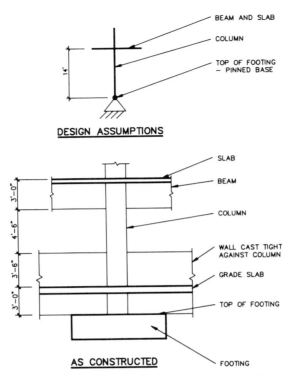

BEAM AND SLAB

COLUMN

TOP OF FOOTING
– PINNED BASE

14'

DESIGN ASSUMPTIONS

SLAB

BEAM

COLUMN

WALL CAST TIGHT
AGAINST COLUMN

GRADE SLAB

TOP OF FOOTING

3'-0"

4'-6"

3'-6"

3'-0"

AS CONSTRUCTED

FOOTING

Figure 5-11.   Assumed vs. actual column-base conditions.

tails further show the walls cast tightly to the columns with wall re-inforcement continuous from the walls into the columns. The resulting unbraced column length is actually only 50–60% of what the designer assumed. The column as shown in Figure 5-11 will probably crack. The point is to design properly and to make sure that the construction mirrors your design.

5.3.3   Lateral Load Resistance

5.3.3.1   Shear Walls. In order to minimize slab cracking in large-plan-area structures, and if temporary isolation joints in shear walls are not practical, arrange the walls so they do not restrain the normal volumetric changes accompanying post-tensioning, temperature changes, shrinkage, and creep. Remember that these effects are more significant in parking structures than in other structures. Ideally, locate the walls at or near the center of rigidity of the structure, whether in the interior

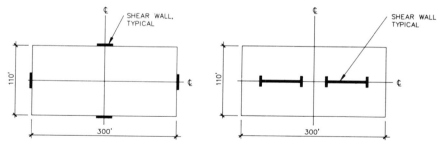

Figure 5-12.   Example of shear wall layout. Keeping shear walls at the exterior
simplifies traffic planning, but may adversely affect structural behavior.

Figure 5-13.   Example of shear wall layout.
Placing shear walls near the center of mass
will simplify structural design, but may adversely affect traffic planning.

or on the perimeter. Interior shear walls may form hiding places—large
formed holes will improve passive security. Coordinate wall location
with isolation joints. Figures 5-12 through 5-14 show example arrangements.

5.3.3.2   Truss Action. Frame action necessary to resist lateral forces
may be lessened by the presence of structurally integral ramps connecting consecutive floors. This same approach may be used with continuously sloping floors. In some configurations you may achieve truss
action by taking the ramps into account, perhaps in one direction only,
but carefully analyze the effect of lateral displacements on intercon-
necting elements (Figure 5-15).

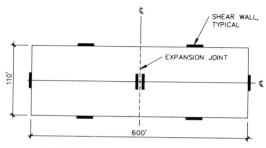

NOTE: PLAN REPRESENTS A 2–BAY STRUCTURE

Figure 5-14.   Example of shear wall layout. For a larger structure, some combination of
Figures 5-12 and 5-13 arrangements may be the best solution.

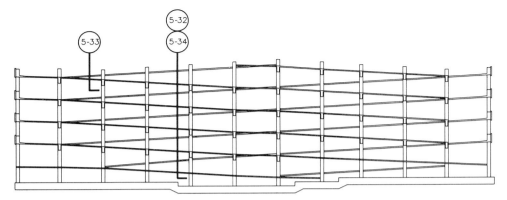

Figure 5-15.   Simplified section showing whole structure as a truss.

5.3.4   Superstructure Systems

Superstructure systems commonly used for parking structures may be classified into the following groups:

Cast-in-place (CIP) concrete
Precast (P/C) concrete
Structural steel

When considering any superstructure system, remember that parking-structure design and construction demands more attention to durability design than is the case for structures protected from the weather.

5.3.4.1   Cast-in-Place Concrete. Cast-in-place concrete structures are typically rigidly framed with monolithically cast slab-to-beam-to-column connections.

5.3.4.1.1   Post-Tensioned CIP Concrete. Figures 5-16 and 5-17 show plan and section views of a typical post-tensioned one-way slab, post-tensioned beam, and conventionally reinforced column-framing system. Post-tensioning a member reduces its size for a given span. The more economical member size produces a smaller total structural weight, reducing moments. Negative moments and associated cracking are further reduced by post-tensioning-induced compressive stresses. This results in a reduction in the exposure of steel reinforcement. *Despite* this advantage, maintain proper concrete cover.

Though post-tensioning will generally reduce cracking, it is not

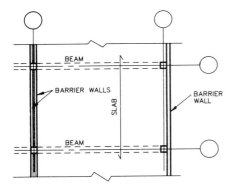

Figure 5-16. Example of post-tensioned system—plan.

necessary, or even desirable, to design post-tensioning to carry all gravity loads. Instead, use it only to reduce the tensile stresses at strategic locations. One concept uses post-tensioning to offset a percentage of the dead load stresses. Too much post-tensioning will increase both elastic shortening and creep-associated problems. Even at the lowest effective level of post-tensioning, these factors are important in design.

Post-tensioning systems contain areas where care must be taken to preserve system durability, especially where reinforcement is near the top surface of the concrete.

Through design, control volume change, creep deflections, and initial camber. Check the initial member camber, elastic deflection, and creep deflection so the final floor profile is consistent with the deck drainage system. Provide adequate concrete cover for reinforcement protection and fireproofing.

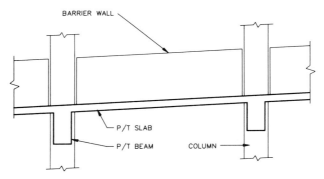

Figure 5-17. Example of post-tensioned system—section.

5.3.4.1.2   Conventionally Reinforced CIP Concrete. Flexural members are typically deeper when designed in conventionally reinforced concrete than they are for post-tensioned concrete members of equal span and load. Because of larger member size and monolithic construction, performance under vehicle-induced vibrations is generally good. Increased member weight leads to increased reinforcement, formwork, and foundation costs. Increased structural depth may lead to taller structures. This system's larger deflections may affect drainage design. We would expect this system to be used with short-span construction as part of a multi-use structure, such as basement parking in an office building.

Analyze for allowable crack width control at negative moment locations, such as beam-column joints. Water penetration at these spots can bring on corrosion. Provide proper rebar cover everywhere. Provide sealant details at construction joints.

5.3.4.2   Precast Concrete. Figures 5-18 through 5-21 show typical precast double-tee framing system plans and sections. For parking structures, precast concrete has two primary advantages; concrete quality is generally good. The speed of assembly reduces on-site construction

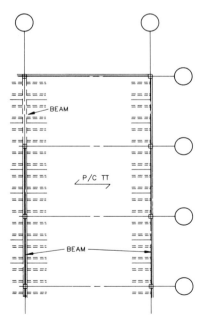

Figure 5-18.   Example of precast system—plan.

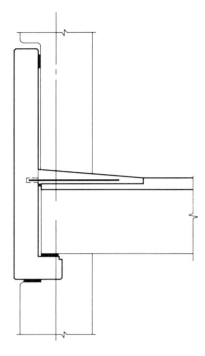

Figure 5-19. Example of precast system—section at exterior.

time and cost. Disadvantages are the installation and maintenance of the large number of connections and sealed joints.

Plant-fabricated members are manufactured with close dimensional tolerances. Embedded item location control is usually critical for achieving quick erection. Coordinate drains and openings to ensure a properly detailed structure. When locating drains, account for member deflections and cambers. Provide for adequate member clearances in design. Pay particular attention to casting and assembly tolerances.

Prohibit forcing units into position during erection; such erection stresses can cause local failures.

Detailing member connections is critical. Review special connections with a local fabricator for ease of construction. Connections are often exposed to water penetration through cracked topping and leaking joints. To prevent cracking and leaking, use properly selected materials. If connections are to be concealed in pockets which are concreted after the connections are made, use a non-shrink nonmetallic grout mix to fill the pockets. Also consider adding fiber reinforcement to the fill mix. Coat the pocket with a bonding agent before filling. Tool a recess around the joint and fill it with sealant to keep water out

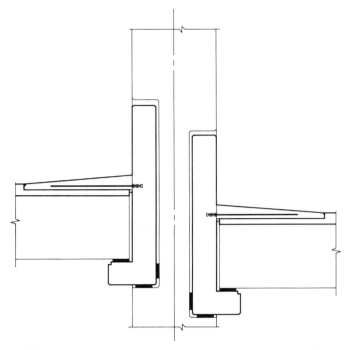

Figure 5-20.   Example of precast system—section through beam at interior.

of the joint between pocket and pocket fill. Consider epoxy-coated or even stainless-steel connections and anchors to reduce metal corrosion and corrosion-caused concrete spalling. Where field welding is required, detail to allow heat-caused expansion of the metal embedments without cracking the adjacent concrete. Use field-applied coatings to protect welds and ferrous metals after welding has been done. Connections need not be concealed if detailed to be clean and simple. Exposed connections, suitably coated against corrosion, are not objectionable; in fact, it may be better to have visible connections which are more easily monitored and maintained.

Connections may become complicated in areas of seismic risk where continuity is essential. Even in areas of low seismic risk, volumetric and temperature changes in large structures may induce large forces and moments in connections.

5.3.4.2.1   Conventionally Reinforced Precast Concrete. Precast concrete with conventional reinforcement is sometimes used for architectural spandrel beams and for shorter span structural members.

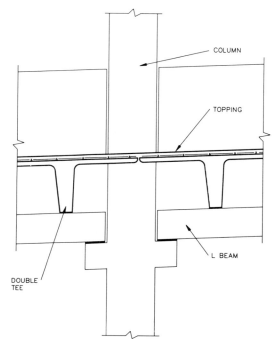

Figure 5-21. Example of precast system—section through double tees at interior.

Structural columns, stair units, walls, short-span slabs, and short ramps may be designed in conventionally reinforced precast concrete. Some precasters do prefer pretensioning these members for ease of handling, so check local precaster preferences to save yourself the trouble of checking a design different from your original. See the discussion for conventionally reinforced CIP concrete.

5.3.4.2.2 Pretensioned Precast Concrete. Properly used, pretensioning provides strength, deflection and crack control, and shallower slabs, joists, and beams. Pretensioning may also be used simply to protect against damage during transportation and erection.

Proper pretensioning effectively closes service-load cracks, reducing water penetration. Pretensioned concrete units have already undergone full elastic shortening before erection, so further elastic shortening may be neglected. However, do not neglect the effects of long-term creep of pretensioned members after erection. Much of the discussion for post-tensioned CIP concrete, as it relates to crack control and stressing levels, applies here also.

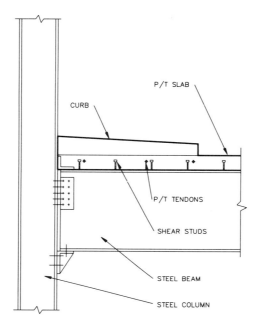

Figure 5-22. Example of structural steel system—section.

5.3.4.2.3 *Post-Tensioned Precast Concrete.* Coupling together an assembly of precast units with post-tensioning may be desirable. Take care to ensure that such stressing does not cause unacceptable changes in the geometry of the structure, either at the time of stressing or in the long term. Individual members also may be post-tensioned. A common scheme is to pretension a large member for self, dead, and construction loads, and to post-tension it for live loads.

5.3.4.3 Structural Steel. Figure 5-22 shows in section a structural steel framing system for parking garages. The plan view would be similar to that for the post-tensioned layout previously shown in Figure 5-16. Structural steel framing has been combined successfully with cast-in-place or precast concrete floors. Corrugated metal deck forms, open-web steel joists, and certain weathering steel connections may not perform well in areas where road salt use is common, rainfall is high, or in coastal environments. Protective coatings are necessary for these systems in such areas; local building codes may require fireproofing.

In many areas, our experience has been that a life-cycle analysis will show that first cost and maintenance cost of a structural steel frame are more expensive than those of a corresponding concrete frame. Consequently, of the over 600 parking structure projects we have designed and which have been carried through to construction, no more than six have had structural steel frames. Reasons for using structural steel might be: the need for as light weight a structure as possible to permit economical foundations despite poor soil, a shortage of skilled concrete tradesmen in the project locale, and local preference.

5.3.5 Structural Components

5.3.5.1 Slabs

5.3.5.1.1 *Thin-Slab CIP Systems.* Thin-slab systems, such as waffle slabs and pan joists usually require less concrete than one-way-slab designs and so may initially appear economically attractive.

These systems use thin slabs, usually 3–4 in. thick, stiffened by a stem-web pattern underneath. The main advantage of these systems is that large spans are achieved with relatively light structural weight.

Waffle slabs and pan joists present specific problems with cracking and reinforcement protection. Small-scale cracking can be expected at the slab periphery due to variations in curing rates, and shrinkage between slabs and joists because of the differences in volume/area ratios. These characteristics lead to stress cracking which might not otherwise occur. In any case, the cracks will accelerate the weathering process when water is present. In a thin slab, it is more likely that cracks will fully penetrate the slab, permitting water to reach both the top and bottom reinforcement and will cause objectionable leaching on the slab underside. These problems may be lessened by careful control of the construction process, particularly curing, and by attention to the arrangement, placement, and protection of the reinforcement. A traffic-bearing protective membrane and epoxy-coated reinforcement should be required protection. Crack control by tooled and sealed-control joints may reduce the section to an unacceptable degree and is not generally practical for thin slabs. However, do provide tooled and sealed joints at every construction joint.

Another system which shares some thin-slab characteristics is that which incorporates cast-in voids to form a floor structure which resembles a thick hollow-core unit. A composite system for which the above statements are also true is one which incorporates precast pretensioned joists spaced at about 7'6" on centers and spanning 40–68 ft.

Hollow-core units with topping share thin-slab characteristics. They are also vulnerable to slab deflections and shear stress failure. In

design, reduce excessive elastic and creep deflections to prevent ponding and poor drainage. Call for weep holes in the downslope ends to
permit drainage.

We have seldom used any of the above-described systems, but
have seen problems with them in our restoration practice. We do not
recommend their use other than in special circumstances. Watch out
for any structural system with thin elements and vulnerable top reinforcement.

One- and two-way slab systems will generally have high tensile
stresses at slab-top fibers at supports. Cracking is likely to be more
visible at these locations, and may penetrate the entire slab. The top
reinforcement in these systems is also vulnerable in any event and will
require protection against corrosion. Deflections are a concern in the
design of longer spans. Careful control of camber and attention to live-
load deflection are essential.

In general, most structural systems can be made to perform adequately in parking structures, if sufficient effort is made in both design
and construction. As discussed above, however, some systems are more
suited for parking structures, while others are better avoided.

5.3.5.1.2 Precast Concrete. Parking structure floors may be made of
solid or hollow-core units. Tees of all kinds, single, double, triple, and
quad, depending on which forms local precasters own have been used.
Hollow-core units and tees may be made composite with cast-in-place
topping or may not require topping. The latter, referred to as "pretopped" or sometimes as "factory topped," have become more common in recent years. See Figures 5-23 and 5-24 for illustrations of
untopped and pretopped double tees.

In both topped and pretopped floors, welded connections between members are required to provide deflection compatibility between adjacent members and to help transfer shear across slab
diaphragms.

If the floor members are topped, there is a change in structural
cross section at the joints between adjacent members. Though the composite thickness of the 2-inch-thick tee flange and the 3-inch topping
is 5 inches, at the joint between flanges only the 3-inch-thick topping
exists. This thinner section comprises a weakened plane which will
crack as the topping concrete cures and shrinks. To make a more durable floor structure, it is important to control cracking and minimize
leaks by specifying tooled and sealed control joints in the topping at
every joint between precast members. Joint depth should be 20–25%
of the concrete thickness to be effective (see Figure 5-25).

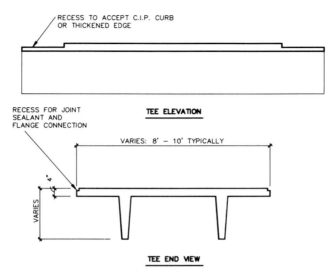

Figure 5-23.   Pretopped double tee.

Avoid sawcutting control joints. If sawing is done too early, the edges of the cut will ravel, leaving uneven edges which are difficult to seal and which may be objectionable in appearance. If the sawing is done too late, the concrete already will have cracked. The sawcut

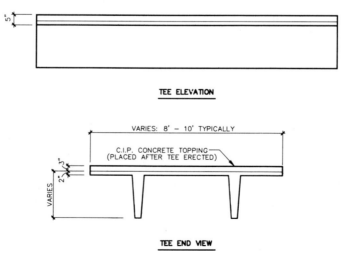

Figure 5-24.   Topped double tee.

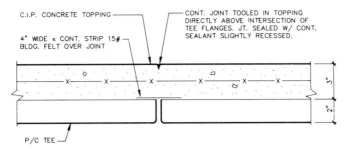

C.I.P. CONCRETE TOPPING

4" WIDE x CONT. STRIP 15#
BLDG. FELT OVER JOINT

CONT. JOINT TOOLED IN TOPPING
DIRECTLY ABOVE INTERSECTION OF
TEE FLANGES. JT. SEALED W/ CONT.
SEALANT SLIGHTLY RECESSED.

P/C TEE

Figure 5-25.   Tooled and sealed joint in topping.

leaves a right-angle edge which is weaker than the rounded edge left by tooling the joint. Even if the sawcut joint is initially sealed successfully, which, because of its narrow width is not easy to do, the edge is likely to crack, rendering the joint sealant useless.

Concrete topping for precast floors is often the last concrete placed. Quality of placing, finishing, and curing may suffer because people are hurrying to complete the job. Also, because the topping usually varies in depth from three inches typically to five inches at a thickened edge at the floor perimeter, one bad truckload means poor topping over a large area.

5.3.5.1.3 *Structural Steel.*

Floors in steel-framed parking structures may be cast-in-place on conventionally reinforced or post-tensioned concrete on composite steel decking, or may be cast on a corrugated metal form. The Steel Deck Institute, however, recommends that steel deck be used only as the form for the concrete slab, not as a composite reinforcement. For corrosive environments, the Institute recommends a traffic-bearing membrane for the concrete and vented decking.

Concrete topping placed on corrugated steel often cracks due to shrinkage. Control this cracking with properly spaced, tooled and sealed construction and control joints. Also use welded-wire fabric and fiber reinforcement. Do not expect post-tensioned temperature reinforcement to be effective in the direction parallel to the steel beams.

5.3.5.2 Beams and Joists.

Cast-in-place beams are usually formed in a T or an L shape, the latter occurring at spandrel beams which may have the vertical leg of the L either upturned or downturned.

Precast members may have many shapes, such as rectangular, trapezoidal, tees, inverted tees, L beams, I beams, etc. (see Figure 5-26), and may be conventionally reinforced, pretensioned, post-tensioned,

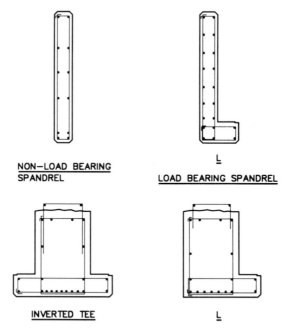

NON—LOAD BEARING
SPANDREL

LOAD BEARING SPANDREL
L

INVERTED TEE

L

Figure 5-26.   Commonly available precast beam sections.

or some combination of two or all three. Precast members are often
made composite with a cast-in-place topping. (For a discussion of torsion in beams, see Section 5.5.3.)

If a perimeter curb is cast perpendicular to and atop a cast-in-
place T beam, it is likely that the curb will crack parallel to the sides
of the beam below. We have never seen problems with reflective cracking over the beam with the slab alone, or with a thickened edge, but
a 6-in. high curb will crack. Prevent these cracks by tooling and sealing
a control joint in the curb directly above and parallel to each beam
face below.

Steel studs are often used with structural steel members to develop composite action between the member and a cast-in-place slab.
Beams and girders may be castellated or cover plated. Joists may be
the open-web type, but we do not recommend them.

5.3.6   Summary of Structural System Selection

As you will see from the discussion above, some structural systems
are more suitable for parking structures than are others. Some systems

will work just as well, but only if additional protection is provided. That additional protection, though, may make the selected system uneconomical.

To help you select a system for your project, see Table 5-2. For every system listed in Table 5-2, consider it a given that the concrete quality is good—a low water-cement ratio, proper air entrainment, proper curing and finishing, and that control joints are properly located, tooled, and sealed.

5.4   VOLUME-CHANGE EFFECTS

Volumetric changes affect frame action in structures, especially those having a large plan area. The results can include the development of high shears and bending moments in the first-story frames at or near the building periphery.

5.4.1   General

Volume change is the change in dimensions of the structural elements due to drying shrinkage, temperature change, elastic shortening, and horizontal creep. The strains and forces resulting from structural restraints have important effects on connections, service load behavior, and ultimate-load capacity. Consider these strains and forces in design. The restraint of volume changes in moment-resisting frames causes tension in the beams and slabs, and moments and deflections in the beams and columns.

5.4.2   Drying Shrinkage

Drying shrinkage is the decrease in concrete volume with time. This decrease is due to changes in the concrete's moisture content and chemistry. These changes are unrelated to externally applied loads. When the concrete shrinkage is restrained enough for shrinking, cracking will provide relief at weak points. For proper durability and serviceability, predict and compensate for drying shrinkage, which is likely to be significant in open parking structures (see ACI 209R for recommended shrinkage values).

5.4.3   Elastic Shortening

In prestressed concrete, axial compressive forces applied to the concrete by prestressing tendons cause elastic shortening. This shortening

TABLE 5-2 Summary of Structural System Selection for Parking Structures

System Type	Minimum Surface Protection for Corrosive Environments	Minimum Internal Protection for Corrosive Environments	Long Clear Spans	Recommend
Cast-in-Place Concrete				
Post-tensioned one-way slab and beam	Sealer	Epoxy-coated rebar in top 3 in. and fully encapsulated P/T	Yes	Yes
Two-way slab	Sealer	As above	No	Depends[1]
Conventionally reinforced waffle slab	Membrane[2]	Epoxy-coated rebar in top 3 in.	No	No
Pan joist	Membrane[2]	As above	No	No
One-way slab and beam	Membrane[2]	As above	No	No
Two-way slab	Membrane[2]	As above	No	No
Precast Concrete				
Prestressed long-span tees topped	Sealer or microsilica topping[3]	Epoxy-coated WWF	Yes	Yes
pretopped	Sealer	As above	Yes	Yes
Short-span tees	Sealer	As above	Can be[4]	No
Prestressed hollow core units with topping	Sealer or microsilica topping[2]	As above	Can be[4]	No
Conventionally reinforced floor elements	Membrane[2]	Epoxy-coated rebar in top 3 in.	No	No

Composite CIP and PC				
Permanent PC form with CIP topping and cast-in voids	Membrane[2]	As above	Can be[4]	No
PC joists with CIP slab	If slab P/T: sealer; if not: membrane[2]	As above	Yes	Depends[1]
PC spread single tees with CIP P/T slab	Sealer	As above	Yes	Yes, but double tees cheaper
PC rectangular beams with CIP P/T slabs	Sealer	As above	Yes	Yes
Structural Steel				
Composite with CIP P/T slabs	Sealer	As above	Can be[4]	Depends[1]
Composite with topped PC tees	Sealer or microsilica topping[3]	Epoxy-coated WWF	Can be[4]	Depends[1]
Composite with topped PC HCU	As above	As above	Can be[4]	Depends[1]

[1]Depends on many circumstances for specific project; cannot make blanket recommendation.
[2]Traffic-bearing membrane
[3]Need sealer on microsilica topping for first year.
[4]Can be either long or short span; for more information, see Chapter 6.

causes some loss of prestressing force which must be accounted for in determining the final effective prestress force. Elastic shortening is additive to drying shrinkage. The Prestressed Concrete Institute and Post-Tensioned Concrete Institute Design Handbooks provide recommendations for predicting elastic shortening and all the types of volume change described in this section.

5.4.4   Creep

Creep is the time-dependent change in dimension of hardened concrete subjected to sustained loads. Concrete continues to deform inelastically over time under sustained loads. Its total magnitude may be several times larger than short-term elastic shortening. Frequently, creep is associated with shrinkage, since both occur simultaneously and provide the same net effect—increased deformation over time.

5.4.5   Temperature

A change in temperature will cause a volume change which will typically affect the entire structure. In addition, sun light will affect local areas such as the roof and edges of lower levels more than it will the rest of the structure. The change can be expansion or contraction, so may be additive or subtractive to the above-discussed volume changes. Unlike drying shrinkage, elastic shortening, and creep, temperature changes are cyclic. The resulting expansion and contraction occur in both daily and seasonal cycles. The structural movements resulting from temperature changes must be a major design consideration.

5.4.6   Comparison of Superstructure Systems

Table 5-3 compares the volume-change characteristics of the major systems covered here. Characteristics exhibited by a structural steel-framed system would be similar to those for the type of floor construction used—conventionally reinforced cast-in-place, precast prestressed, or a cast-in-place post-tensioned concrete.

5.4.7   Control Measures

5.4.7.1   Overall Structure. The degree of fixity of a column base has a significant effect on the size of the forces and moments caused by volume-change restraint. The assumption of a fully-fixed column base in the analysis of the structure may result in significant overestimation

TABLE 5-3   Effect of Volume Changes on Parking Structures

Volume-Change Type in Floor Elements	Construction Type		
	Cast-in-Place Conventionally Reinforced Concrete	Precast Pretensioned Concrete	Cast-in-Place Post-Tensioned Concrete
Elastic shortening	None	None	Significant
Shrinkage	Significant	Some	Significant
Creep	Some	Some	Significant
Temperature	Some	Some	Some

of the restraint forces. Assuming a pinned-column base may have an opposite effect. The degree of fixity used in the volume-change analysis should be consistent with that used in the analysis of the column loads, the determination of column slenderness, and construction document details.

A change in the center of rigidity or column stiffnesses will change the restraint forces, moments, and deflections. The areas of a structure to be treated with extra care for volume change are:

a. Any level with direct exposure to the sun and the columns directly below that level
b. The first supported level and the columns directly below it
c. the southern (in the northern hemisphere) and western faces

Creep- and drying-shrinkage effects take place gradually. The effect of the shortening on shears and moments at a support is lessened because of both creep and micro-cracking of the member and its support. These volume-change shortenings can be provided for by using the concept of equivalent shortening in the *Prestressed Concrete Institute Design Handbook.*

5.4.7.2   Design Measures. In dealing with the volume-change forces, consider:

a. Parking structures have large plan areas. This characteristic will result in significant secondary stresses due to temperature change, shrinkage, and creep. Place isolation joints to permit separate segments of the structural frame to expand and contract without adversely affecting the structure's integrity or serviceability. Table 5-4 gives some *general* guidance in spacing permanent and temporary isolation joints in a structure. Dividing the structure into smaller areas with isolation joints

TABLE 5-4   Example Guidelines for Isolation-Joint Spacing

Number of Isolation Joints	Cast-in-Place Post-Tensioned Concrete	Precast Concrete Prestressed in Direction of Consideration	Precast Concrete No Prestress in Direction of Consideration
0	200 ft maximum (no pour strip) 275 ft maximum (one pour strip at center)	225 ft maximum	300 ft maximum
1 at center plus or minus	400 ft maximum (no additional pour strips) 550 ft maximum (two additional pour strips at $\frac{1}{4}$ points)	450 ft maximum	600 ft maximum
2 at $\frac{1}{3}$ points plus or minus	600 ft maximum (no additional pour strips) 825 ft maximum (three additional pour strips at $\frac{1}{6}$ points)	675 ft maximum	900 ft maximum

may be complicated by the presence of inter-floor-connecting ramps.

b. Isolating the structural frame from stiff elements—walls, elevator cores, stair cores. Not allowing the superstructure to move freely with respect to the substructure is a trap for the unwary. Figures 5-27 and 5-28 show two details which will be helpful.

c. Reducing the rigidity of certain members or connections, using pinned (or partially pinned) connections at column foundations, or using longer unbraced column lengths. Unbraced column lengths may be increased without increasing structure height, particularly between the grade and first supported levels.

Leaving a sealed gap between slab on the grade and the column should result in an unbraced length beginning at the top of the foundation rather than at the top of the grade slab. This measure may also be combined with lowering the top of the foundation to further increase the unbraced length. Using only soft connections at the first supported level may double the unbraced length of precast columns.

To temporarily reduce column rigidity at any location

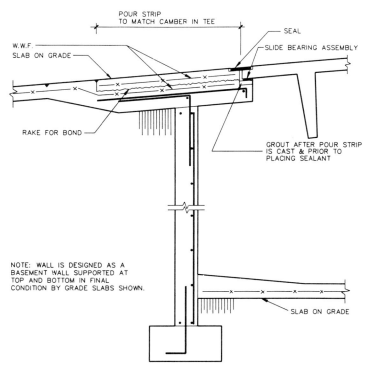

Figure 5-27.   Example of detail which permits relative movement between foundation and superstructure.

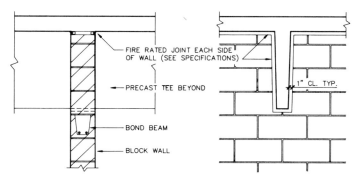

Figure 5-28.   Example of detail which permits relative movement between wall and superstructure.

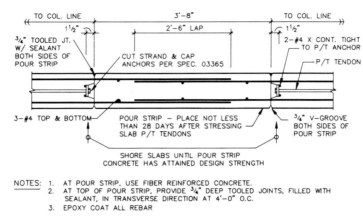

NOTES: 1. AT POUR STRIP, USE FIBER REINFORCED CONCRETE.
2. AT TOP OF POUR STRIP, PROVIDE ¾" DEEP TOOLED JOINTS, FILLED WITH SEALANT, IN TRANSVERSE DIRECTION AT 4'–0" O.C.
3. EPOXY COAT ALL REBAR

Figure 5-29.   Pour strip detail (temporary isolation joint).

consider a temporary hinge. Block out the column form, leaving a length of concrete supported only by the vertical reinforcement; later fill the gap. If you use this method, be sure to check the capacity of the vertical reinforcement alone versus the column loads expected before the gap is filled and full column capacity is restored.

d. Installing temporary open "pour-strip" joints which are closed before construction is complete (see Figure 5-29). These temporary contraction joints allow the dissipation of early-age volume-change effects, such as elastic shortening, temperature movement, and shrinkage, which occur before placing the pour strip. Give close attention to pour-strip concrete quality. Use tooled and sealed construction joints and protected reinforcement. Control joints transverse the pour-strip length at spacing equal to the strip width, and fibrous reinforcement may be necessary. In particularly hostile environments, consider a traffic-bearing coating overlapping the pour-strip-edge construction joints.

e. Frames with unequal column height and column stiffnesses, differential thermal response of members, and inelastic behavior cause further difficulties in predicting isolation-joint movement. Computer modeling of parking structure frames can be effective in predicting volume-change-induced moments, forces, and movements, provided that the model is not oversimplified.

5.5   PROBLEM AREAS

5.5.1   Beam-Column Joints

Columns in parking structures are often subjected to unusual forces compared with those in other buildings. Effects of the prestressing system, relatively high joint moments and shears associated with long spans, and effects of volume change all contribute to highly stressed joints.

Exterior columns and beams typically will have high joint moments, requiring special consideration of the anchorage of the beam-top reinforcement. In columns, the shear within the joint caused by the beam negative moment can exceed the shear capacity of the concrete alone. Ties are required within the joint. (Factors affecting the behavior of these joints are discussed in ACI 426 and 352.) Shear in the columns may require increased tie reinforcement throughout the column height. Where column vertical bars lap, development of those bars and the corresponding column-tie requirements both need evaluation. Typical post-tensioned beam-column details are shown in Figures 5-30 through 5-33.

In cast-in-place post-tensioned structures, shortening of the first-supported level beams due to elastic shortening, creep, and shrinkage will induce tension in the beam bottoms. Similar but lesser effects will occur at the upper levels.

Precast concrete beam-column joints require special attention. Joints in precast concrete structures are subjected to repeated movement due to cyclic volume change and vehicular traffic.

On the roof levels of parking structures of all types, the sun heats the top side of the structural system while the underside remains cooler. The result is a daily cycle of camber and deflection. Rotations and forces at member ends can distress both simple-span and rigid-frame construction. Special detailing may be necessary.

5.5.2   Variable-Height Columns

The typical method for accessing the successive levels of a multilevel structure is via continuous sloping ramps. These ramps may comprise entire floors and may be used for both parking and through traffic (refer to Figures 5-15 and 5-34).

The presence of integrated interior or exterior ramps will have a significant effect on the behavior of the structure. Internal ramps interrupt the floor diaphragms and complicate their analysis. High mo-

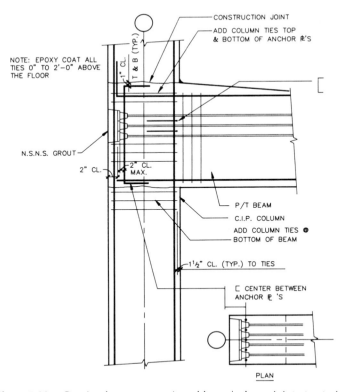

Figure 5-30.   Cast-in-place post-tensioned beam/column joint at exterior.

ments and shears are induced in columns adjacent to ramps where monolithic beams enter opposite sides of the columns at varying elevations (Figure 5-34).

5.5.3   Torsion

Avoid torsion if you can. Spandrel beams at slab edges, built integrally with the floor slab, are subjected not only to transverse loads, but to a torsional moment per unit length, equal to the restraining moment at the slab's edge. ACI 318-83, Chapter 11, addresses design requirements with respect to torsion requirements in combination with shear and bending for non-prestressed members. Typically, monolithically cast spandrel beams, whether prestressed or not, are easily reinforced to

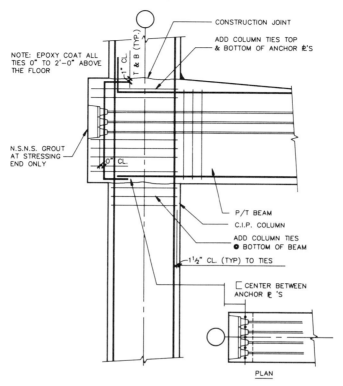

Figure 5-31. Cast-in-place post-tensioned beam/column joint at exterior. This detail alleviates reinforcement congestion at the post-tensioning anchorages by moving them outside the column reinforcement. It is also suited for use when the column cross section is not rectangular.

meet minimum code requirements, but design them to minimize cracking.

Precast spandrel beam design is one of the most complex elements in parking structures. Figure 5-35 shows the major loads on one such typical beam. Industry practices and published design procedures vary in fundamental aspects. ACI 318 does not address combined shear and torsion in prestressed beams, although PCI Research and Development Project No. 5 does address a number of issues concerning the behavior and design of precast spandrel beams. Pocketed beams, such as those shown in Figure 5-36, are one way to reduce torsion in spandrel beams. PCI R&D Project No. 5 considers both L and pocketed beams.

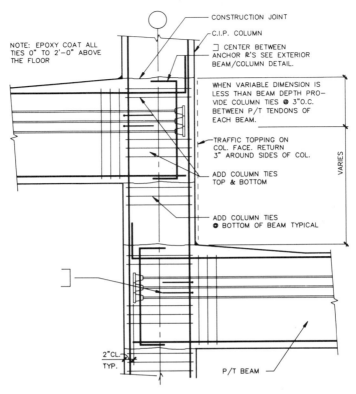

NOTE: EPOXY COAT ALL
TIES 0" TO 2'-0" ABOVE
THE FLOOR

CONSTRUCTION JOINT

C.I.P. COLUMN

⊐ CENTER BETWEEN
ANCHOR ℝ'S SEE EXTERIOR
BEAM/COLUMN DETAIL.

WHEN VARIABLE DIMENSION IS
LESS THAN BEAM DEPTH PRO-
VIDE COLUMN TIES @ 3"O.C.
BETWEEN P/T TENDONS OF
EACH BEAM.

TRAFFIC TOPPING ON
COL. FACE. RETURN
3" AROUND SIDES OF COL.

ADD COLUMN TIES
TOP & BOTTOM

ADD COLUMN TIES
@ BOTTOM OF BEAM TYPICAL

VARIES

2"CL.
TYP.

P/T BEAM

Figure 5-32. Cast-in-place post-tensioned beam/column joint at interior. Ramped floors result in beam/column relationships like this and the one shown in Figure 5-23.

5.5.4 Stair and Elevator Shafts

Shafts interrupt the regular pattern of framing (Figures 5-37 and 5-38), and may cause differential deflections in the adjacent structure, causing localized cracking. For instance, one beam or tee may end at the wall of a shaft while the adjacent one continues. The effects of dead-load deflection may be minimized by prestressing; however, differential deflections due to live-load will occur both in beams and in the connecting floor structure. Local differential cambers and movements parallel to the members may also be a problem. These may be handled with a reinforced, cast-in-place topping, or by accepting the movements and installing an isolation joint between the two members.

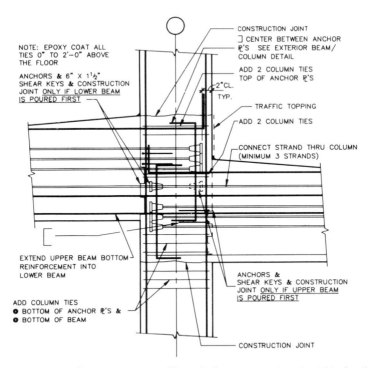

NOTE: EPOXY COAT ALL
TIES 0" TO 2'-0" ABOVE
THE FLOOR

ANCHORS & 6" X 1½"
SHEAR KEYS & CONSTRUCTION
JOINT ONLY IF LOWER BEAM
IS POURED FIRST

CONSTRUCTION JOINT
◻ CENTER BETWEEN ANCHOR
℞'S   SEE EXTERIOR BEAM/
COLUMN DETAIL

ADD 2 COLUMN TIES
TOP OF ANCHOR ℞'S

2"CL.
TYP.

TRAFFIC TOPPING

ADD 2 COLUMN TIES

CONNECT STRAND THRU COLUMN
(MINIMUM 3 STRANDS)

EXTEND UPPER BEAM BOTTOM
REINFORCEMENT INTO
LOWER BEAM

ANCHORS &
SHEAR KEYS & CONSTRUCTION
JOINT ONLY IF UPPER BEAM
IS POURED FIRST

ADD COLUMN TIES
● BOTTOM OF ANCHOR ℞'S &
● BOTTOM OF BEAM

CONSTRUCTION JOINT

Figure 5-33.   Cast-in-place post-tensioned beam/column joint at interior. This detail would apply at a point near where two ramped floors meet.

5.4.5   Expansion Joints

The best expansion (isolation) joint is one formed by a complete separation of one side of the joint from the other. Avoid sliding joints. (See Section 8.3 for more discussion.)

5.6   SPECIFIC LOAD CONDITIONS

5.6.1   Live Loads

Building codes commonly require a uniformly distributed load of 50 psf and a 2000-lb concentrated wheel load distributed over a 20-sq-in. area anywhere on a floor, with additional load for snow (see next section) on the top (roof) level. Most building codes allow for the reduc-

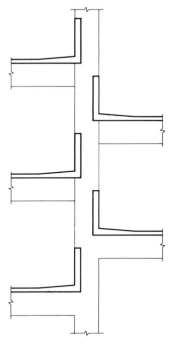

Figure 5-34. Section of ramped floors at center column line.

tion of live loads for members supporting tributary slab areas. In some parts of North America, roof-level parking requires combining parking live loads with roof snow loads.

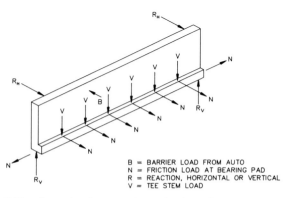

B = BARRIER LOAD FROM AUTO
N = FRICTION LOAD AT BEARING PAD
R = REACTION, HORIZONTAL OR VERTICAL
V = TEE STEM LOAD

Figure 5-35. Example of precast beam loads and reactions at exterior.

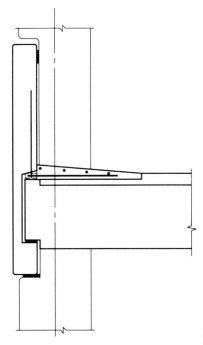

Figure 5-36. Example of precast pocketed beam.

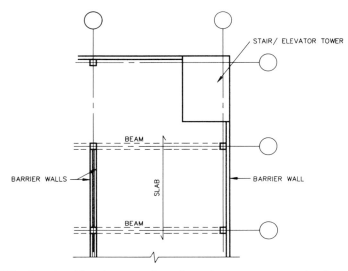

Figure 5-37. Structural framing around a stair or elevator tower—cast-in-place construction.

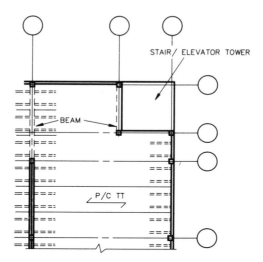

Figure 5-38. Structural framing around a stair or elevator tower—precast construction.

5.6.2   Snow/Live-Load Combination

Some building codes require adding roof loads (usually snow) in addition to the normal parking load. Some designers believe that for the design of principal members, this requirement is too restrictive. Combinations of snow and live loads which might be encountered in building codes are:

a. Full snow load over entire span with live load reduced by code
b. Full snow load over entire span with live load in stalls only
c. Full snow load over entire span with live load (reduced by code) in stalls only
d. Full snow load over entire span with auto load in stalls only (no impact load or reducible)

An appeal may be made in advance of construction to the local building department to reduce the requirements. The following is an example of option d above for combining snow and parking live loads to produce a realistic prediction of required capacity of a top level, which we recommend.

The option-d approach to combining live loads assumes autos in the parking stalls only. The autos, however, are considered immobile

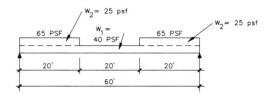

Figure 5-39. Snow and live load diagram.

and therefore no impact is considered (that is, 25 psf live load non-reducible). For this example, we will assume a building code snow load of 40 psf over the entire 60-ft simple span, and a 1-ft tributary width as shown in Figure 5-39.

$$\text{Service load moment} = w_1 1^2/8 + w_2 1_2/2$$

$$= (40 \times 60 \times 60)/8 + (25 \times 20 \times 20)/2$$

$$= 23{,}000 \text{ ft lbs}$$

$$\text{equivalent load} = (23{,}000 \times 8)/1^2$$

$$= (23{,}000 \times 8)/60^2 = 51.1 \text{ psf}$$

5.6.3 Wind Loads

Design and construct every parking structure and its component elements to resist the equivalent wind pressures given in governing building codes. Model building codes have methods with which to calculate wind pressures, using basic wind speed, importance factor, exposure factor, and projected areas. In most cases it is unrealistic to use anything less than the full building face area for the projected area. Do not subtract the open areas of the face; consider the face solid unless you make a more rigorous analysis.

5.6.4 Seismic Loads

ACI 318 Appendix A and the Structural Engineers Association of California's (SEAOC) "Recommended Lateral Force Requirements and Commentary," are excellent sources of information for use with the applicable building code.

High concentrations of diaphragm stress commonly exist at discontinuities in the slab, such as at the juncture of ramps with level areas of the supported slab. The presence of precast slab joints at such locations creates a high probability of cracking along the joint lines.

The presence of a ramp system zigzagging down the height of the building complicates the lateral force analysis. The ramp slabs, whether or not precast, must be able to carry their share of in-plane tension and compression.

5.6.5   Car Bumper Barrier Loads

Not all building codes deal with lateral load requirements for car bumper barriers at edges of floors. Requirements vary from none, to the National Parking Association's (NPA) recommendation given below, to Houston's 12,000 lb. Designing for these loads will restrain a slow-moving vehicle. These requirements are in addition to other building code requirements for handrails or similar barriers. The Parking Consultants' Council of the NPA recommends a factored concentrated lateral load (strength design load) of 10,000 lb at 18 in. above the driving surface. At least one code requires the load be applied 27 in. above the floor—50% more than the NPA requirement. Barrier-load-resisting reinforcement is additive to that required by other loads.

A typical curb, 6 in. high, or a precast wheel stop, will not stop other than a relatively slow-moving car, and should never be considered as a barrier. A faster-moving car will jump the curb or wheel stop and will hit the bumper barrier beyond with barely diminished force. Curbs do add to driver comfort with the facility, though. A curb of proper width will ensure that the car's rubber tires hit the curb before its chrome and steel bumpers hit the wall beyond.

Some building codes require a barrier with greater impact resistance at locations like the perimeter wall at the bottom of a ramp, especially if there are no parking places directly in front of that wall.

5.7   SUMMARY

In Chapter 5 we reviewed a number of factors which might influence the structural design of a parking structure, such as cost, schedule, and local building codes.

We also looked at items, such as the drainage of rainwater runoff and expansion joint sealants, which, while not an integral part of structural design, certainly must be taken into account if the structure is to work.

We considered many possible structural systems regarding the relative suitability of each for parking structures. We also noted which systems will need additional protection in corrosive environments.

Since parking structures are usually large enough and exposed enough to the elements for volume-change-induced forces to become significant in structural design and performance, we examined the effects of shrinkage, elastic shortening, creep, and temperature.

There are some areas of design which are unique to parking structures, such as variable height columns, which we have also discussed. Finally, we examined the types of loads encountered in parking structure design.

Chapter 6
Durability

ANTHONY P. CHREST

6.1 INTRODUCTION

You will have seen from Chapter 5 that with parking structures, structural design and durability design go hand-in-hand. One depends on and directly affects the other. For organizational purposes, this chapter will address construction material use which will improve the life of your structure.

Parking structures deteriorate at a rate more rapid than that of other building types, simply because more things attack them. Exposure conditions, even in climates like those found in southern North America, can be severe, and require proper protection.

The cost of protection systems varies widely. Some measures are almost free. Others are uneconomical except as a last resort, or because the project involves more than just the parking structure. For example, consider an office building with parking below. The structure of the parking area supports that of the office area above. If the parking structure deteriorates to the point where it is unsafe, the entire building would have to be condemned; this has already happened.

The discussion of why some concrete deteriorates and some does not appears in Chapter 9. There you will also find an explanation of why it is a good idea to keep road salt and water out of your concrete.

We will classify protection systems as internal or built-in and external or applied. With a few exceptions, internal measures are relatively inexpensive, while external systems are usually more expensive. In Sections 6.2 and 6.3, the alternatives presented range from least to most expensive. Use any system with proper care, or it will be a waste of money. Another caution—do not expect the impossible from a product; each one has limitations and none will work miracles.

For example, we were consulted on a project in New England

because the protective sealer was not working. A review of the construction records showed that the concrete quality was poor and that floor deflections exceeded accepted limits. The result was cracks in the floors. No sealer can bridge cracks in concrete. The cheap fix obviously desired was not successful, not because the sealer was of poor quality, but because of the ignorance of its capabilities.

As a final introductory note, when considering one element of durable design, do not forget to look at its impact on the total design. For example, increasing reinforcement cover may increase cracking, unless crack widths and tensile stress levels are checked and adjusted if required.

6.2 BUILT-IN PROTECTION SYSTEMS

6.2.1 Drainage

While not an ingredient of concrete, we must certainly build in proper drainage. If we do not allow chloride-carrying water to collect anywhere in a structure, we reduce considerably the opportunities for chlorides to attack the structure. Minimum pitch for drainage should be one percent, with two percent preferred. Size the drains to handle anticipated runoff volume. Include a sediment bucket (to reduce chances of clogging the piping) and corrosion protection.

Do not be misled by the brevity of treatment of drainage here. It is your first line of defense against corrosion. For more discussion of drainage in new structures, see Chapter 5. For remedial measures to drainage in an existing structure, see Chapter 10.

6.2.2 Concrete

Look at the basic ingredients of a good quality concrete first—cement, water, and aggregate. As basic as they are, these materials are not always selected properly, so take nothing for granted.

6.2.2.1 Cement. Any Type I cement conforming to ASTM C150 should be acceptable, provided that local experience records show that it will perform well. Also, depending on the project area or project requirements, special-use cements such as sulfate-resistant cement, high-early-strength cement, or quick-setting cement may be needed. Different brands of cement will produce different results, so it is wise to use just one brand throughout the project. Be more careful about cement selection when you are designing for 28-day yield strengths exceeding 7000 psi.

6.2.2.2 Water. Water must conform to ASTM C94, which basically calls for water of drinkable quality. Chloride content must be such that total chloride content of the mix does not exceed the limits given in the next section.

6.2.2.3 Aggregate. Both coarse and fine aggregate must conform to ASTM C33. For coarse aggregate, use crushed and graded limestone, or a local equivalent which you will accept. For fine aggregate, use natural sand having the preferred grading shown for normal-weight aggregate in ACI 302.1R, Table 4.2.1.

Aggregate gradation is important. Too many small particles will increase the water requirement, leading to a high water-cement ratio and consequent shrinkage cracking. There will be more on that later in this chapter. Gradation may have to be different for pumped concrete, rather than that placed by some other means (see ACI 304).

Another important but easily overlooked characteristic of an aggregate is its chloride ion content. Check this quantity through the laboratory making the trial concrete mixes. According to ACI 318, the total mix water-soluble chloride ion content, including all ingredients, should not exceed 0.06% chloride ions by weight of cement for prestressed concrete. The corresponding figure for reinforced concrete is 0.15%. Until ASTM comes up with a test procedure, use that given in Federal Highway Administration Report No. FHWA-RD-77-85, "Sampling and Testing for Chloride Ion in Concrete."

6.2.3 Additives

6.2.3.1 General. In this context, we will use the term "additives" to include compounds added in relatively small amounts to the ingredients already mentioned. In Section 6.2.4 we will address ingredients like fly ash and microsilica, which replace some part of the cement.

Prohibited additives are calcium chloride, thiocyanates or additives containing more than 1% chloride ions by weight of additive. At this writing there is some disagreement over whether or not thiocyanates are harmful, but our opinion is that we will continue to avoid them. Additionally, each additive must not contribute more than 5 ppm, by weight of chloride ions to total concrete ingredients. We arrived at these limits after discussion with additive manufacturers. Interestingly, we found that chloride percentages of additives, even from the same manufacturer, varied all over the map. Quality control of chloride ion content by the manufacturer needs improvement.

The engineer must approve all additives in writing before use. Use all additives according to the manufacturers' instructions. Prudent practice should require all additives to be furnished by the same manufacturer. Require a statement of additive compatibility from the manufacturer (see Section 6.2.4.3).

6.2.3.2 Air Entraining Agents (AEA). These materials must conform to ASTM C260. AEAs give the concrete resistance to freezing and thawing. Without an AEA, concrete in a cold climate will deteriorate quickly, as highway engineers discovered early in this century. Concrete mixed and placed using conventional practices will have 1–3% entrapped air. Entrapped air is air trapped in relatively large pockets distributed randomly in the mixture. Entrapped air does not help much with freeze-thaw resistance. Entrained air is distributed uniformly throughout the mixture in millions of tiny bubbles. The bubbles relieve the pressure generated by water becoming ice, thus preserving the concrete. To be effective in climates where freeze-thaw damage may occur, air entrainment should be in the 5–7% range in the placed mix for $\frac{3}{4}$-in. nominal size aggregate. (See ACI 301 for entrained air requirements for other aggregate sizes.) Transportation handling such as pumping, over-consolidation, and over-finishing will all reduce the air content. How do you know how much entrained air is in the hardened concrete? See Chapter 7. A fringe benefit of air entrainment is that it improves the workability of concrete. Air-entrained concrete segregates less easily and handles better than its non-air-entrained counterpart.

6.2.3.3 Water-Reducing Additives (WRDAs). In most areas use these additives to reduce the water content of a mix while retaining workability and slump. The result is a lower water-cement ratio, which leads to stronger and more durable concrete without adding cement.

Normal range WRDAs should pass ASTM C494, Type A. High range WRDAs, usually called superplasticizers, drastically reduce the water needed for a mix. Or, they dramatically increase slump, or both, depending on dosage amount and water volume. Dosage must be kept within the manufacturer's limits. High range WRDAs should meet ASTM C494, Type F or G. Use must not change the specified requirements for:

Maximum allowable water-cement ratios
Minimum allowable concrete strength

Minimum allowable air concrete
Minimum allowable cement content

6.2.3.4   Corrosion Inhibitors. Certain proprietary products containing calcium nitrite as their main ingredient slow corrosion of unprotected conventional steel reinforcement. Calcium nitrite reacts with ferrous ions to protect uncoated reinforcement. With continued addition of chloride ions from outside sources, the calcium nitrite supply ends and chloride ion corrosion starts. As with most other forms of protection described here, corrosion is delayed, not stopped. These products, while effective, are relatively more expensive than the measures listed above. Whether or not they will be cost-effective for your project is something only value engineering and life-cycle cost analysis will tell you.

6.2.4   Admixtures

6.2.4.1   General. The admixtures described here are used in super-structure concrete to replace some of the cement in the original mix. Except for fly ash, cost considerations tend to limit their use to certain elements within a parking structure. We do not use fly ash in post-tensioned concrete because we need high early strength. Though the eventual strength of fly ash concrete will exceed that of a comparable straight Portland cement concrete, the slower strength gain of the former makes it unsuitable for the post-tensioned application. Silica fume may be used in beam, column, and slab concrete and in slab toppings. Because of its relatively high cost, latex-modified concrete is commonly used only in slab toppings. Both silica fume concrete and latex-modified concrete require careful attention to proper finishing and curing procedures.

6.2.4.2   Fly Ash. According to ACI 116, fly ash is the finely divided residue resulting from the combustion of ground or powdered coal. It is a material with cementitious properties used as a partial replacement for cement in some mixes. When properly used, it will improve workability and final strength. It may improve impermeability; however, fly ash-rich concrete will gain strength more slowly than a comparable mix with none, even though final strength may be greater. For that reason, we do not specify fly ash in concrete to be post-tensioned. We do not permit it to replace more than 25% of the cement, by weight of cement in a mix on a pound-for-pound basis in other concrete.

Fly ash should meet ASTM C618, class C or F; test according to

ASTM C311. Use trial mixes to be sure that the proposed fly ash does not cause variation in specified strength or entrained air content by more than the specified tolerances. Choose the appropriate class, C or F, for the intended use. For example, sulphate resistance varies greatly from C to F. Finally, by our practice, fly ash carbon content should be 4% or less.

6.2.4.3   Silica Fume. Silica fume, also called microsilica, is another finely divided by-product of industry which has properties helpful to concrete. When added to a concrete mix in the right proportions, the silica particles fill some of the spaces between cement particles and react chemically with the cement. The result is a concrete with much improved strength, impermeability, and electrical resistivity. These properties make the concrete considerably more durable than a comparable mix without silica fume. Finishing and curing procedures require careful attention.

By the time the silica fume concrete is one year old, the impermeability of the concrete reaches a value comparable to that of concrete coated with a protective sealer. (See the section under Exterior Protection for more on sealers.) You need not seal silica fume concrete of a proper mix design after it is over a year old. Because it does take a year for the full impermeability to develop, initial protection will be needed for silica fume concrete. There then will be no first-cost savings for not sealing silica fume concrete, but maintenance costs will be lower. To reduce first cost, though, consider using a sealer with a shorter expected life such as linseed oil, instead of a more expensive, longer-lasting one that will be more than is needed. The first coat of sealer will be the last. Comparable concrete without silica fume will require sealer application every one to three years, depending upon traffic loads.

Proprietary silica fume products may contain a retarder. Require a statement from the manufacturer that his product is compatible—will not react deleteriously—with the other additives.

6.2.4.4   Latex. Latex is a water emulsion of synthetic rubber obtained by polymerization. It is used in place of water to produce latex-modified concrete (LMC). LMC shares many of the properties of silica fume concrete, though for different reasons, since strength does not improve. LMC is relatively more expensive than either ordinary Portland Cement concrete or silica fume concrete. For that reason, it is commonly used only as a topping for new or rehabilitation work.

Finishing can be troublesome because the fresh surface tears easily. Curing must also be done with care to prevent excessive surface cracking. Choose your contractor with care. Make sure your project is not his first LMC job.

6.2.4.5 Other Admixtures. There are a number of polymer admixtures, such as methyl methacrylate, which will produce improved concrete. Characteristics usually include improved impermeability, high early strength, and thin layer installation, as thin as $\frac{3}{4}$ in. in some products. These products are all relatively expensive and are used for repairs, not new structures.

6.2.4.6 Summary. Table 6-1 summarizes concrete ingredients.

6.2.5 Reinforcement

6.2.5.1 Cover. ACI 318 recommends increased cover over reinforcement in corrosive environments such as those found in parking structures. ACI is referring to reinforcement which is not epoxy coated.

TABLE 6-1 Concrete Ingredients

Ingredient	Reference	Requirement	Reason for Use
Cement	ASTM C 150	≥ 6 sacks	Strength
Water	ASTM C 94	w/c ≤ 0.40	Durability
Aggregate	ASTM C 33	Varies	Strength, durability, pumpability
Chlorides	FHWA-RD-77-85	≤ 0.06% for P/S, otherwise ≤ 0.15%	Durability
Thiocyanates		Zero	Durability
Entrained air	ASTM C 260	Varies with climate	Freeze-thaw resistance, workability
Water Reducers	ASTM C 494, type A	Varies	Durability, workability
Superplasticizers	ASTM C 494, type F or G	Slump ≤ 6″	Durability, workability
Accelerators	ASTM C 494, type C	Varies, but chloride free	Cold weather concreting
Retarders	ASTM C 494, type B	Varies	Hot weather concreting
Corrosion inhibitors		Varies	Durability
Fly ash	ASTM C 618	≤ 25%	Economy, handling, low heat
Silica fume	ACI 226	≤ 7%	Strength, durability, impermeability
Latex	ACI 548R	Varies	Impermeability

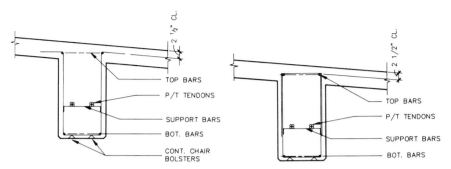

Figure 6-1. Cast-in-place post-tensioned beam cross section—open stirrups.

Figure 6-2. Cast-in-place post-tensioned beam cross section—closed stirrups.

Increase cover, over code requirements, to allow for cover loss through traffic abrasion. Our practice in floor structures is to specify a minimum of $2\frac{1}{2}$-in. cover over all reinforcement which is not epoxy coated. We prefer, however, to epoxy coat all reinforcement within 3 in. of the floor surface, and provide $1\frac{1}{2}$-in. cover, minimum, or 3 bar diameters, whichever is greater. Figures 6-1 and 6-2 show examples involving sloping floors. This discussion applies to floors in areas where pavements are salted.

6.2.5.2 Epoxy-Coated Reinforcement. This heading refers to welded wire fabric, conventional reinforcing bars, and high-strength prestressing strands. Welded wire fabric and reinforcing bars may be epoxy coated to reduce the opportunities for corrosion. You can epoxy coat the strand reinforcement used in prestressed concrete. Present practice in most places is to epoxy coat only the strand reinforcement used in pretensioned precast concrete when additional protection is needed, if fire protection of the epoxy coating is not a problem. Protect the strand reinforcement used in post-tensioned concrete with a coating of grease and a plastic sheath. Epoxy-coated strand reinforcement for post-tensioning use is available if exposure conditions warrant. See Chapter 7 for more information on post-tensioning strand protection.

In our opinion, galvanized reinforcement does not work long term in the corrosive environments commonly found in parking structures. Galvanizing is sacrificial protection; therefore, in a corrosive environment, it too corrodes. The corrosion products occupy more space than do the uncorroded galvanizing. The resultant pressure exerted on the surrounding concrete will spall or crack it. Laboratory and field test results conflict, but we do not specify galvanized reinforcement.

Protected reinforcement is worth the investment in climates where pavements are salted in winter. Note that testing has indicated that epoxy-coated conventional reinforcement needs longer anchorage and development lengths than uncoated reinforcement. We expect the ACI 318-89 code to reflect this fact. Epoxy-coated pre-tensioning strand reinforcement needs a grit embedded in the epoxy coating to develop bond.

6.2.5.3   Prestressed Reinforcement. Prestressing high strength steel reinforcement puts concrete structural elements into compression before it is loaded by its own weight and service loads. Prestressing a floor slab or beam will result in that member's remaining in compression or at a low level of tension under self and service loads. No or little tensile stress in the concrete means little cracking of the kind usually found in conventionally reinforced concrete.

In most precast concrete parking structures, the pretensioned precast elements are of simple-span construction, which means that the top fibers of those elements are always under compression. Further, the pretensioned reinforcement is below the centroid of the concrete section. In the case of a typical 24-in.-deep double tee, that reinforcement is likely to be protected by a foot or more of concrete cover above it. So long as there are no cracks in the precast concrete through which moisture can penetrate, the pretensioned reinforcement will be adequately protected from corrosion by the concrete alone. Also, since the entire length of the top fibers of the member are under compression, any cracks which do form will be kept closed by that compression.

In most cast-in-place post-tensioned concrete parking structures, the post-tensioned concrete elements are of continuous span construction, which means that the member top fibers will be in tension at supports and in compression at midspans. The post-tensioned reinforcement must therefore be close to the member top surface at the supports. Not only is the protective concrete cover for the reinforcement less at the supports, but if there is sufficiently high tensile stress in the negative moment area at a support, flexural cracks may open, providing a path for surface moisture from the floor to reach the post-tensioned reinforcement. It is necessary, then, to provide additional protection for the post-tensioned reinforcement.

The first layer of protection around the post-tensioned reinforcement strand is a coating of corrosion-inhibiting grease, which also lubricates the strand to reduce friction when the strand is tensioned. The second layer of protection is a continuous, seamless plastic sheathing extruded around the greased strand. The grease must completely fill the annular space between the strand and the sheathing. This sheathing keeps moisture away from the strand. At the ends of the strand, the anchor assemblies are completely encapsulated and the joint between strand sheathing and anchor encapsulation is made watertight.

Stressing pockets are very susceptible to water penetration. The joint between the pocket and the pocket fill material may form a path for water penetration; provide sealants or bonding agents at pockets. Also provide a sealed, tooled joint and epoxy-coated conventional reinforcement at each construction joint to minimize water penetration through slabs and to provide structural redundancy should tendon integrity be breached. For more information, see Chapter 7.

6.2.5.4 Fiber Reinforcement. Fiber reinforcement may be steel or plastic; we specify a polypropylene product. The fibers, when mixed uniformly throughout the concrete in the manufacturer's recommended proportions, improve its crack-resistance during finishing and curing. We use them for concrete where, because of location or place in the construction sequence, such cracks can be frequent. Examples are the cast-in-place pourstrips at the ends of pretopped double tees and at temporary isolation joints. These are long, narrow placements which tend to crack due to plastic shrinkage and drying shrinkage-induced stress, due to neglect of proper curing. Further, concrete at these locations is commonly placed late in the project when schedules are pressing. With the best of intentions, workmanship quality may drop. If the standard quality control procedures are not working, you may wish to minimize cracking at these places by using fiber-reinforced concrete. An evaporation retarder may also be effective, but neither measure will replace simply following proper placing and curing practices.

6.2.6 Construction Practices

6.2.6.1 Mixing, Transporting, and Placing Concrete. It is hard to know where to stop explaining when concrete is the subject; there are so many pitfalls for the unprepared. ACI has a number of publications which will help you guard against the many mistakes which will lower concrete quality. For a start, read ACI 304 through 306.1. Next, set up a preconcreting meeting with the owner, general contractor, ready-mixed concrete supplier, concrete pumping contractor if there is one, finishing contractor, forming contractor, testing agency, design professionals, and anyone else whose work will affect concrete quality.

Define procedures and agree on them. The outcome of this meeting (it may take more than one) should be a written procedure outlining every major step in the process. It is in these meetings where every forseeable problem must be examined and provided for, *not* when construction is underway. It must cover producing, delivering, placing, finishing, testing, and curing each type of concrete on your project. Everyone involved must sign off on the written procedure. There is more on these topics in later sections of this chapter and in Chapter 8.

6.2.6.2   Formwork for Concrete. Aside from suggesting that formwork be built tightly enough to prevent paste loss from the hardened concrete, which will result in honeycombing repairs, and that formwork be kept clean (cigarette butts, paper, and sawdust are not acceptable admixtures in concrete), we recommend ACI SP-4 on formwork; the subject matter is well covered there.

6.2.6.3   Consolidation. Properly done, consolidation reduces trapped air voids, improving the end-product concrete. Most of the time consolidation is done with vibrating screeds for thin slabs and overlays. Internal vibrators are used for thicker slabs and other members. Too little vibration will produce voids. The visible voids cost money to repair. The hidden voids will reduce strength and durability. Too much vibration will drive entrained air out of the plastic concrete and will bring too much paste to the surface. Both these effects will make the concrete less durable, though the loss of entrained air in the final product will be important only in colder climates where freeze-thaw damage is a concern. Premature surface deterioration because of too much cement paste at the floor surface is a concern in any climate. Good inspection will help prevent over vibration. With the help of ACI 309, education of the people running the vibrators will be more thorough.

6.2.6.4   Finishing. Finishing is another operation which can help or hurt the final product. As with many things, it is easier to do wrong than right. In the mistaken belief that more is better, finishers will overwork the concrete surface, bringing paste and fines to the top and driving out entrained air. The result is a weakened surface susceptible to scaling.

The best finish for floors with auto and pedestrian traffic consists of these operations—screed to the specified elevations and profiles. We recommend a vibrating screed for slabs to reduce the internal vibration needed. Wait until bleed water has evaporated or remove it. Bullfloat, then final finish with a light broom perpendicular to the direction of traffic. The term, "light broom finish," in this context means a textured finish achieved with a soft broom where the amplitude between adjacent grooves and ridges is between $\frac{1}{32}$ and $\frac{1}{16}$ in. The surface must not be too smooth. A typical sidewalk finish is too smooth. If the finish is too rough, it will be a tripping hazard. Be careful not to drag aggregate from the surface.

If a traffic-bearing membrane is to be installed on the slab, be sure to check with the membrane manufacturer to obtain his finishing requirements; then specify them. Include specification wording to the effect that the finish shall conform to the membrane manufacturer's requirements wherever that membrane is to be installed.

Some finishers like to add water to the surface as they work. This practice produces a thin surface layer which is weak and permeable. Again, inspection will help prevent this practice. Education via the preconcrete meeting mentioned earlier is better and will work even when the inspector is not around. Keep the same finishing foreman throughout the project to maintain a consistent quality of finish.

6.2.6.5   Curing. Curing is the important last step in the concreting process, but is sometimes ignored or done carelessly. Using a curing compound is often the preferred method. Evidence shows that curing compounds do not perform well on rough surfaces like the light broom finish discussed above. We prefer to specify a wet cure; it is proven to produce the best results. The cost is higher than other methods, but the results are worth it.

Pay special attention to ACI 305 and 306.1 for concreting in hot and cold weather, respectively. Some practices to avoid should be well known, but every so often they appear again, so we will mention them. In cold weather, heaters may be needed to maintain proper curing temperature in the concrete. Be sure to vent the combustion products of the heaters to the outside air, not to the concrete surface. If the combustion products come in contact with curing concrete, carbonation occurs. Carbonation produces dusting of the surface which is impossible to stop once it starts. As cold weather approaches in northern climates, do not seal concrete; it will not continue to lose moisture as it should and damage will result when the water in the concrete freezes. Also, sealers are not effective on high-moisture concrete.

In hot weather, especially on windy days, plastic shrinkage cracking is a real danger. Windbreaks, fogging, and application of an evaporation retarder will reduce the chances of it occurring.

6.2.6.6   Summary. Table 6-2 summarizes construction practices.

TABLE 6-2 Construction Practices

Activity	References
Mixing	ACI 304
Transporting	ACI 304
Placing	ACI 304, ACI 304.2R
Formwork	ACI 347, ACI SP-4
Consolidation	ACI 309, ACI 309.1R, ACI 309.2R
Finishing	ACI 302.1R
Curing and protection	ACI 305R, ACI 306.1, ACI 308

6.3 EXTERIOR PROTECTION SYSTEMS

6.3.1 General

Systems under this heading will range from least to most expensive. Remember, though, that you cannot use any external protection to make a silk purse out of a sow's ear. In other words, if the underlying concrete is not of good quality, no external protection system, no matter how expensive, will work for long. To emphasize the point one more time—spend your money on good concrete—then you will not have to throw it away later on applied protection.

6.3.2 Sealants

There is sometimes confusion between the terms "sealant" and "sealer." A sealant is a viscous material applied in fluid form, hardening somewhat to provide a long-term flexible seal which adheres to the surrounding concrete. In the next section, we will cover sealers.

In parking structures use sealants to keep water out of joints. These joints may be gaps inches wide between members. They may be grooves tooled into the plastic concrete to provide weakened planes for crack control. Whatever the joint width, a good sealant will keep water and water-borne salts out of the joint, extending its serviceability. Water and water-borne salts attack both the concrete and any unprotected ferrous metals in the joint.

Always specify that the edges of joints to be sealed be tooled, not sawn, then be ground to remove concrete laitance. Next, the joint should be primed. Both practices will greatly improve the bond between sealant and concrete.

Though some engineers and architects prefer a two-part sealant over a one part, there are pluses and minuses for each. We have never had a problem with the one-part sealants which we specify. We have had problems occasionally with installer mistakes in proportioning or mixing two-part sealants. See Chapters 5 and 9 for more on proper joint design and sealants.

6.3.3   Sealers

Sealers are protective coatings applied over a concrete surface to prevent water and water-borne salts from penetrating that surface. A good sealer for parking structure use penetrates the concrete surface, but allows vapor to escape. A good sealer may extend the service life of a sound concrete surface, but will not bridge cracks.

There are both good and worthless sealers available. To tell them apart, see Chapter 7 for an extensive discussion.

6.3.4   Membranes

We classify waterproof membrane systems (sometimes called traffic toppings) for parking structure pavements into two kinds—those applied over the pavement as a traffic-bearing surface and those which must be protected by a wearing course. Figures 6-3 and 6-4 show example installations of each kind. Note that the protected membrane installation has more components than the traffic-bearing type. The protection and drainage systems for the protected type are often difficult to install correctly. Protected systems are more expensive than traffic-bearing systems. Finding a leak in a protected membrane can be difficult. Repairing that leak means removing some part of the protection layer, fixing the leak, and replacing the protection layer. If the protection layer is made of brick or concrete pavers, the removal and repair operations are not particularly difficult, but if the layer is concrete, removal and repair operations are more difficult. On the other hand, protected membranes are not subjected to traffic wear or to deterioration from the ultraviolet rays in sunlight.

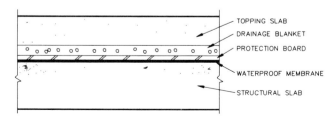

Figure 6-3.   Non-traffic bearing membrane system.

Figure 6-4. Traffic-bearing membrane system.

Traffic-bearing membranes typically come in four wear grades. The first is for areas in which wear is low, such as in parking spaces or pedestrian areas. The second is for flat, straight driving aisles. The third is for areas where high wear is likely. Examples are turns, steeper slopes, ramps, and spots where stopping and starting will occur, as at gates or ticket dispensers. The fourth grade is not really a grade at all. It can be any of the other three grades, except that it has a light-colored top coating. It should also contain an aliphatic compound, to increase its resistance to ultraviolet light. This grade is, of course, for areas exposed to sunlight. Traffic-bearing membranes are relatively easy to repair. Service life varies with abrasion wear received. Turns or areas where frequent starts and stops occur may need repairs every 2–3 years Parking spaces may need repairs only after 10 years or so. Note that at this writing, there are no data available which relate membrane wear to traffic volume.

Like good sealers, good membranes keep water and water-borne salts out of the underlying concrete while letting vapor out. Unlike good sealers, good membranes will bridge narrow cracks which may develop in the underlying concrete after the membrane is in place. (For specification information, see Chapter 7.)

6.3.5   Overlays

6.3.5.1   General. Except for cast-in-place overlays on precast concrete, overlays are not often used in construction of new parking structures. Their use, particularly of the more exotic materials, is more common in restoration work. For a further discussion, see Chapter 10. For the sake of completeness, we include them here also.

Topping on precast concrete is commonly used to provide a level, even wearing surface which compensates for the unevenness between the tops of the precast members caused by differential camber. At the same time, the topping reinforcement may also serve to transfer lateral loads to the structural framing. Further, the topping can be used to conceal connections between precast members. Keep in mind that the joints between the precast pieces beneath a topping will reflect through that topping. Tool and seal the topping above all joints in the precast.

On a few recent projects, owners requested that we specify a noncomposite sacrificial topping over already pretopped precast double tees.

Several different types of concrete overlays will be presented here. Though they will all fulfill the functions listed above, they do differ in durability and in expense, and are listed in order of increasing degree of expense. These overlays also vary in minimum recommended thickness. The densities of the overlays listed below are all around 145 lb/cu ft. So, the thinner the overlay can be applied, the lighter it will be. Extra structure weight drives up the cost of construction in both new and restoration work.

6.3.5.2 Portland Cement Concrete. This economical alternative has been used successfully as an overlay many times. Where high impermeability, low self-weight and a thin overlay are requirements, this material would not be a solution. A 3-in. thick overlay is a practical minimum. If of normal weight (145 lb/cu ft) concrete, it will weigh 35–40 lb/sq ft. If thickness and self-weight are acceptable, you could use a sealer or membrane on the new overlay to improve impermeability. The cost of such a combination will probably exceed the cost of one of the other overlays listed below.

6.3.5.3 Low Slump, Dense Concrete (Iowa Method). This alternative has the same attributes as that above. However, its lower water-cement ratio (0.32) makes it more durable and less permeable. It has been used extensively on new and repaired highway bridge decks.

6.3.5.4 Silica Fume Concrete. Density is a little higher than that of the two alternatives above. Its much higher strength makes a $1\frac{1}{2}$- to 2-in. thickness practical. Impermeability is quite high. See Section 6.2.4.3 for more information on this material.

6.3.5.5 Latex-Modified Concrete (LMC). LMC is similar in performance to silica fume concrete, but significantly higher in cost and of lower strength. Thickness may be less. See Section 6.2.4.4 for more information.

6.3.5.6 Other Overlays. Combinations of materials and protection are not uncommon. Fiber-reinforced overlays and overlays with covered or exposed membranes are not unusual. See Section 6.2.5.4 for more information on fiber reinforcement.

TABLE 6-3   Exterior Protection Systems

System	Usual Application	Relative Cost[1]
Sealants	Waterproof control, construction, and isolation joints	Lower
Sealers	Keep out chlorides	Lower
Membranes	Keep out water and chlorides; bridge cracks	Higher
traffic bearing	Parking structures	Lower
non-traffic bearing	Pedestrian plazas	Higher
Overlays	Durability, protection	
Portland cement concrete		Lower
low slump, dense concrete		Lower
silica fume concrete		Higher
latex-modified concrete		Higher

[1]Costs are relative only within a system category.

6.3.6   Summary

Table 6-3 summarizes exterior protection systems.

6.4   CATHODIC PROTECTION

Cathodic protection may be internal (built in during initial construction) or it may be external (retrofitted during restoration construction).

Cathodic protection works by putting energy in the form of elec-

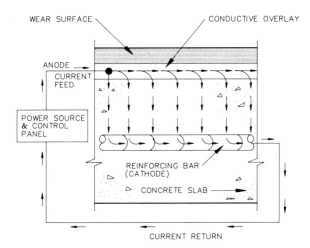

Figure 6-5.   Example of cathodic protection system schematic.

TABLE 6-4   Durability Measures

	Basic Protection	Added Protection
Surface	Finishing, curing, sealer	Traffic topping
Concrete	Mix design, low w/c ratio, entrained-air superplasticizer, etc.	Silica fume concrete
Reinforcement	Cover, epoxy coating	Calcium nitrite cathodic protection
Design	Drainage, prestressing, volume change	

trical current into the concrete to be protected. The introduced energy prevents corrosion in the steel reinforcement. Corrosion of metal is, put simply, a loss of energy from that metal. Feeding in more energy prevents that corrosion. Cathodic protection is the only protective measure which prevents corrosion from starting. If corrosion has already started before cathodic protection is introduced, it is the only protective measure which will stop corrosion. All other measures described in this chapter are but delaying actions, though some are very effective; they will slow corrosion, but not stop it. See Figure 6-5 for a diagram showing how one cathodic protection system works.

Why are not all parking structures cathodically protected? First, such protection may not be necessary, as in mild climates where there is no road salt, salt air, or salt spray to attack the structure. Second, the cost of a typical installation is prohibitive unless replacement cost of the structure would also be prohibitive. An example of such a case could be a parking structure which comprises the lower levels of a high-rise building. Third, the cathodic protection process generates hydrogen gas which may cause hydrogen embrittlement in high strength prestressed reinforcement. Hydrogen embrittlement can cause abrupt failure of metal. It is not a concern with conventional steel-deformed bar reinforcement, but do not use it in prestressed structures, even (using present technology at this writing) with encapsulated prestressed systems.

6.5   SUMMARY

Table 6-4 gives a summary of recommended combinations of protection measures.

Chapter 7
Specifications

ANTHONY P. CHREST

7.1 INTRODUCTION

This chapter is directed primarily at engineers and specification writers, though it may interest others. Its scope will be limited to discussing sections of specification Divisions 3 and 7 as they relate to parking structures. These sections cover technical areas of concrete, reinforcement, and waterproofing systems and are the heart of parking structure specifications. They represent areas of continually changing technology and are often the basis of disputes, so should be written with care and strictly enforced. Unfortunately, enforcement is often lacking. Determined long-term education of field staff is the only practical way to improve enforcement.

7.2 COMMUNICATION

Make it your business to talk to suppliers of all types. Ask them to review your master specifications and discuss their views with you. You will soon know who is trying to be genuinely helpful and who is just pushing a product or service. These reviews will improve your specifications. More importantly, they will build good communication, respect, and trust between the supplier and you. If suppliers know that you are approachable, they just may be able to help you prevent some of the inevitable problems that arise during construction. Supplier familiarity with your specifications will produce lower bid prices.

Some people will be reluctant to invite these review discussions. They will feel uncomfortable because they think such action will suggest a lack of knowledge. Such a view is short-sighted and self-defeating in today's competitive market. Think about how you react when a stranger asks you for directions. Your first reaction is likely one of pleasure at being asked to share your knowledge. Everyone likes to be thought helpful, so do not waste time worrying about how a supplier will take your request; just do it.

7.3   PERFORMANCE SPECIFICATIONS

A performance specification states *what* is wanted, not *how* to get it. As an example, a performance specification for concrete might list required strength, water-cement ratio, air-entrainment range, maximum slump, and minimum cement content. It will not list required ingredient proportions or give directions as to how to achieve the results.

A prescription specification states both the what *and* the how. In the example above, the prescription specification would require the same results, but would continue with details as to how to proportion, mix, transport, place, consolidate, finish, and cure the concrete. Most specifications today combine elements of both prescription and performance.

As much as possible, we use a performance specification approach. Prescription specifications can lead to headaches for the designer, especially in product areas where new developments are rapid. Within reason, we really do not care what a product is made of, so long as it meets all our requirements. A good performance specification must contain requirements which are measurable quantitatively by standard testing methods such as those of ASTM. Pass/fail limits must be set for each test required. If a product is not listed in a specification section, the product vendor should be able to find all the performance requirements for his product category in that section.

Make sure that your specifications contain the flexibility to deal with local or regional conditions—more reason to communicate as suggested above.

If there is an industry-accepted requirement for a product, such as ASTM C 260 for air-entraining agents for concrete, it is sufficient to list only that requirement. If there are no industry-accepted requirements for a product type, then listing acceptable products by name, followed by the corresponding performance specification, reduces questions during bidding and construction and limits time spent with prospective vendors.

We make changes to our master specifications as needed, but

rarely make exceptions to them. The exceptions are only for a trial of a new product. Exceptions may help one vendor, but they are unfair to everyone else. When we explain to a vendor that an exception for him on the job at hand might lead to an exception for his competitor on the next, he will not persist in a request for an exception.

7.4   SPECIFICATION PRODUCTION

Computer-Assisted Specifying (CAS) has been available from a few vendors for a few years. The National Institute of Building Standards, the military, and larger firms such as Sweet's now provide CAS systems. Reexamine your present specification production system in light of these developments. There may be little reason to maintain your current methods.

7.5   COPYRIGHT

We copyright our master specification. A copyright notice appears on the last page of the table of contents in the project manual and on the first and last page of every specification section. Some designers unfortunately have the habit of appropriating the specifications of others, mistakes and all, for their own use. Copyrighting your specifications will not stop others from using your specifications, but will strengthen your position should you ever wish to make an issue of it.

7.6   DIVISION 3

7.6.1   Section 03300, Cast-in-Place Concrete

In many bidding situations, bidders may neglect to read your concrete requirements in full. For example, concrete specified with a 28-day strength of 4000 psi *and* a water-cement ratio of 0.40 will cost appreciably more than that specified with the strength requirement alone. If the bidder sees the first requirement but not the second, his profit margin may be reduced enough to cause him, and you, problems as the project progresses.

Example:
    Required: 4000 psi compressive strength at 28 days
                    0.40 water/cement ratio
                    3 in. slump

Water demand: concrete generally requires approximately 275
                lb (33 gals) of water per cu yd to achieve a 3 in.
                slump.
Cement factor: water/cement = 0.40
                cement = 275 lb/0.40 = 687.5 lb/cu yd
Problem: If the strength requirement is shown as 4000 psi at 28
           days, the ready-mix concrete producer often bids on
           this information, but misses the maximum water/ce-
           ment ratio which governs. A cement factor of 550 lb/
           cu yd is usual for 4000 psi compression strength. The
           137.5 lb of cement per cu yd increase (687.5 − 550)
           necessitated by the water/cement ratio is a significant
           extra cost to the ready-mix producer if he did not in-
           clude it in the bid.
Result: The mix design may be "adjusted" to meet the specifi-
           cation. The cement factor included in the bid may be
           used with a fictitious water content in order to conform
           to the maximum water/cement ratio of 0.40.
Listed water content = cement factor × 0.40
                     = 550 lb × 0.40
                     = 220 lb/cy (26.4 gals)

Concrete cannot be mixed at this water content. If you include a
short caution under the "Scope of Work" heading in Section 03300,
Cast-in-Place Concrete, it might prevent this problem.

   Example:
   Work Included: In accordance with Contract Documents, pro-
                   vide all materials, labor, equipment, and ser-
                   vices necessary to furnish and install cast-in-
                   place concrete.
   CONCRETE SUPPLIER: THIS SECTION CONTAINS REQUIRE-
                       MENTS FOR HIGH-STRENGTH AIR-
                       ENTRAINED CONCRETE WITH LOW
                       WATER-CEMENT RATIO AND SU-
                       PERPLASTICIZER.

7.6.1.1 Entrained Air. Correct entrained-air content in concrete improves workability and protects the hardened concrete from freeze/
thaw damage. Other common additives will improve workability, but
few others will provide freeze/thaw protection—none as inexpen-
sively.

Entrained-air content requirements to provide adequate protection vary widely depending on climate. Requirements in North America typically vary from a range or 5–7% in northern latitudes to 4% or less farther south. State or Provincial Department of Transportation requirements for entrained air for bridge construction are reliable guidelines if you have no experience in an area.

If requirements are set, the next question is, where should the air content be measured? If the concrete leaves the ready-mix plant with x% air, chances are that the in-place concrete will have $\frac{1}{2}$x% air or less. Transportation, pumping, consolidation, and finishing all tend to remove entrained air from plastic concrete, particularly the latter two.

Check air content at the truck. Also check it after screeding. A sample specification excerpt is included in Appendix 7-1 at the end of this chapter, along with a commentary. The commentary will be in italics in all appendixes.

7.6.1.2 Finishing. Improper finishing is usually over finishing. The best finishing is that accomplished with the least effort.

For a parking structure, the finish itself should be rough enough to provide traction for car tires and to prevent slipperiness when wet. It must not be so rough as to become a tripping hazard for pedestrians.

We suggest you specify a light broom finish. A sample specification is included in Appendix 7-2 at the end of this chapter, including a commentary.

7.6.2 Section 03365, Post-Tensioned Concrete

Unbonded tendons, with hardware and protection, should be treated as an integral system and looked at as a whole. For that reason, we have reproduced the entire section in Appendix 7-3 at the end of this chapter, along with a commentary.

7.6.3 Section 03400, Precast Concrete

Whether you use a performance specification, with some or all of the piece design assigned to the precaster, or you provide full structural design services, the success of any project with sizeable quantities of precast elements is in the precaster's hands. While even the best precaster cannot produce a good project from a poor design, a poor precaster can certainly spoil a good design. Your best defense is to work with precasters whom you have come to trust. If you are working in a locale new to you, use well-written prequalification requirements to exclude questionable performers. Requiring PCI plant certification is recommended. Keep in mind, though, that neither PCI plant certification nor any industry standard will protect you from dishonesty by a vendor. Sometimes your only defense may be to refuse to work with a vendor who has cheated or tried to cheat on a previous project, even if it means losing the current project.

7.7 DIVISION 7

7.7.1 Section 07100, Waterproofing Systems

This section contains example requirements for:

protective concrete sealers
traffic toppings
expansion (isolation) joint sealants
concrete control joint sealants

and is included in its entirety in Appendix 7-4 at the end of this chapter. As you read this appendix and included commentary, you may wonder if the constant safeguards and checking are necessary. There certainly are honest people and firms in the construction business, but it has been our experience that safeguards and checking are quite necessary, particularly in the protective concrete sealer market. Competition is fierce; the market is crowded; corners do get cut; outright dishonesty does exist. Be on your guard; give no one a second chance to cheat you or your client.

7.8 KEEPING CURRENT

Treat your master specification as a living document. We update some part of ours every month. What you see in this chapter was current when it was written. By the time this book is published, several months will have elapsed. We guarantee that some part of the quoted sections will have been superseded during the intervening months, because of new experiences, changing technology, and new product research.

7.9 SUMMARY

Specifications are one of the two principal means to communicate project requirements from the engineer to the builder. As such, specifications must be clear, complete, and fair. We have discussed some of the likely trouble spots in a typical parking structure specification and have included excerpts and sections from our master specification. Finally, no matter how watertight the specification, without enforcement at the job site, the specification will be worthless.

Appendixes to Chapter 7

APPENDIX 7-1
Sample Specification Excerpt for Field Quality Control by the Testing Agency

3.04 FIELD QUALITY CONTROL BY TESTING AGENCY (Reference: ACI 301, CHAPTER 10)
  A. Air Content:
    1. Sample freshly-mixed concrete in accordance with ASTM C172 and conduct one air content test in accordance with ASTM C231 or ASTM C173 for each truck of ready-mix, air-entrained concrete delivered to Project.
    2. Sample fresh concrete immediately following placement and screeding and conduct air content tests in accordance with ASTM C231 or ASTM C173 at rate of one for every 10 truck loads of ready-mix, air-entrained concrete delivered to Project.
    3. Core and test hardened concrete topping for air content in accordance with ASTM C457 at rate of one core per 15,000 sq ft of topping or structural slab.
    4. If concrete consistently meets requirements of Specification and concrete mix design and placement procedures remain unchanged, Engineer may waive requirement for further testing of hardened concrete.
    5. Contractor shall patch holes resulting from concrete coring. Use patching materials meeting Specification requirements.
  Our experience with restoration of deteriorated concrete indicates that lack of specified entrained-air content is often the reason for the deterioration, or was the opening wedge for problems. The only method available for determining the entrained-air content of the hardened concrete is to take core samples and check. The text above also gives suggested testing.
  Suggested quality control requirements for entrained air content:
  B. Air Content:
    1. Of freshly mixed concrete by pressure method, ASTM C231, or volumetric method ASTM C173

2. Of hardened concrete by microscopic determination, including parameters of air-void system ASTM C457.

State only the percentage of entrained air required. Suggested additional requirements:

3. Hardened concrete shall have an air-void spacing factor of 0.0080 in. maximum.

4. Specific surface (surface area of air voids) shall be 600 in.^{2} prcu in. of air-void volume, or greater.

5. Number of air voids/in. shall be $1\frac{1}{2}$–2 times numerical value of entrained-air content percentage, as determined by ASTM C457.

Is entrained-air content less than that specified? ACI 301, Chapter 17 is specific on acceptance requirements for concrete strength, but not for entrained-air content. The wording below is similar in intent to Section 17.2 of ACI 301:

C. Acceptance of Structure (ACI 301, Chapter 18):

1. Acceptance of completed concrete Work will be according to provisions of ACI 301, Chapter 18. Entrained-air content of hardened concrete acceptable so long as averages of all sets of three consecutive test results equal or exceed specified entrained-air content and no individual test falls below specified minimum by more than $\frac{1}{2}$%.

Example:

If 7% $\pm$ $1\frac{1}{2}$% is specified, then $5\frac{1}{2}$% is specified minimum and 5% is limit below which no individual test may fall.

If entrained air content is found to be below the specified limit, then negotiated corrective measures must follow at the builder's expense. Corrective measures range from replacement of the concrete, to installation of a protective topping or membrane, to an extended warranty.

APPENDIX 7-2
Sample Specification Excerpt of Finishing Concrete Flatwork

3.03 INSTALLATION

A. Placing (*text omitted here*)

B. Finishing (reference ACI 301, Chapters 10 and 11):

1. Formed Surfaces: ACI 301, Article 10.4

2. Flatwork in Parking and Drive Areas (BROOM Finish, ACI 301, 11.7.4.):

a. Begin bull floating after bleeding of water through surface of concrete has been completed and water sheen has disappeared from surface of concrete and concrete has stiffened sufficiently to allow operation (ACI 302.1R, Article 7.2.3).

b. Immediately after bull floating, concrete shall be given coarse transverse-scored texture by drawing broom across surface. Texture shall be as accepted by Engineer from sample panels.

c. Finishing tolerance: ACI 301, Paragraph 11.9: Class B tolerance. In addition, floor surface shall not vary more than $\pm\frac{3}{4}$ in. from elevation noted on Drawings anywhere on floor surface.

d. Before installation of flatwork and after submittal, review, and approval of concrete mix design, Contractor shall fabricate two acceptable test panels simulating finishing techniques and final appearance to be expected and used on Project. Test panels shall be minimum of 4 ft × 4 ft in area and shall be reinforced and cast to thickness of typical parking and drive-area wearing surface in Project. (Maximum thickness of test panels need not exceed 6 in.) Test panels shall be cast from concrete supplied by similar concrete batch, both immediately after addition of superplasticizer or water-reducing admixture, and at maximum allowed time for use of admixture supplemented concrete in accordance with Specifications. Intent of test panels is to simulate both high and low workability mixes, with approximate slump at time of casting of test panels to be 6 and 3 in., respectively. Contractor shall finish panels following requirements of paragraphs above, and shall adjust finishing techniques to duplicate appearance of concrete surface of each panel. Finished panels (one or both) may be rejected by Engineer, in which case Contractor shall repeat procedure on rejected panel(s) until Engineer acceptance is obtained. Accepted test panels shall serve as basis for acceptance/rejection of final finished surfaces of all flatwork.

e. Finish all concrete slabs to proper elevations to insure that all surface moisture will drain freely to floor drains, and that no puddle areas exist. Contractor shall bear cost of any corrections to provide for positive drainage.

3. Flatwork in Stair Towers and Enclosed Finished Areas (Floated Finish, ACI 301, 11.7.2):

a. Give slab floated finish. Texture shall be as accepted by Engineer from sample panels.

b. Finishing tolerance: ACI 301, Paragraph 11.9: Class A tolerance. In addition, floor surface shall not vary more than $\pm\frac{3}{4}$ in. from elevation noted on Drawings anywhere on floor surface.

APPENDIX 7-3
Section 03365, Post-Tensioned Concrete

PART 1   GENERAL

1.01   WORK INCLUDED

A. In accordance with contract Documents, provide all materials, labor, equipment, and supervision to fabricate and install all post-tensioning Work. Support bars shall conform to Section 03200.

1.02 RELATED WORK

A. Work in other Sections related to post-tensioned concrete:
1. 03100 Concrete Formwork
2. 03200 Concrete Reinforcement
3. 03250 Concrete Accessories
4. 03300 Cast-in-Place Concrete

1.03 SYSTEM DESCRIPTION

A. Unbonded post-tensioning system described here is intended to perform in corrosive environment without long-term corrosion or other distress. Post-tensioning strand, couplers, intermediate and end anchorages shall be completely protected with watertight system. Tendon grease shall be as specified, with corrosion inhibitors. End anchors shall be protected against long-term corrosion. (See Part 2 of this section for detailed requirements.) Post-tensioning system is expected to function satisfactorily for 40 yr.

The purpose of this paragraph is to signal clearly, at the beginning, what result we are shooting for in this section. We also cover this item in the prebid and preconstruction conferences.

1.04 QUALITY ASSURANCE

A. Work shall conform to requirements of ACI 301, ACI 318, and CRSI MSP-2 except where more stringent requirements are shown on Drawings or specified in this Section.

Our requirements for certain tolerances and reinforcement cover are sometimes more stringent.

B. Welders and welding procedures shall conform to requirements of AWS D1.1 and AWS D1.4. Except where shown on Drawings, welding is prohibited unless approved in writing by Engineer.

Heat from welding can severely weaken the high strength steel in post-tensioning tendons.

C. Provide post-tensioning strand systems supplied by PTI certified manufacturers and licensees only, conforming to all material and installation requirements of ACI 301, ACI 318, and approved by International Conference of Building Officials (Uniform Building Code).

These requirements tend to get rid of less reliable suppliers who cannot spare the time or expense to meet these requirements. Always demand the paperwork to back up vendors' claims, though. Also keep in mind that no amount of certification will guarantee honesty.

D. Comply with ACI 301, Article 15.2.3.

E. Post-tensioning Work shall be performed by an organization that has successfully performed at least five previous post-tensioning installations similar to one involved in this Contract.

You do not want anyone learning on your project.

F. All post-tensioning Work shall be under immediate control of post-tensioning contractor's superintendent experienced in this type of Work. Superintendent shall exercise close check and rigid control of all operations as necessary for full compliance with all requirements. Post-tensioning contractor's superintendent assigned to Project shall have supervised five prior projects of similar magnitude and design, and shall be present during all placing and post-tensioning operations. Superintendent shall be approved by Engineer. Superintendent's failure to insure full compliance with Specification will result in his removal from Project.

Do not be reluctant to invoke this clause if reasonable efforts to obtain required performance have failed.

1.05   REFERENCES
   A. American Concrete Institute (ACI):
      1. ACI 301, "Specifications for Structural Concrete for Buildings"
      2. ACI 318, "Building Code Requirements for Reinforced Concrete"
      3. ACI 347, "Recommended Practice for Concrete Formwork"
      4. ACI 423.3R, "Recommendations for Concrete Members Prestressed with Tendons"
   B. American Society for Testing and Materials (ASTM):
      1. ASTM A416, "Specification for Uncoated Seven-Wire Stress-Relieved Strand for Prestressed Concrete," including supplement, "Low-Relaxation Strand"
      2. ASTM E328, "Recommended Practice for Stress-Relaxation Tests for Materials and Structures"
   C. American Welding Society (AWS):
      1. AWS D1.1, "Structural Welding Code—Steel"
      2. AWS D1.4, "Structural Welding Code—Reinforcing Steel"
   D. Concrete Reinforcing Steel Institute (CRSI):
      1. CRSI MSP-2, "Manual of Standard Practice"
   E. Post-Tensioning Institute (PTI):
      1. PTI, "Guide Specifications for Post-Tensioning Materials"
      2. PTI, "Performance Specification for Corrosion-Preventive Coating"
      3. PTI, "Specification for Unbonded Single-Strand Tendons"

1.06   SUBMITTALS
   A. Make submittals in accordance with requirements of Division 1 and as specified in this Section.
   B. Because Work of Sections 03200 and 03365 are interdependent, Contractor shall have both suppliers review each other's Shop Drawings and note any potential interferences. Contractor shall then review 03200 and 03365 Shop Drawings against each other and inform Engineer of any potential interferences.

Otherwise, interferences will be discovered only in the field, where corrections are costly.

   C. Shop Drawings. Include:
      1. Numbers and arrangement of post-tensioning tendons
      2. Methods of maintaining tendon alignment
      3. Type of post-tensioning sheathing
      4. Type and chemical analysis of grease
      5. Type, material, and thickness of post-tensioning sheathing repair tape
      6. Detailing of anchorage devices

7. Other incidental features
8. Superintendent qualifications
D. Submit following information with Drawing Submittal:
   1. Sealed calculations, prepared under supervision of qualified, professional, registered structural engineer in state Project is to be constructed, of losses due to anchorage seating, elastic shortening, creep, shrinkage, relaxation, friction, and wobble, used to determine tendon sizes and number.

The construction documents state the required final, effective, post-tensioning forces in structural members such as beams and slabs. To obtain the values of the initial forces in each member, and therefore the quantities of post-tensioned reinforcement to be supplied, the post-tensioning contractor's engineer must make assumptions, based on his experience, for the losses mentioned here. He is then able to calculate the initial forces required, and from those, the number of tendons needed in each member. The engineer's seal requirement ensures only minimum qualifications on that engineer's part, but it does have legal liability implications for him.

   2. When low-relaxation strand is used, Contractor shall submit additional sealed calculations to satisfy requirements of design effective prestress force and design moment strength for all post-tensioned sections as indicated above. If additional reinforcement is required due to change of strand type, reinforcement shall be furnished at no additional cost to Owner and shall be coordinated between post-tensioning contractor and Contractor before Drawings and calculations are submitted to Engineer.

Use of low-relaxation strand instead of stress-relieved strand results in less post-tensioning reinforcement for a given final effective force. With the reduced area of very high-strength reinforcement comes the need for even more assurance that the calculations are correct. Further, if the post-tensioning supplier decides to change strand type, the owner should not have to pay for more than the structure originally specified.

   3. Post-tensioning tendon and end anchorage sizes.
   4. Sample of complete tendon and anchorage protection system including live, dead, and intermediate anchorage assemblies
   5. Satisfactory evidence of compliance with requirements for PTI certification and ICBO approval
E. After review, Shop drawings and data shall not be changed, nor shall construction operations be revised unless resubmitted for approval of Engineer. Engineer's review of details and construction operations will not relieve Contractor of responsibility for completing Work successfully in accordance with Specifications and within Contract Time.
F. Submit following to Engineer for review before beginning construction:
   1. Post-tensioning experience record of contractor who is to perform post-tensioning Work
   2. Certified Mill Test Reports for each coil or pack of strand, containing, as a minimum, following test information:
      a. Heat number and identification

b. Standard chemical analysis for heat of steel
c. Ultimate tensile strength
d. Yield strength at 1% extension
e. Elongation at failure
f. Modulus of elasticity
g. Diameter and net area of strand
h. Type of material (stress-relieved or low relaxation).

Follow up and make sure that you receive everything specified.

3. Relaxation losses for low-relaxation type material shall be based on relaxation tests of representative samples for period of 1000 hr, when tested at 70°F, and stressed initially to not less than 70% of minimum guaranteed breaking strength of strand. Tests shall be in accordance with ASTM E328 and ASTM A416.

4. Low-relaxation strand shall be provided with verification that proposed supplier has adequate safeguards to ensure that only low-relaxation strand is furnished.

5. Evidence of satisfactory performance on similar projects in United States, if strand is manufactured outside the country. (See also 2.01A.)

6. Results of tests required by ACI 301, Article 15.2.3

7. Current certification of welders

Again, especially for 7 and 8, follow up to make sure that certifications are current. Contractor response will provide a measure of his competence and quality control.

8. Current PTI certification of tendon supplier

9. Certified calibration curve for each jack

G. Keep post-tensioning records and submit to Engineer. Record on each report:

1. Calculated elongation, based upon elastic modulus and cross-sectional area of tendons used

2. Actual field elongation of each tendon

3. Calculated gauge pressure and jacking force applied to each tendon

4. Actual gauge pressures and jacking forces applied to each tendon

5. Required concrete strength at time of jacking

6. Actual concrete strength at time of jacking

7. Range of allowable elongations for jacking force

H. Submit copies of actual field records to Engineer promptly upon completion of each member or slab tensioning run. *Question neat, clean records with all elongations the same or nearly so. They may be office-generated. Also, if actual elongations are not as predicted, check the actual modulus of elasticity against the value used in the elongation prediction calculations. A difference in the two is often the source of the discrepancy. At time of stressing first member of each type, check individual tendon to establish procedure for ensuring uniform results.*

I. Certify that stressing process and records have been reviewed, and that forces specified have been provided. *Follow up.*

J. If it appears that design post-tensioning stresses are not being attained, a recheck may be ordered by Engineer. *Follow up.*

K. Do not cut or cover tendon ends until Contractor receives Engineer's written review of post-tensioning records. *Cover in preconstruction conference.*

1.07 DELIVERY, STORAGE AND HANDLING

A. Assign all post-tensioning tendons within every group or in same member a heat number and tag accordingly.

B. Strand shall be packaged at source in manner which prevents physical damage to strand during transportation and storage and which positively protects strand from moisture and corrosion during transit and storage. *Cover this requirement emphatically in the preconstruction conference. Moisture accumulating inside the tendon sheathing is the first step toward tendon corrosion.*

C. Remove and replace at no cost to owner wires or strands which are broken or show severe fabrication defects. *Enforce with no exceptions.*

D. Do not store materials on slabs to be prestressed before final prestress of slabs is accomplished. At no time shall weight of stored material placed on slab area, after prestressing is completed and concrete has reached specified 28-day strength, exceed total design load of that slab area. Between time final post-tensioning is accomplished and time concrete has reached specified 28-day strength, weight of stored material placed on slab area shall not exceed half total design load of that slab area. *Cover in preconstruction conference.*

1.08 SEQUENCING

A. See Drawings for construction and tensioning sequence. *Because of the relatively large volume changes occurring in the typical cast-in-place post-tensioned concrete parking structure, the casting sequence will induce axial forces, shears, and moments in the structure which may be both unexpected and significant. Although a construction sequence is part of the contractor's "means and methods," you need to advise him. Most contractors will readily accept such advice. Besides, if cracking develops during construction due to an ill-chosen casting sequence, you will have to solve it anyway. Prevent the problems by showing a casting sequence on a small isometric drawing of the superstructure.*

1.09 ALTERNATES

A. High-strength bar and button-headed wire systems, bonded or unbonded, may be used instead of unbonded wire strand systems, only with prior written approval of Engineer. *The great majority of parking structure post-tensioning is done with unbonded monostrand systems. The reason is cost. For further reference, the PTI Manual has a chapter which covers most systems currently available in North America.*

PART 2 PRODUCTS

2.01 MATERIALS

A. Post-tensioning tendons, ASTM A416: Seven-wire, uncoated low-relaxation steel strand, Grade 270 with minimum ultimate strength of 270,000 psi, unbonded system. All strand shall be manufactured by single source.

If manufactured outside United States, strand shall be subject to Engineer's acceptance. Acceptance will be based on Engineer's review of evidence of satisfactory performance of strand in United States over past 5 yr. Engineer's decision will be final. *Strand comes from all over the world and properties may vary widely. Do not hesitate to exercise your option. Some foreign-made strand is not acceptable.*

B. Sheathing: Tendon sheathing for unbonded single-strand tendons shall be made of material with following properties:
   1. Sufficient strength to withstand damage during fabrication, transport, installation, concrete placement, and tensioning
   2. Watertightness over entire sheathing length
   3. Chemical stability, without embrittlement or softening over anticipated exposure temperature range and service life of structure
   4. Non-reactive with concrete, steel, and tendon corrosion-preventive coating
   5. Color shall contrast with black grease so that sheathing tears will be readily visible. *The concrete itself is the first line of defense against tendon corrosion. Sheathing and grease are the second and third lines, respectively. Do not compromise your defenses.* Sheathing shall be seamless and extruded. Sheathing thickness shall not be less than 0.040 in., ±0.003 in., for medium- or high-density polypropylene. Sheathing shall have inside diameter at least 0.010 in. greater than maximum diameter of strand. Sheathing shall be connected to all stressing, intermediate, and fixed anchorages in watertight fashion, providing complete encapsulation of prestressing steel, all couplers, intermediate, and end anchorages, including wedge cavities.

C. Tape (*trade names deleted*):

D. Tendon grease: Lithium-based, containing corrosion inhibitors, wetting agents, and less than fifty parts per million of chlorides, sulphides, or nitrates:
   1. Approved greases (*trade names deleted*):
   2. Grease shall completely fill void between tendon and sheathing.
   3. Minimum weight of grease on tendon strand shall be 2.5 lb per 100 ft of 0.5-in. diameter strand, and 3.0 lb per 100 ft of 0.6-in. diameter strand.
   4. Grease shall meet all requirements of PTI "Performance Specification for Corrosion Preventive Coating."

The post-tensioning reinforcement installation on your project is a sophisticated, high-technology product. Unfortunately, there are a number of small suppliers who do not perform well, yet manage to stay in business year after year. If your prequalification requirements have not kept them from bidding, and one becomes the successful post-tensioning bidder, your only hope of obtaining satisfactory performance is to enforce, firmly, fairly, and completely, all of your specification. If you have to refuse the first few payment requests, or even fire him to make yourself heard, do it.

E. Couplings: In accordance with ACI 301, Article 15.2.2 where indicated on Drawings or specified by Engineer.

F. Anchorages: In accordance with ACI 301, Article 15.2.2; design slab anchors for transfer at 2500 psi concrete strength; design beam anchors for transfer at 3000 psi concrete strength; size bearing plates in accordance with ACI 301 unless certified test reports are submitted proving acceptable deviation. Anchorage system must meet all requirements below and those of PTI Guide Specifications for aggressive environments. Also see 1.04C. All anchorage systems must be approved at least 14 days in advance of Bid date. All anchor plates shall be epoxy coated in accordance with coatings specified in Section 03200 or protected by other means to conform to this Section.

G. Tendon anchorages and couplings shall be designed to develop static and dynamic strength requirements of Sections 3.1.6(a) and 3.1.8(1) and (2) of PTI "Guide Specifications for Post-Tensioning Materials." Castings shall be nonporous and free of sand, blow holes, voids, and other defects. Provision shall be made for plastic cap which fits tightly and seals barrel end on stressing side of anchor. Bearing side of anchor casting shall have provision for plastic sleeve which shall prevent moisture leaks into anchor casting or tendon sheathing. For wedge-type anchorages, wedge grippers shall be designed to preclude premature failure of prestressing steel due to notch or pinching effects under static and dynamic test load conditions stipulated under paragraph (a), for both stress relieved and low relaxation prestressing-steel materials. *The intent of the entire Part 2 of this specification section is to obtain a post-tensioned reinforcement protection system which will guard the high-strength strand against corrosion. This section and the others in Division 3 are probably the most detailed and stringent in our entire master specification. These characteristics reflect their importance in providing the necessary durability to the concrete structure. If the structure is not durable, its service life will not outlast the term of the financing package.*

H. Tendon anchor plates and wedges shall be same as those which passed static and dynamic test in accordance with Part 1.

I. Blockouts: Plastic pocket former shall be used at stressing end to provide minimum recess of 2 in. to anchor casting and minimum of $2\frac{1}{2}$ in. in width to allow access to cut off excess strand without damage to wedges and anchor casting. At intermediate stressing ends, prevent moisture leaks into anchor casting or tendon sheathing.

J. Anchor Cap:
Protection of tendon ends is vital. Do not compromise here either.
   1. Stressing Ends: Plastic cap shall fit tightly, covering stressing end of barrel and wedges, and shall be fitted with sealing device. Cap shall allow minimum $1\frac{1}{4}$-in. protrusion of strand beyond wedges.
   2. Intermediate Stressing Ends: Plastic cap similar to above shall be used with exception that cap shall be open to allow passage of strand with minimum $\frac{3}{4}$-in. chimney extension of cap.
   3. Coating Material: Wedge area and plastic cap shall be completely filled with same grease used along length of strand.

K. Sleeve: Plastic sleeve shall be used on bearing side of anchor casting which will prevent moisture leaks into anchor casting or tendon sheathing. Plastic sleeve shall be 10 in. long, minimum. *Wrapping exposed tendons or anchor assemblies with duct tape or any other type is not acceptable; duct tape leaks. Any tape will provide less protection than that specified in this section.*

L. Intermediate Anchorage Sheathing: At intermediate stressing anchorages, exposed stressing length shall be protected with corrosion preventive coating (same as above), covered with plastic sheathing, adequately taped along its length, and taped to tendon sheathing and chimney. Tape shall be that specified in paragraph 2.01C.

M. Acceptable protection systems (*trade names deleted*): *Review the literature and samples with care. If the system does not have the credentials required in paragraph 1.04C, we suggest you reject it until you have those credentials in your hands and are satisfied that they are genuine. In our opinion, there are only three systems of which we are aware that are acceptable. One of these is quite expensive so only two are ever bid. Competition will bring more qualified systems into the marketplace; we are considering two more systems. Figure 7-1 illustrates a generic system of the type which we would accept.*

N. Design Forces and Stresses:

1. Effective post-tensioning forces, after all losses have occurred, are shown on Drawings.

2. Maximum tensile stress in post-tensioning tendons due to jacking forces shall not exceed 80% of specified tensile strength or 94% of specified yield strength of post-tensioning tendon, whichever is smaller, but not greater than maximum value recommended by manufacturer of post-tensioning tendons. *Except for the 80% provision, this requirement is straight out of ACI 318, Article 18.5.1(b). We prefer the 80% requirement instead of the 85% that ACI 318 allows because it is a little more conservative. Also, see ACI 318 commentary for Article 18.5.*

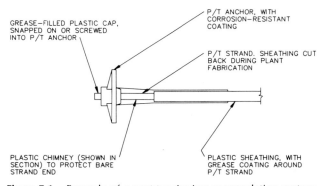

GREASE—FILLED PLASTIC CAP, SNAPPED ON OR SCREWED INTO P/T ANCHOR

P/T ANCHOR, WITH CORROSION—RESISTANT COATING

P/T STRAND. SHEATHING CUT BACK DURING PLANT FABRICATION

PLASTIC CHIMNEY (SHOWN IN SECTION) TO PROTECT BARE STRAND END

PLASTIC SHEATHING, WITH GREASE COATING AROUND P/T STRAND

Figure 7-1. Example of a post-tensioning encapsulation system.

3. Maximum tensile stress in post-tensioning tendons immediately after anchorage shall not exceed 70% of specified tensile strength.
4. Allowable slip of strand at anchorage shall not exceed $\frac{1}{4}$ in. as long as elongations are within $\pm 5\%$. *Should the anchorage slip exceed $\frac{1}{4}$ in., there are two alternatives available. If a lower effective force per strand will be adequate for the member in question, you may be able to accept the higher slip, assuming that the force corresponding to the actual slip is acceptable. If the higher slip is not acceptable, try restressing and reseating the high-slip strand with new wedges. If the unacceptable slip persists, more expensive correction will be your only alternative.*
5. Design effective prestress force shown on Drawings and design moment strength of all post-tensioned sections are based on effective stress of 173,000 psi in prestressed reinforcement after allowance for all prestress losses. *This information tells the post-tensioning contractor's engineer the basis for your design, should he wish to change strand type.*

PART 3   EXECUTION

3.01   PREPARATION
  A. Furnish necessary information, materials, accessories, and other items for prestressing and attaining effective post-tensioning forces, after all losses have occurred, as shown on Drawings and specified in this Section.
  B. Maintain post-tensioning equipment in safe, working condition.
  C. Maintain spare jack on site during post-tensioning operations.
  D. Satisfactorily protect post-tensioning tendons from rust or other corrosion before placement. Provide sufficient protection for exposed prestressing steel at ends of members to prevent deterioration by rust or corrosion.
3.02   INSTALLATION
  A. Placement:
    1. Place tendons with parabolic profile in vertical plane conforming to control points shown on Drawings unless otherwise specified:
       a. Dimensions locating profile are given to center of gravity of tendons.
       b. Low points are at midspan unless noted otherwise.
       c. Where tendons interfere with one another, contact Engineer before relocating tendons.
    2. Space tendons evenly within slab to achieve required effective prestress as shown on Drawings:
       a. Slight deviations in spacing are permitted to avoid specifically located openings and inserts. *But, keep tendon profiles smooth in both planes. Do not permit sudden changes in direction or unplanned curvature reversal. Distress will result.*

    b. Maximum main slab tendon spacing is six times slab thickness but not more than 36 in. unless otherwise noted on Drawings.

3. Straighten strands to produce equal stress in all tendons that are to be stressed simultaneously and to insure proper positioning of anchors.
4. Run tendons parallel to grid lines unless otherwise noted.
5. Run tendons full length within pour without splices or couplers.
6. Install horizontal and vertical spacers or chairs to hold tendons in required position and to conform to specified profile. Conform to Section 03200. *Section 03200 has further requirements for reinforcement supports.*
7. Space tendon support chairs at 48 in. maximum.
8. Tendons and anchorages shall be supported firmly to prevent displacement during subsequent operations. *Keep alert for problems during concrete placement. Chairs tip, displace, or crush easily.*
9. Place tendons and anchorages to both horizontal and vertical tolerances for corresponding horizontal and vertical member dimensions:
    a. 8 in. and less: plus or minus $\frac{1}{8}$ in.
    b. 8 to 24 in.: plus or minus $\frac{3}{8}$ in.
    c. Greater than 24 in.: plus or minus $\frac{1}{2}$ in.
    d. Deviations in horizontal plane which may be necessary to avoid openings or inserts shall have radius of curvature of not less than 21 ft. *Also see the commentary after paragraph 3.02 A2a.*

B. Sheathing Inspection and Repair:
1. After installing tendons in forms and before concrete casting, inspect sheathing for possible damage.
2. Repair damaged areas by restoring grease coating in damaged area, and repairing sheathing.
3. Sheathing repair procedure:
    Do not compromise here, or all your previous work will be wasted.
    a. Restore tendon grease coating in damaged area.
    b. Coat with grease outside of sheathing the length of damaged area, plus 2 in. beyond each end of damage. Example: If sheathing tear is 6 in. long, then greased area will be 10 in. long, centered on tear.
    c. Place piece of longitudinally slit sheathing around greased tendon. Slit shall be on side of tendon opposite tear. Length of slit sheathing shall overlap greased area by 2 in. at each end. Example: If greased area is 10 in. long, then sheathing will be 14 in. long, centered on tear.
    d. Tape entire length of slit sheathing, spirally wrapping tape around sheathing to provide at least two layers of tape. Taping shall overlap slit sheathing by 2 in. at each end. Before taping, sheathing shall be dry and free of grease. Example: If slit sheathing is 14 in. long, then taped area will be 18 in. long, centered on tear.

C. Tensioning:
1. Take safety precautions to prevent workers from standing behind or above jacks during tensioning. *It is not at all uncommon for an anchor and a short length of tendon to explode upward or outward during tensioning.*
2. It is imperative that concrete slabs to be prestressed reach 2500 psi compressive strength in 96 hr or less. Tensioning should commence as soon as concrete strength reaches 2500 psi. If within 96 hr (including Saturdays, Sundays, and holidays) after concrete placement commenced, strength has not reached this limit, $\frac{1}{2}$ stress shall be applied to each wire or strand and full stress applied when concrete reaches 2500 psi. No exceptions to this requirement will be permitted. *None means none. Proper scheduling will eliminate most conflicts. Cover this requirement in the prebid and preconstruction conferences. The reason for the requirement is that the post-tensioning reinforcement is usually the only continuous reinforcement in the slab. In fact, at some locations, it may be the only reinforcement. If more than 96 hr pass without compression being introduced into the concrete by the stressing operation, the concrete will crack. Cracks are avenues for future leaks. Later stressing may or may not close those cracks. In any event, the enhanced durability which the post-tensioning system would have provided will be compromised to an unknown extent if the time limit requirement is not met.*
3. No tensioning will be permitted unless post-tensioning tendons are free and unbonded in enclosure.
4. Stress all post-tensioning tendons by means of hydraulic jacks equipped with accurate reading, calibrated hydraulic pressure gauges to permit stress in post-tensioning steel to be computed at any time:
   a. Provide certified calibration curve with each jack.
   b. If deviation greater than 5% occurs between measured elongation and computed elongation for given jack-gauge pressure, immediately recalibrate gauges.
   c. If after recalibration, computed and measured elongation for given gauge pressure still deviates by more than 5%, cease tensioning operations until cause of deviation is found and corrected.
5. Anchor prestressing steel at initial force that will result in effective forces, after all losses occur, not less than those shown on Drawings.
D. Cutting Tendons After Stressing:
1. Do not cut tendons or cover pockets until elongation records are reviewed by Engineer.
2. Clean tendons, anchorages, and pockets of grease with non-corrosive solvent before removal of excess length of post-tensioning tendons.
3. Do not damage tendon, anchorage, or concrete during removal of excess length of tendon. *Make sure the man with the cutting torch knows what he is doing.*

3.03 FIELD QUALITY CONTROL
  A. Tendons shall not be exposed to weather more than seven calendar days. *Intent: minimize chances for moisture to enter sheathing.*
  B. Tendon sheathing damaged over more than 10% of length shall be rejected. Damaged length need not be continuous.
  C. Before concrete placement around sheathing, all tendon damage shall be repaired to watertight condition. Repairs shall be approved by Engineer. *You really need to observe the entire placement. Watch what is happening and arrange to have a competent, conscientious ironworker assigned to you to make repairs and replace chairs at your direction.*
  D. Inspect sheathing for unrepaired damage, for watertight seal between sheathing and anchor, and for correct installation of anchors, before concrete is placed around tendons. *Same comment as for C above.*

3.04 PROTECTION *Watch this being done.*
  A. After removing excess length of tendon, exposed end of tendon and chucks shall be made watertight by covering with approved grease-filled tendon cap, or by other approved methods.
  B. After sealing exposed end of tendon and chucks, and before grouting tendon pocket, coat pocket surfaces with bonding agent.
  C. Grout tendon pockets solid with non-shrink, non-stain, chloride-free grout as specified in Section 03300.

3.05 EXTRA STOCK
  A. Maintain on site adequate supplies of repair tape and tendon grease to make repairs.
  B. Maintain spare jack on site during post-tensioning operations. *Make sure it is there.*

APPENDIX 7-4
Section 07100, Waterproofing Systems

PART 1   GENERAL

1.01 WORK INCLUDED
  A. Single installer shall be responsible for providing complete sealant and waterproofing system designed to minimize occurrence of common sealant, waterproofing, and concrete deterioration problems. All measures called for in these Specifications will be rigorously enforced. *The intent of this provision is to permit only those contractors capable of doing all the work to bid it. Fragmentation of responsibility leads to quality problems.*
  B. In accordance with Contract Documents, furnish all labor, materials, equipment, and installation of appropriate system shown and detailed on Drawings and described in this Section:
    1. Protective concrete sealer on these surfaces: *Edit as appropriate for your project.*
       a. Supported concrete floor and roof surfaces including curbs, walks, islands, and pour strips

      b. Concrete stairs and landings

      c. Slab-on-grade within parking facility, including curbs, walks, and islands

      d. Approach drives and adjoining sidewalks within construction limits

   2. Traffic topping: Fluid applied, waterproofing, traffic-bearing elastomeric membrane with integral-wearing surface

   3. Expansion-joint sealant system

   4. Concrete-control-joint sealant system

1.02 RELATED WORK

  A. Work in other sections related to Waterproofing System:

    1. 03300 Cast-in-Place Concrete

    2. 03370 Concrete Curing

    3. 03400 Precast Concrete

    4. 07920 Caulking and Sealants

    5. 09912 Pavement Marking

  B. Materials shall be compatible with materials of related Work with which they come into contact, and with materials covered by this Section—*especially curing compounds. Sodium-silicate-based curing materials are prohibited in Section 03370, Concrete Curing because:*

    All curing compounds tested in accordance with ASTM C309 are intended to be membrane-forming compounds. This essentially means that resin-based compounds capable of forming a film serve as the primary mechanism for water retention and concrete curing. Sodium silicate assists in water retention by chemically reacting with the calcium hydroxide in the concrete and reducing the concrete porosity by forming calcium silicate gels. The water retention mechanisms are not similar.

    Sodium silicate is primarily used as a floor hardener. Some manufacturers have tried to use sodium silicate as a sealer; however, sodium silicate has performed poorly as a sealer when tested in accordance with NCHRP 244.

    Although some sodium silicate formulations can meet the requirements of ASTM C309, they take advantage of a relatively weak ASTM test. The ASTM committee is presently considering a ballot for more stringent criteria for the ASTM C309 test. This step may eliminate marginally adequate sodium silicate curing materials.

1.03 QUALITY ASSURANCE

  A. Testing Agency: Independent testing laboratory employed by Owner and approved by Engineer.

  B. Prequalification of Bidders:

    1. At prebid conference, submit proposed Section 07100 subcontractor qualifications. Engineer shall notify General Contractor whether or not subcontractor is acceptable within 48 hr of prebid conference close.

    2. Prequalification criteria, all in writing:

      a. Evidence of compliance with 1.04.C below, and with 1.01.A above.

b. Evidence of acceptable previous work on WALKER-designed projects. If none, so state.

c. Engineer retains absolutely, right to reject any prequalification statement.

C. General:

1. Provide written certification by each system manufacturer to Engineer that system installer is approved applicator. *Paragraphs 1 and 2 guard against applicator inexperience and ensure that the manufacturer is involved in the application quality-control process.*

2. All Work under Section 07100 shall be performed by organizations which have successfully performed at least five verifiable years of installations similar to those involved in this contract. System installer shall submit listing of five or more prior installations in climate similar to that for this Project.

3. Final selection of Section 07100 installer shall be subject to approval of Engineer. Engineer retains right to reject any installer. *This is our last defense to prevent an unqualified installer from getting the work.*

4. All Section 07100 Work shall be under immediate control of person experienced in this type Work. Exercise close check and rigid control of all operations as necessary for full compliance with all requirements. Contractor's superintendent assigned to Project shall have supervised five prior projects of similar magnitude and design, and shall be present during all operations. Superintendent shall be approved by Engineer. Engineer retains right to remove superintendent from project if he fails to ensure full compliance with Specification. *Do not be shy about taking necessary action here either.*

5. Submit certification that products and installation comply with applicable EPA and OSHA requirements regarding health and safety hazards.

D. Protective Concrete Sealer:

1. Manufacturer: Provide complete system of compatible materials designed to produce protective sealer.

2. Manufacturer: Furnish sample of materials to be used on Project to Owner for spectrographic analysis. *Paragraphs 2 and 3 guard against substitution of some other material for that specified. Unfortunately, we have found that such measures are necessary. In the preconstruction meeting, let the sealer installer know about these safeguards; then be sure to follow up.*

3. General Contractor: Furnish additional samples—one for each 100 gal of materials used—to owner as sealer is installed. Designate lot and batch number from which each sample is taken and identify bay and tier where sampled batch is applied. All samples will be sent to Testing Agency for spectrographic comparison with original sample to verify material installed on Project is same as that submitted initially by manufacturer as standard.

4. Sealer shall be applied to project at same rate used to pass NCHRP 244 test. However, Section 07100, Part 3 must also be met. *This sen-*

tence refers to the skid-resistance requirement. If both requirements cannot be met simultaneously with single application rate, sealer will be rejected. *This paragraph guards against the specified product being applied at a lesser rate than that used to meet the specification requirements. Also, the NCHRP 244 test has no provision for excluding materials which will reduce skid resistance, making the sealed surface slippery.*

    5. Testing agency shall take three cores—minimum in each instance—from high traffic-wear areas, and trial sections referenced under Section 07100, Part 3, to test for sealer effectiveness in accordance with ASTM C642 and modified as follows: concrete core samples shall be taken 14 days after application of sealer. Report water absorption through top and bottom surfaces of core only. Sealer effectiveness as determined by comparison of water absorption through sealed top surface and core bottom surface shall be at least 85%. *This is the last check to ensure that the Owner is getting what was specified.*

E. Traffic Topping:
    1. Manufacturer: Review concrete finish specification and confirm in writing to Engineer that finishes as specified are acceptable for system to be installed; *otherwise, if there are problems later, the topping manufacturer can and will blame the finish for those problems.*
    2. Confirmation: Send to Engineer one month before placement of any concrete which will receive traffic topping.

F. Expansion-Joint Sealant System:
    1. Manufacturer: Review and approve all details before construction. Confirm in writing to Engineer. *Follow up.*
    2. Coordinate services with related Work, including layout of joint system and approval of methods for providing joints.
    3. Inspect site to ensure proper joint configuration in field.
    4. Testing Agency shall check Shore A hardness in accordance with ASTM D2240. *Too soft or sticky a seal will create a tripping hazard.*

G. Concrete Control-Joint Sealant System:
    1. Review and approve joint details before construction.
    2. Coordinate layout of joint system and approve methods for providing joints with precast concrete and concrete contractors.
    3. Inspect site and precast plant before precast production to ensure proper joint configuration. *Follow up. Same comment as for E.1 above.*

1.04 REFERENCES
A. American Society for Testing and Materials (ASTM):
    1. ASTM C642, "Test for Specific Gravity, Absorption and Voids in Hardened Concrete"
    2. ASTM D2240, "Test for Indentation Hardness of Rubber and Plastics by Means of a Durometer"
    3. ASTM E119, "Fire Tests of Building Construction and Materials"

1.05 SUBMITTALS
A. Make submittals in accordance with requirements of Division 1 and as specified in this section.

B. General: *Be sure to get these. Act quickly and decisively if submittals are unacceptable.*
   1. Section 07100 contractor's experience record for past 5 yr
   2. Superintendent qualifications
   3. Evidence of applicator's being licensed by manufacturer
   4. Reviewed Shop Drawings to all others whose Work is related
   5. Five copies of snow removal guidelines for areas covered by guarantee
C. Protective Concrete Sealer System:
   1. Written computations to Engineer of material quantities to be applied to concrete surfaces at least two weeks before sealer application. *Follow up to see that calculated quantities are applied.*
   2. Proposals for alternate application methods to Engineer at least 2 wk before sealer application
   3. Alternate materials, with full text of test reports required in Part 2, at least 3 mo before application.
D. Traffic Topping:
   1. Written computations to Engineer of material quantities (by components) to be applied to concrete surface at least 2 wk before application of traffic topping. *Follow up to see that calculated quantities are applied.*
   2. Proposed method of preparation of concrete surface. *Do not permit acid etching. This specification covers this prohibition, but acid etching sometimes slips in. More on this subject later in this specification.*
   3. Proposed method and details for treatment of cracks on concrete surface.
   4. Product samples.
   5. Quality-Control Procedures: System manufacturer shall submit written quality-control plan to Engineer for approval one (1) month prior to construction for application procedures which specifically address following:
      a. Surface preparation acceptance criteria
      b. Method of application of coats
      c. Primer type and application rate
      d. For all coats, wet mils required to obtain specified dry thickness
      e. Number and type of coats
      f. Quality-control plan for assured specified uniform membrane thickness that utilizes grid system of sufficiently small size to designate coverage area of not more than 5 gal at specified thickness. In addition, employ wet mil gauge to continuously monitor thickness during application. Average specified wet mil thickness shall be maintained within grid during application with minimum thickness of not less than 80% of average acceptable thickness. Immediately apply more material to any area not maintaining these standards.
      g. Amount of aggregate required with each coat
      h. Maximum and minimum allowable times between coats

   i. Temperature, humidity, and other weather constraints
   j. Final cure time before resumption of parking and/or paint striping
   k. Any other special instructions required to ensure proper installation

E. Expansion-Joint Sealant System:
  1. Material samples
  2. Installation plans and large-scale details
  3. Field samples of premolded joint sealant. Width, thickness, and durometer hardness of sealant shall be checked by Testing Agency. *Follow up. One problem which may arise is that the sealant thickness and concrete blockout depth are unequal.*
  4. Other information required to define joint placement or installation

F. Concrete Control-Joint Sealant System:
  1. Material samples
  2. Installation plans and large scale details
  3. Any other information necessary to show placement of control-joint system

1.06 DELIVERY, STORAGE, AND HANDLING

A. Deliver all materials to site in original, unopened containers, bearing following information:
  1. Name of product
  2. Name of manufacturer
  3. Date of preparation
  4. Lot or batch number
    Watch for new labels applied over old ones. We have seen it happen.

B. Store materials under cover and protect from weather. Replace packages or materials showing any signs of damage with new material at no additional cost to Owner.

C. Do not store material on slabs to be post-tensioned before final post-tensioning of slabs is accomplished. At no time shall weight of stored material being placed on slab area, after post-tensioning is completed and concrete has reached specified 28-day strength, exceed total design load of slab area. Between time final post-tensioning is accomplished and time concrete has reached specified 28-day strength, weight of stored material placed on slab area shall not exceed half total design load of slab area. *Review this requirement at the preconstruction meeting.*

1.07 WARRANTY

A. This article applies to all materials covered in Section 07100 EXCEPT Protective Concrete Sealer. *Even though warranties are offered by most sealer manufacturers, and many have used warranties of up to 10-yr duration as marketing tools, our opinion is that 1) Warranties are simply a gamble by the manufacturer that his product will hold up under a variety of conditions. 2) Warranties increase the sealer initial cost without adding any real compensating benefits. 3) Warranty statements contain so many legalisms and exceptions that they are virtually worthless.*

An exception is one manufacturer's warranty which includes benchmark testing and annual follow-up testing and maintenance performed under a separate agreement. Instead of warranties, we rely on our testing and installation requirements. Vendors also know that one dishonest performance will bar them permanently from our master specification.

B. System Manufacturer: Furnish Owner with written total responsibility guarantee from manufacturer that system will be free of defects, water penetration, and chemical damage related to system design, workmanship, or material deficiency, consisting of:
   1. Any adhesive or cohesive failures
   2. Spalling surfaces (applies to traffic topping only)
   3. Weathering
   4. Surface crazing (does not apply to traffic topping protection course)
   5. Abrasion or tear failure resulting from normal traffic use
C. If material surface shows any of defect listed above, supply labor and material to repair all defective areas and to repaint all damaged line stripes.
D. Guarantee period shall be 3 yr commencing with date of acceptance of Work. *Three years is fairly standard in the industry. Anything longer may cost disproportionately more, but is usually available.*
E. Perform any repair under this guarantee at no cost to Owner.
F. Before construction, provide Engineer with sample of final guarantee. Guarantee shall be provided by manufacturer.
G. Snow plows, vandalism, abnormally abrasive maintenance equipment, spinning studded-snow tires, truck traffic, and construction traffic are not normal traffic use and are exempted from warranty.
H. State in bid additional cost of providing 5-yr total guarantee in lieu of 3-yr guarantee as defined above.

Many of the above provisions and precautions may seem extreme, but we have been given reason to institute every one.

PART 2   PRODUCTS

2.01   MATERIALS
A. Protective Concrete Sealer:
   1. Acceptable protective concrete sealers are listed below:
      a. Class S (*trade names deleted*):
      b. Class H (*trade names deleted*): *The S in Class S stands for Standard. The H in Class H stands for Heavy Duty. Class H product-acceptance requirements are more stringent, so performance is expected to be better and service life longer than for Class S. Class H products are also more expensive than Class S products. We developed the two categories so we could choose the appropriate performance level while comparing prices on an apples-to-apples basis. Which class to choose for each application may be a matter of economics or preference, but in general, if there is opportunity*

for choice, choose the Class H product for higher traffic areas, or for the entire structure in colder climates, if affordable.

    2. Supplier shall furnish application rate at which following tests were passed:

We instituted the following provisions as one of the first performance specifications for sealers. So far it has worked well. Requirements given here for the 244 tests are more stringent than those used in the original NCHRP study.

       a. NCHRP 244 tests:

         1) Four-Inch Cube Series II: Upper limits of average weight gain and chloride content at completion of cube test series shall be limited to: Class S: 21% of weight gain and 20% of net chloride content of untreated control cubes; Class H: 15% of weight gain and 13% of net chloride content of untreated control cubes.

         2) Southern Climate: Upper limit of average chloride content at end of 24-wk test shall be limited to: Class S: 10% of net chloride content of untreated control slabs; Class H: 3% of net chloride content of untreated control cubes.

       b. ASTM C 672 test: In Part B, acceptable rating for Class S shall be "one" and for Class H shall be "one minus."

    3. Proposed substitutions: Contact Engineer. *We only permit a substitution if the proposed substitution meets all the testing requirements given above. Since the tests are not cheap and take several months to run, the proposed substitution seldom has the required credentials. We prefer to consider a product for inclusion in our master specification, where it is then eligible for all projects, rather than accepting it under time pressure for one project. This same philosophy holds true for all product categories.*

  B. Traffic Topping (*sometimes called Traffic-Bearing Membrane*):

    1. Acceptable toppings:

       a. Class C (*trade names deleted*):

       b. Class E (*trade names deleted*): *The reasons for choosing a C to indicate the standard grade of traffic topping and an E to designate the heavy duty grade of products are forgotten. However, the commentary for sealer class designations following 2.01.A.1.b above applies here as well, with the following exceptions: It is not uncommon to use both classes of topping from the same manufacturer on the same project. The Class E product might be applied at high wear areas, such as entrances and exits, and the Class C product might be used for areas of lighter wear, such as parking stalls. Class E products will be more expensive than Class C products and may not fit into the project budget. For that reason, since predicting bids is not an exact science, both Class E traffic toppings and Class H sealers are often designated "add alternates" so one or both may be accepted, if feasible.*

         Proposed substitutions: Contact Engineer to obtain copy of document PM-2-88, which contains requirements for proposed substitutions. *The commentary directly above for paragraph A.3 applies here also. Though the requirements for substitutions are not listed here because of their length, they are on the PM-2-88 document and are written on a performance basis similar in concept to the sealer requirements above.*

C. Expansion-Joint Systems: *See Chapter 5 for figures.*
   1. Premolded sealant system where shown and as detailed on Drawings:
      a. Acceptable materials (*trade names deleted*):
      b. All premolding: By material manufacturer at factory
      c. All sealant strips: Fabricated in one piece, continuous from end to end of joint
      d. Premolded sealant systems: Shore A hardness:
         1) 25 minimum between −15°F and +120°F
         2) 20 minimum between 121°F and 150°F
         3) 25 minimum at installation
   2. Other expansion-joint systems where shown and as detailed on Drawings (*trade names deleted*): *There are no standard tests for expansion-joint sealant systems at this writing. Requests for substitutions are handled by a single person who speaks for the entire firm. Decisions are necessarily subjective, based on experience. Before accepting a product into our master specification, we give a small trial installation, a 1- or 2-yr in-service performance test on a few selected projects. The owner of the selected project must give his permission for the trial and usually receives a price break. He always receives a guarantee that an originally specified product will be installed free if the trial product does not perform.*

D. Concrete Control-Joint Sealant System: *See Chapter 5 figures.*
   1. Provide complete system of compatible materials designed by manufacturer to produce waterproof, traffic-bearing control joints as detailed on Drawings.
   2. Compounds used for sealants shall not stain masonry or concrete. Aluminum-pigmented compounds not acceptable.
   3. Color of sealants shall match adjacent surfaces.
   4. Bond breakers and fillers: As recommended by system manufacturer.
   5. Primers: As recommended by sealant manufacturer, used in accordance with manufacturer's printed instructions.

E. Sealant for Horizontal (Non-Cove) Joints:
   1. Acceptable control-joint sealants (*trade names deleted*):

F. Sealant for Vertical and Cove Joints:
   1. Acceptable sealants (*trade names deleted*):

G. Fire-rated joint-sealing systems are shown and detailed on Drawings: *Usually used to seal flange-to-flange joints between pretopped precast double tees where a fire-rated joint sealant is required by the local building code.*
   1. Fire-rated joint-sealing systems shall prevent passage of flame or hot gases and stop transmission of heat in accordance with ASTM E119 and be two-stage, 2-hr-rated joint system.

2. Backup filler: Approved materials are:
   a. Polyethylene closed-cell foam backup
   b. Ceramic-fiber blanket insulation

PART 3   EXECUTION

3.01   INSPECTION
A. General
1. Inspect surfaces to receive Work and report immediately in writing to Engineer any deficiencies in surface which render it unsuitable for proper execution of Work.
2. Coordinate and verify that related Work meets following requirements: *Make sure that these requirements are met or later problems will only bring finger pointing, not solutions.*
   a. Concrete surfaces are finished as acceptable for system to be installed.
   b. Curing compounds used on concrete surfaces are compatible with Work to be installed.
   c. Concrete surfaces have completed proper curing period for system selected.
3. Acid etching: Prohibited. *The acid, even when diluted by wash water, can be an environmental hazard. Also, if it is not completely removed, the acid will remain active beneath whatever is applied over it and may cause problems later on.*

3.02   PREPARATION   *Proper surface preparation is the key to problem-free installations.*
A. Protective concrete sealer:
1. All control-joint and expansion-joint Work shall be complete and accepted by Owner before beginning protective sealer surface preparation and application.
2. This contractor shall be responsible for repair or replacement of all sealant materials damaged by surface preparation operations.
3. Clean surfaces to be treated as specified by sealer manufacturer before sealer application.
4. Equipment used during floor-slab cleaning shall not exceed height limitation of facility and shall not exceed 3000 lb axle load or vehicle gross weight of 6000 lb.
5. Water jet pressure of waterblasting equipment, if used, shall have constant pressure between 8000 and 10,000 psi. Keep water jet from previously installed sealant materials. *High-pressure water jets can be a safety hazard and can do significant damage to existing construction work.*
B. Traffic Topping:
1. Remove all laitance and surface contaminants, including oil, grease, and dirt in accordance with manufacturer's recommendations. If wet preparation system is employed, substrate shall be allowed to dry as required by manufacturer of traffic topping.

2. Before applying materials, apply system to small area to assure that it will adhere and dry properly and to evaluate appearance.

3. All cracks on concrete surface shall be prepared in accordance with manufacturer's recommendations.

C. Expansion-Joint Sealant System:

1. Coordinate expansion-joint sealing system with other related Work before installation of expansion-joint sealant.

2. Check adhesion to substrates and recommend appropriate preparatory measures.

3. Proceed with expansion-joint sealant only after unsatisfactory conditions have been corrected in manner acceptable to installer.

4. Clean joints thoroughly in accordance with manufacturer's instructions to remove all laitance, unsound concrete, and curing compounds which may interfere with adhesion.

5. Cease installation of sealants under adverse weather conditions, or limitations for installation.

6. Prepare for installation of extruded expansion-joint systems in accordance with manufacturer's recommendations.

D. Concrete Control-Joint Sealant System:

1. Correct unsatisfactory conditions in manner acceptable to installer before installing sealant system.

2. Grind joint edges smooth and straight before sealing. All surfaces to receive sealant shall be dry and thoroughly cleaned of all loose particles, laitance, dirt, dust, oil, grease, or other foreign matter. Obtain written approval of method from system manufacturer before beginning cleaning.

3. Check preparation of substrate for adhesion of sealant.

4. Prime and seal joints according to manufacturer's directions and protect as required until sealant is fully cured.

5. Cease installation of sealants under adverse weather conditions, or when temperatures are outside manufacturer's recommended limitations for installation.

3.03   INSTALLATION/APPLICATION

A. General:

1. Do all work in strict accordance with manufacturer's written instructions and specifications including, but not limited to, moisture content of substrate, atmospheric conditions (including relative humidity and temperature), coverages, mil thicknesses, and texture, as shown on drawings.

B. Protective Concrete Sealer:

1. Submit manufacturer's recommended application rates in writing before start of sealer application. Section 07100, Part 1 states minimum rate.

2. All concrete to be treated shall be cured above 50°F for at least 14 days before applying sealer.

3. All concrete to be treated shall be air dried for at least 72 hr (following

surface wetting) at temperatures above 50°F immediately before applying protective sealer system.

4. Ambient and concrete temperatures shall be 50°F or higher during application of protective sealer, but temperature, humidity, and wind velocity shall be within manufacturer's specified limits to prevent solvent flash-off. *Important. If the solvent flashes off, you may have no sealer left on the concrete.*

5. Install three trial sections of sealer to verify treated surface is not glazing as result of sealer application. If application of sealer causes glazing at trial section, Contractor shall contact sealer manufacturer to obtain written recommendations for solving problem. Contractor shall not proceed with sealer application following trial section applications until directed to do so in writing by Engineer. *This provision protects against having a slippery surface. Such a surface is usually caused by insufficient surface preparation, by too heavy an application rate, or by too impermeable a surface for the sealer to penetrate.*

6. Clean all surfaces affected by sealer-material overspray and repair all damage caused by sealer-material overspray to adjacent construction or property at no cost to Owner.

7. Unsatisfactory test results reported under Section 07100, Part 1 shall be grounds for rejection of sealer and sealer application or sealer reapplication at no additional cost to Owner.

C. Traffic Topping:

1. Manufacturer's technical representative, acceptable to Engineer, shall be on site during installation.

2. Do not apply coating material until concrete has been air dried at temperatures at or above 40°F for at least 30 days after curing period specified in Section 03370.

3. All adjacent vertical surfaces shall be coated with traffic topping minimum of 4 in. above coated horizontal surface. Requirement includes columns, walls, curbs, and islands.

4. Complete all Work under this Section before painting line stripes.

D. Expansion-Joint Sealant System:

1. During months when historic mean daily temperature at Project is 20°F or colder than annual mean daily temperature, install premolded sealant on temporary basis to prevent hot weather buckling. Permanent installation shall be done in summer when Engineer directs.

2. Install extruded expansion-joint system in accordance with manufacturer's instructions.

E. Concrete Control-Joint Sealant System:

1. Completely fill joint without sagging or smearing onto adjacent surfaces.

2. Fill horizontal joints slightly recessed to avoid direct contact with wheel traffic.

Chapter 8
Construction

ANTHONY P. CHREST

8.1 INTRODUCTION

This chapter is for owners, engineers, and builders. The construction process involves all of you, or should. When it does not, problems arise. All too often relations between owner, engineer, and builder become adversarial shortly after construction begins. Then they stay that way for the rest of the project. Instead of working *with* each other, the team members waste most of their energy working *on* each other. This sorry state does not have to be. Imagine how much less time a project would take if even half the disputes and "cover your tail" paperwork could be cut; it can be done; it has been done. Trust is the key; however, trust is not free; it must be earned. Earn trust by fair treatment, clear communications in every step, and clear communications *before* every step. All this will take unaccustomed effort. It also initially may require behavior which makes you uncomfortable, but the resulting savings in time and hassle will be worth it. Give it a chance.

8.2 COMMUNICATION

8.2.1 Plans and Specifications

The lifeblood of the construction process is communication—clear, concise, and constant. It must begin well before construction begins—during plans and specification preparation in the engineer's office. At this point the owner and engineer must communicate. The owner must make clear his intentions for his project. The engineer must achieve that intent while keeping the project within the owner's total budget. Trade-offs during this process are likely.

210

Repetition through different kinds of communication will make the message clearer. It is usually best to follow up spoken messages with written confirmation of one party's understanding of what was said. Invite corrections before a stated deadline. If the deadline passes with no corrections received, the ground rules must say that the written message stands. Memories are short, especially in disputes where what really happened conflicts with what we would like to have happened.

During the preparation of plans and specifications, engineer and owner should review the documents together, signing off as they go. This practice, if followed, will bring the project from start to finish with few detours. Keep in mind though, that some arguments will happen and may even make for a better project.

Drawings must clearly give the prospective builder enough information both to estimate a construction cost and to build the project.

Quality control during drawing production is important. Drawings for a recent project, for which we were consulted as an expert witness, showed an obvious conflict between column vertical bars and slab dowels, yet the conflict was not noticed until construction, when the dowels pulled out of the columns. When questioned, the response from the "Engineer of Record" was, "The draftsman showed it that way." The response betrayed both lack of quality control and lack of knowledge of what was on the drawings.

Enough has been said elsewhere on the subject of unclear documents. Unfortunately, there are still engineers who take on document preparation at a fee so low as to prevent doing a complete job. What results is a project with many problems, an unhappy owner, and lawsuits. Any savings resulting from the low design fee will be wiped out by added costs during construction. A higher design fee may produce a lower total cost for a project. So, if an engineer quotes you—an owner—a fee which seems too good to be true, it probably is.

8.2.2   Understandable Specifications

Too many design professionals believe that a good specification is full of legal-sounding words and phrases. A review of proprietary master specifications such as the PSAE "Masterspec" or the Construction Specifications Institute "Masterspec" shows that the briefer, the better. Specifications must be clear, concise, and no longer than necessary. Look for ways to shorten your specifications while still getting your message across.

8.2.3   Bidding or Negotiating

Common bidding procedures often work against good communications during this critical phase of a project's life. Adversarial relationships may begin here because information that one party, usually the bidder, wants is not available.

Use a prebid conference during the middle of the bidding period to emphasize to bidders what you want them to understand, particularly those items unique to parking structures or to your specification. (See the commentary to Chapter 7 appendixes for some examples.) Though a prebid conference is common practice our experience has been that meaty questions are seldom asked there. Bidders, being competitors for the project, will not say too much for fear of giving away a real or imagined edge. The result, while the motives are certainly understandable, is little worthwhile communication.

Another problem with bid jobs is there may be little control over who may bid and so, who gets the work. Unless prequalification procedures are permitted and used, the owner may have no control over to whom he gives his hard-earned money. Even then the successful contractor may not be the owner's first choice. So, bidding may result in a forced marriage. How many of those work?

Step one in successful construction price negotiating is owner selection, with engineer advice, of a reputable builder with demonstrated success in parking structure construction. Check the prospective builder's financial stability too, even (and especially) if he is a friend or has worked with you before. In negotiating, the parties must talk. Make available the opportunity for give and take on both sides. All three sides must win the negotiation. Fairness is the key. Mutual trust and a spirit of working *with*, not *on*, one another will begin here. The negotiated price may end up higher than a bid price might have been; however, the final negotiated project price will not, because there is better understanding by all parties of what is in the project, which will result in fewer change orders during construction. Also, the owner's increased control over who does the construction work stacks the deck for just such an outcome.

8.2.4   Preconstruction

Before the initial euphoria of signing the construction contract wears thin, begin preconstruction planning. The owner, engineer, and builder must meet with all the major players in the construction process:

Concrete supplier
Concrete placer

Concrete finisher
Concrete curer
Concrete inspector
Concrete tester
Precaster supplier
Precast erector
Others

"Others" includes contractors, suppliers, and professionals who are involved in almost any multilevel building construction. Examples are excavators, elevator contractors, roofers, soil engineers, and painters, to name a few. We are concentrating on matters more particular to parking structures. Since concrete is the single-most important and expensive component of most parking structures, we emphasize it here. It may be desirable to hold more than one preconstruction meeting for different parties and/or during different phases of construction. Example: Why discuss painting until the builder schedules that work?

The outcome of a preconcrete meeting, one of the preconstruction meetings to be held, must be a written procedure which spells out who will do exactly what and when. This written procedure must be reviewed, discussed, and signed by every party involved in the work for the items listed:

Mix design(s)
Mix design(s) submission
Mix design(s) approval
Concrete proportioning and mixing
Concrete transportation
Concrete placement method
Consolidation procedures
Finishing procedures
Weather protection procedures
Curing procedures
Testing procedures, locations, and frequency
Testing personnel qualifications
Inspection requirements
Inspector qualifications
Test and inspection reporting
Repair criteria and procedures
Conflict resolution
Acceptance criteria
Criteria for reduced payment
Suppliers' quality control

Subcontractors' quality control
General contractor's quality control
Designer's site quality control

Every party must keep a copy of the agreed-upon procedure on hand, particularly when he is at the job site. The above list may seem too long, but every item listed has been disputed among some of the parties involved in construction. Several examples of common disputes are in the following paragraphs. You can avoid them all by reaching agreement on the list above before construction begins, not after the problem occurs.

On several projects a year, we used to receive a frantic call from the builder, "We are ready to pour footings. The truck is here. Is the mix design okay?" To which we would reply, "What mix design?" because we had not received it. The builder, "Well, can we pour anyway?" We always say no, making sure that the problem will never again arise on that project.

Another dispute may arise just after the first compression test-cylinder reports come in, showing that some concrete is under strength. Next, the builder alleges that the testing agency's site person took, stored, handled, and transported the test cylinders wrong. Besides, the person was just some kid who did not know which end was up anyway. The testing agency stoutly denies all these ridiculous claims. The important point, whether or not the concrete is really under strength, is left hanging.

The project specification clearly states that provisions must be ready to protect fresh concrete from rain and snow. The first time it rains we suddenly discover that no protection of any kind is ready. Once more, blame is tossed back and forth while the real problem is obscured. The unacceptable concrete surface will still be there, too.

Opportunities for mistakes abound during the finishing process. Since the slab finish is exposed to the weather, it has to be durable. The best finish is therefore one which can be produced with the least working of the surface possible. It also must be roughened somewhat so it is not too slippery for tires or shoes. The best way to define the finish: take the finishing crew foreman to another project; show him what you want for appearance; spell out the finishing requirements in the project specification and review them with the foreman also. For these requirements, see Chapters 6 and 7. Take the time to explain why you want what you do. We have found that few engineers do take time to give explanations for what they want. When we do, it really makes an impression. The people doing the work are appreciative of our concerns and become very cooperative.

If you see a concrete finisher with a bucket of water doing anything other than cleaning his tools with it, get rid of the bucket. Some
finishers do not feel they have done a good job unless they add water
to the surface as they work. This practice does make finishing easier,
but it results in a weakened surface which will deteriorate quickly.
Avoid this potential problem too, with good use of a preconcrete meeting.

If the project includes precast concrete, a similar meeting with a
similar result is necessary for:

Precast concrete mix design(s)
Precast concrete mix design(s) submission
Precast concrete mix design(s) approval
Reinforcement and accessories submission
Reinforcement and accessories approval
Shop drawing submissions, including schedule
Shop drawing approvals
Design requirements (if applicable)
Design submission (if applicable)
Design approval (if applicable)
Manufacturing tolerances
Testing procedures
Inspection procedures
Testing personnel qualifications
Inspector qualifications
Test and inspection reporting
Precaster's quality control
Fabrication, curing, storage, and handling procedures
Erection plan
Shipping plan
Handling procedures at site
Erection procedures
Erection tolerances
Erector's quality control
Engineer's site quality control
Repair criteria and procedures
Acceptance criteria
Criteria for reduced payment

The following examples discuss areas of potential dispute. You
can avoid them by good communication in a preconstruction meeting.

Most precasters would prefer that the engineer not check their
shop drawings. The checking process delays the start of production

even if all drawings are correct. More delays occur if the engineer finds mistakes or makes changes. These delays cut into the already tight schedule, completely filling the time between precast concrete contract signatures and erection of the first piece. Yet, if no one checks the shop drawings, serious mistakes will carry through to fabrication, erection, and the final structure. We know of two recent structures where members were made with insufficient reinforcement. The result in both cases was partial dismantling of the structures and replacement of the defective pieces.

We now have a policy of requesting from the precaster a list of what we call the basic piece marks. We check the shop drawings for the basic piece marks carefully. The precaster must follow corrections made on the basic piece-mark shop drawing for the applicable associated piece marks. An example of a basic piece mark is T1, a typical 10-ft-by-2-ft, 55-ft double tee. Another is B4, a typical 8-in. thick, 6-ft deep, 30-ft L beam. An associated piece mark for T1 might be a double tee mark T1a. It is the same in all respects including reinforcement, as T1, but has a blockout for a floor drain at one corner. An associated piece mark for B4 could be an L beam mark B4c, like B4, but 6 in. shorter.

Improper storage practices in the precaster's holding yard can result in damaged pieces. Your experience may suggest that you caution the precaster about support points and stacking heights, even though these matters are not really your responsibility.

Other potential trouble spots are erection tolerances which affect connections, and the sequence of erection and connections. The Prestressed Concrete Institute has adopted some erection tolerances, but you may have to adopt others to fit your project requirements. If you do, you are also probably aware of who the precast bidders are likely to be for your project. Before you add new tolerances, check with the prospective bidders to get a consensus of what is realistic. Then try to work from that information. Most erectors will want to make permanent connections as late in the erection process as they can. Doing otherwise interrupts and delays erection, costing expensive crane time. However, C-clamps and other temporary connections are not as strong as permanent ones. Frame stability may be less than it should. With all the serious construction collapses during the past five years the last thing anybody needs is a collapse waiting to happen. You should agree on some middle ground so permanent connection installation does not lag much behind erection.

Another preconstruction meeting must be held for those involved with concrete reinforcement, especially if post-tensioned (P/T) reinforcement will be used:

Materials submissions
Materials approvals
Shop drawing submissions
Shop drawing approvals
Design requirements
Placement tolerances
Stressing procedures
Stressing record keeping
Stressing record submission
Stressing record approval
Tendon cutoff procedures
Stressing pocket fill and seal procedures
Repair criteria and procedures
Acceptance criteria
Contractors' quality control
Inspection procedures
Inspector qualifications
Engineer's site quality control

Note the number of items in the list having to do with stressing the P/T tendons. That is because there are usually problems. Too often the stressing records submitted are too neat and the recorded elongations are exactly the same. We do not trust such reports; they are usually generated in the office and have little to do with actual elongations. Another point to convey: we do not allow the post-tensioner to cut off tendons and fill the stressing pockets until he resolves all tendon-elongation discrepancies.

8.2.5 Construction

Whether you are the owner, builder, or engineer, if you laid the groundwork properly during the preconstruction phase, you will have fewer communications problems during the construction phase. Do not relax, though; answer questions from the other two major parties quickly and completely; concentrate on building and maintaining trust. This business of trust is not easy. It requires setting your own ego second to the success of the project. Whether or not you are covered against any eventuality is not all that important. What is important is the completion of the project according to all requirements—including being on schedule and under budget.

A few years ago, to sort out some ongoing construction disputes, the architectural, engineering, and construction management firms met. At the end of an inconclusive meeting, the architect's principal announced that he was leaving before the general contractor was to come in for the next meeting. His comment, ''I do not want to talk to any contractors.'' The project finished with poor quality, well behind schedule, and considerably over budget. The lawsuits are still going on and show no signs of ending soon. But, the principal's ego was protected.

Many of us approach construction problems by ignoring them in the hope that they will go away or get better. They do not.

8.3   BUILDABLE DETAILS

Designers do not contribute to construction problems by requiring details that have to be built like a watch. Our firm requires all design engineers to spend at least six months on a construction site so they can see first hand what details work and which do not.

One situation that may arise is the case when a detail which works well with precast concrete is translated into a cast-in-place detail; it may have little chance of proper execution. Consider the detail shown in Figure 8-1. This detail is not uncommon in precast parking structures. It shows a beam-column sliding joint. A few years ago a contractor called us in to help him in a construction dispute. The plans showed this detail, but in cast-in-place concrete. Note the construction sequence required.

First, the column must be formed and placed to the top of the haunch. Next, form and place the portion of the column above the haunch, then strip the column form. Install the lower slide-bearing

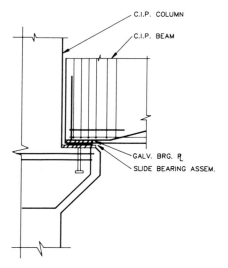

Figure 8-1.   Beam slide bearing.

assembly so that it will not be displaced during subsequent steps; form
the beam. The end of the beam bottom form must be left open to allow
the upper slide-bearing assembly to bear directly on the lower assembly. Install the upper bearing assembly so that it too will not be displaced during later steps. Seal between the upper bearing assembly
and the beam form so that the cement paste will not leak out, leaving
voids. Install the congested beam end and main reinforcement. Place
the beam concrete.

   Figure 8-2 illustrates a similar detail for a slab bearing and sliding
on a beam. While not as difficult to build as the detail shown in Figure
8-1, it is easier to build it wrong than right.

   Rather than use details like those shown in Figures 8-1 and 8-2,
we recommend using only details which are easy to build in the particular structural system. With but a few specific exceptions, we recommend avoiding slide bearings unless no other solution works.
Avoiding slide bearings may result in having to add a column to replace a group of beam slide bearings, but the extra initial cost will be
saved in lower maintenance costs. The only condition where we use
slide bearings is illustrated in Figure 5-27. There the vertical load on
the bearing is low, producing low friction forces on the bearing surfaces. The anticipated movement is small because of the location of
the joint in the structure and the fact that only one side of the joint is
free to move.

   We were engaged as expert witnesses on a project built five years
ago. One of the owner's complaints was that a large number of post-
tensioning tendons had little or no cover and were becoming exposed.
The contractor had been instructed by the structural engineer to maintain a top cover of between one half and three quarters of an inch.
Conscientious workmen can maintain a vertical tolerance of plus or

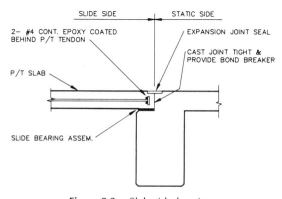

Figure 8-2. Slab side bearing.

minus four-tenths of an inch in placing reinforcement. Within that tolerance, the tendons would be placed with covers ranging from one tenth to one and fifteen hundredths inches. The lower figure is just about what the engineer got. Why is the contractor being criticized?

8.4 CONSTRUCTION SEQUENCE

The task of building a structure with floors that continually slope carries with it a few problems which may not be immediately apparent to one who has never built one before. Obviously, the floors are not discrete segments, but form a continuous ribbon of concrete from bottom to top and back again. While the engineer will suggest a construction sequence and may give certain constraints, the builder must decide for himself what the best arrangement of construction joints and concrete pours will be. Figure 8-3 gives an example of a suggested construction sequence. The circled numbers indicate the order of concrete placement. (See Figures 8-4, 5-27, and 5-29 for sections cut on Figure 8-3. Also see Chapter 7 for further discussion.)

8.5 SITE VISITS

8.5.1 General

Engineer, set the tone for all future visits with your first one. Answer all questions promptly and completely. If you do not know an answer, say so instead of trying to come up with a statement that you may have to retract later. Do get a response back to the questioner quickly if you

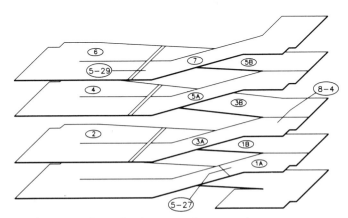

Figure 8-3. Example of suggested construction sequence.

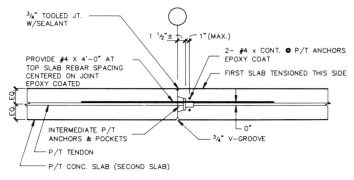

Figure 8-4. Section at construction joint.

cannot answer at the site. *Listen* to the project superintendent. You can learn a lot from him, if you are willing; he may have been in construction before you were born.

Do your best to maintain good relations with the builder's people, but hold your ground on important requirements. We believe that the best way to approach a construction project is to have clear requirements in the contract documents. If, on a project, a requirement cannot be fully satisfied after the construction contract has been signed, there must be a trade-off to compensate the owner for getting less than the contract stated. The trade-off could be a credit to the owner, a reduction in the originally agreed-upon construction time, or enrichment in some other aspect of the project. Whatever the eventual agreement on the particular requirement, record it. If the owner does not receive fair value, you, the engineer, have not performed.

If the contract documents give you the right to remove a contractor's or subcontractor's superintendent for nonperformance or incompetence, and such turns out to be true, do not hesitate to assert your rights. Life is too short to continue to buy trouble.

8.5.2 Before Supported Cast-in-Place Concrete Floor Placement

Rarely will a concrete parking structure floor receive a floor covering later. The only protection for the structural reinforcement is the concrete itself. Proper cover is therefore vital. Check to see that all reinforcement—conventional and post-tensioned—is placed correctly. Ensure that post-tensioning tendons are free from kinks. Check for cover and alignment of tendon anchors. See that tendon sheathing is continuous and intact from anchor to anchor. Get all sheathing tears repaired according to the procedure outlined in the Appendix to Chapter 7.

Ironworkers often tie sheathed tendons to support chairs or bars so tightly that the sheathing is crimped. The crimp provides a spot for moisture already inside the sheathing to condense. This condensation is the first step toward tendon corrosion. Get all too-tight ties redone. Better yet, work with the foreman and his crew before they start. Explain to them what you want and why. You will likely find them pleased with the time you took and quite cooperative. If the foreman does not cooperate, get one that will.

Likewise, work with the concrete finisher and his crew to be sure that they understand that you want the absolute minimum working of the concrete floor surface necessary to achieve the desired finish—screed, bullfloat, broom, or float. (See the Appendix to Chapter 7 for detailed requirements for concrete finish.) Be certain that everyone involved in the concreting operation understands that *no* water may be added to the concrete surface during finishing.

8.5.3   During Supported Cast-in-Place Concrete Floor Placement

Do not permit vibrators to be used to move concrete, even if it is superplasticized. Guard against over vibration which will drive out entrained air and bring paste to the surface. Both results reduce durability; under vibration results in honeycombing.

The typical placement (pour) in a parking structure is rectangular in plan. Place the concrete beginning at one of the shorter ends of the rectangle and bring the pour face forward toward the opposite shorter end in the direction parallel to the two longer sides. Always keep the pour face parallel to the shorter sides, which will result in the open or forward face of the pour being kept as short as possible. You should not have to advise a builder to reduce risk by maintaining a short pour face. For the good of the project, we have had to do this.

Watch for post-tensioned tendon sheathing tears, dislodged reinforcement, and crushed support chairs. Get these items repaired or replaced before they are covered by the concrete. In long-span construction with relatively shallow members, there is little tolerance for misplaced or damaged reinforcement.

8.5.4   Field Observation Guidelines

8.5.4.1   Cast-in-Place Post-Tensioned Structure. We prefer to have a project resident on site full time during all cast-in-place post-tensioned concrete construction to help guard the owner against deviations from the contract documents. If the project's budget renders a full-time resident infeasible, then we would like to have him on site at least the day before each slab and beam pour to review conventional and post-tensioned concrete placement. We would also want the resident on site during the pour itself to guard against reinforcement being displaced and post-tensioning system protection being damaged. He should also review all the concreting, finishing, and curing operations, as well as concrete quality-control testing. One person can usually handle these tasks satisfactorily for each pour. If two or more pours are occurring simultaneously, more people will be needed.

Appendix 8-1 at the end of this chapter contains a guideline, which could easily be converted into a checklist for the resident, for reviewing the items which must be checked for cast-in-place post-tensioned concrete prior to each pour.

8.5.4.2   Precast Structure. Field observations for precast concrete structures are usually confined to concern with erection and connection of the precast pieces.

There will be some cast-in-place concrete work, for instance, at the ends of the double tees for a pretopped double tee structure, or over the entire tee for untopped double tees. Treat that work as you would any other cast-in-place work, but recall the cautions about topping work in Chapter 5.

Watch for conformance to specified erection and connection sequences. Appendix 8-2 at the end of this chapter contains a guideline which could be converted into a checklist for the resident's use each day or each site visit.

8.6   PRECAST CONCRETE PLANT VISITS

Visit the plant supplying your project during fabrication of the first few pieces of each member category. Check for proper placement of reinforcement and cast-in assemblies such as connection hardware. If welding of conventional reinforcement is detailed, spot check the welds. Review concrete test results. Visit the storage yard to see if members are stacked in a way which will not damage them. Revisit the plant at appropriate intervals.

Appendix 8-3 at the end of this chapter contains a checklist to use for each visit to the precast concrete fabrication plant.

8.7   SUMMARY

We have talked about the overriding need to establish good communication among members of the project team during the construction phase. Concentrate on prevention of common problems before they occur. The president of a large construction company recently told us, "You are going to have problems on any job." The implication was that problems cannot be avoided. With good communications, many problems—admittedly not all—can and will be avoided. Along that line, we discussed using buildable details to help ensure proper execution of contract documents. A well-planned construction sequence, covered here and in Chapter 7, is another preventive measure. Finally, we offered some help with site and precast plant visits during construction. During construction, the engineer's most important task is to guard against unplanned departures from the intent of the contract documents. What was designed must be built.

8.8 TRANSITION

Sometimes, providing a maintenance program for the new parking structure is part of the agreement between the owner and engineer for professional services. Sophisticated owners will recognize the prudence of funding a maintenance budget and providing programmed maintenance from the beginning. In other cases, maintenance is not considered until the need becomes obvious. Chapter 9 reviews parking structure maintenance needs.

Appendixes to Chapter 8

APPENDIX 8-1
Guidelines for Field Observation for a Cast-in-Place Post-Tensioned Concrete Parking Structure

A. P/T Tendons—General
   1. Sheathing thickness
   2. Tears repaired
   3. Proper amount of grease
   4. Encapsulation system
B. Slabs
   1. Main tendons:
      a. Number
      b. Drape profile
   2. Temperature tendons:
      a. Number

        b. Location at center of slab

        c. Straight placement

        d. Properly supported so that they do not affect main tendons' profile

    3. Plates at ends:

        a. Correct position

        b. Truly vertical

        c. #4 bars behind plates

        d. $1\frac{1}{2}$ in. minimum/2 in. maximum cover for dead end

        e. Epoxy coating

    4. Conventional reinforcement:

        a. At slab midspan

        b. At slab supports

        c. Epoxy coated?

        d. Epoxy-coated chairs, tie wires?

    5. Expansion joint

    6. Bumper wall reinforcement placement/size

    7. Stainless steel, plastic-tipped or plastic chairs for slabs with bottom exposed.

C. Beams

    1. Stirrup number and spacing

    2. Conventional reinforcement

        a. Top

        b. Bottom

    3. Hook at ends of top and bottom reinforcement to outside face of column in line with column reinforcement.

    4. Tendons:

        a. Number

        b. Location at midspan

        c. Location at ends

        d. Spot check at $\frac{1}{4}$ and $\frac{3}{4}$ points

    5. P/T plates:

        a. Correct location

        b. Extra ties at each side of plate

        c. Rebars behind plates

    6. Expansion joint

    7. Reinforcement from adjacent beam in place if required

    8. Other

D. Columns

    1. Vertical reinforcement

        a. Number

        b. Location

        c. Splices

    2. Ties:

        a. Number

        b. Location

E. Miscellaneous
　1. Provisions for future construction
　2. Provisions for precast facia connections
　3. Provisions for expansion joint—structural, electrical, and mechanical
　4. Joints at stair towers
　5. Electrical boxes and conduits in proper position
　6. Forms to be clean—free of cut reinforcement, tie wire, and debris before pour
　7. Cylinders for P/T concrete—leave open on site and cure with the members
　8. Undue slab and frame restraint
　9. Check drain number and locations

F. Possible Problem Areas
　1. Forms shall not be removed before tensioning.
　2. Stressing shall be performed according to specifications.
　3. Vertical elevations to be checked before and after pouring and prestressing.
　4. Ensure that removal of forms at pour strips is according to notes on structural drawings.
　5. Contractor should be aware that the structure is going to shorten upon prestressing. (May be accomplished by the project manager at preconstruction meetings.)
　6. Use a reference point on the ground to check plumbness and elevations for all levels. *Do not use the previous floor as a reference.*
　7. Check edges of structure to ensure coverage for reinforcement and anchor plates.
　8. Slab plates shall be vertical and shall not project above pour.
　9. Tendons shall be straight in a horizontal position.
　10. Be sure bottom of reinforcement cage has proper concrete cover.
　11. If in your judgment, pour should not be allowed, inform project manager immediately.

APPENDIX 8-2
Guidelines for Field Observation for a Precast Concrete Structure

A. Connections
　1. Column/Column
　　a. Column base grout
　　b. Column/column splice
　2. Beam/Column
　　a. Exterior L beam
　　　1) Coil rods/reinforcement

           2) Pockets grouted
           3) Bearing pads
        b. Interior L beam
           1) Coil rods/reinforcement
           2) Pockets grouted
           3) Bearing pads
        c. Interior inverted tee beam
           1) Coil rods/reinforcement
           2) Bearing pads
        d. End spandrel
           1) Coil rods/reinforcement
           2) Pockets grouted
           3) Bearing pads
   3. Tee/Beam
        a. Exterior/interior "L" beam
           1) Coil rods
           2) Weld plates
           3) Bearing pads
        b. End spandrel beam coil rods
   4. Tee/Tee
        a. Flange connections
        b. WWF in topping/pour strip
        c. Reinforcement at crossovers
        d. Reinforcement at pourstrips/topping
   5. Wall Panel/Tee
        a. Wall panel/beam
        b. Wall panel/column
B. P/C Members
   1. Finish
        a. Columns
        b. Tees
        c. Beams
        d. Walls
C. Miscellaneous
   1. Provisions for future construction
   2. Provisions for light standards at top tier
   3. Provisions for expansion-joint structural, electrical, and mechanical
   4. Joints at stair towers
   5. Electrical boxes and conduits in proper position
   6. Undue column and frame restraint

APPENDIX 8-3
Checklist for Precast Concrete Plant Visit

A. Pre-Visit Review
   1. Reviewer has obtained permission to observe operations.

      2. One authority in plant has been designated to receive reviewer's report.

      3. Reviewer has not interfered with plant operations.

      4. Report of review has been given to the designated authority.

      5. A copy of this review report has been sent to the plant.

B. Pretensioning Strand

      1. Strand is clean and free of dirt or from oil.

      2. Strand is new and free of broken or nicked wires.

      3. Strand is located per plans.

      4. Strand mill report verifies size and strength.

C. Tensioning

      1. Strand vises are clean.

      2. Strand jack has been calibrated within last year. (Report is available.)

      3. An initial stress of approximately 5% of the final is applied.

      4. Final stress is checked by measuring elongation and jack-gauge reading.

      5. If strand is depressed during stressing, frictionless holddown devices are used.

      6. Strand is depressed at the proper locations.

D. Concrete

      1. Mix design is approved.

      2. All concrete admixtures are approved.

      3. Admixtures containing calcium chloride are not allowed.

      4. All admixtures are properly measured.

      5. Admixtures applied to mix at proper time and per manufacturer's recommendation.

      6. Aggregate used is of correct size.

      7. Various aggregates are properly segregated.

      8. Aggregates are stored and handled in a manner which will keep them clean.

      9. Moisture content of aggregate is determined twice daily.

     10. Gradation of aggregate is checked weekly.

     11. Mill reports are available for cement.

     12. Cement is protected from moisture during storage.

     13. Scales for measuring cement, aggregate, and water have been calibrated in the last four months.

     14. All scales are in proper working order.

     15. Water in aggregate is accounted for in measuring water.

     16. Cement is not allowed to free fall into mixer.

     17. Mixer is free of hardened concrete.

     18. Mixer blades are not excessively worn or missing.

     19. Concrete is mixed for the length of time equipment manufacturer recommends.

E. Placing Concrete

      1. Concrete is not allowed to segregate in transporting from mixer to form.

2. Concrete temperature is checked and maintained between 50° and 80°F.
3. Concrete is properly vibrated by either internal or external vibration.
4. No cold joints are allowed to form in adjacent layers of concrete.
5. If placed in layers, lower layer is still plastic when the upper layer is placed and the layers are consolidated.
6. Workmen do not move embedded items or reinforcing while placing concrete.
7. If inserts and plates are put in concrete after it is placed, the concrete is still plastic and concrete around them is properly consolidated.
8. All forms for pockets, blockouts, and ledges allow air to escape during concrete placement.
9. If ambient air temperature is below 40°F, form is preheated to above 40°F prior to placing concrete.

F. Finishing Concrete
1. Concrete surfaces are screeded to correct dimensions prior to applying finish.
2. Surfaces receiving form finish or manual finish are as specified in plans and specifications.
3. If broom finish is specified, striations are made in specified direction.
4. If rough finish is specified, a minimum of $\frac{1}{4}$ in. amplitude is maintained.
5. If steel trowel finish is specified, final finishing operation does not start until all surface water has evaporated and surface cannot be easily dented with a finger.

G. Curing Concrete
1. Concrete is covered to prevent loss of moisture.
2. If ambient temperature is below 50°F, form is heated or insulated to prevent concrete temperature from dropping below 50°F.
3. If heat is used to accelerate cure, continuous recording thermometers are used to monitor temperatures.
4. At no time is concrete temperature allowed to go below 50° or above 175°F.

H. Concrete Testing
1. Cylinders are made in accordance with specifications.
2. Air content is checked in accordance with specifications.
3. Slump is checked in accordance with specifications.
4. Cylinder testing machine was calibrated within the last six months.
5. Cylinders are properly capped prior to testing.
6. Operator applies load to cylinder at a uniform rate.
7. Two cylinders are cured on the form in the same manner as the concrete and break at or above 3500 psi prior to detensioning strand or stripping product.
8. Twenty-eight-day cylinders are tested and concrete strength is in accordance with the specifications.

I. Detensioning
   1. Concrete strength has reached 3500 psi strength as required by design prior to detensioning.
   2. Strands are detensioned by slowly heating the strand.
   3. Strands are heated simultaneously from both ends.
   4. Strands are detensioned in a symmetrical pattern.
J. Quality Control
   1. Plant does have a designated quality-control department.
   2. Quality-control department supervises the tensioning, detensioning, and all concrete testing.
   3. The quality-control department checks dimensions on all products.
   4. The quality-control department maintains records of all stressing, testing, and mill reports.
K. Storage
   1. Dunnage is placed only at pick points.
   2. Product is handled only at pick points.
   3. Product is stored on a level plane.
   4. Product is marked so it can be identified by date cast and location in the building.
   5. If product is stacked more than three high, plant engineer has calculations showing he is not exceeding allowable concrete strength.
   6. If product is stored at below 32°F all inserts and sleeves are sealed to prevent ice forming in them.
L. Precast Double Tee Inspection
   Tee Forms
   1. Forms are clean and free of pits, bends, bowing, and uneven joints.
   2. Form will provide an approved finish.
   3. Form oil is applied properly per manufacturer's recommendations and no puddles are left in form.
   4. Form is of correct configuration and dimensions.
   5. All skews and blockouts are correctly positioned.
   Tee Reinforcement
   1. Shear reinforcement in stems provided per drawings.
   2. Shear reinforcement held with nonferrous chairs which provide specified concrete cover.
   3. End-bearing plate in place and held in proper position.
   4. Bearing plates have received the proper finish and have the correct reinforcement welded to them.
   5. Proper preheat used in welding reinforcement.
   6. Mill reports verifying size and strength of all reinforcement are available.
   7. Flange welded wire fabric (WWF) located per plans and held with nonferrous chairs which provide specified concrete cover.
   8. Extra layer of flange WWF at 2-in. flange.
   9. WWF from 4-in. flange extends proper distance into 2-in. blockout.
   10. Mill reports available showing proper strength and size for all WWF.

11. Flange WWF extends through floor drain holes.
12. Additional reinforcement around all flange holes.
13. Flange connectors made per plan.
14. Flange connectors properly spaced and held securely.
15. Lifting devices in 2-in. flange area only.
16. All reinforcement held securely so that correct cover is maintained.
17. Flange WWF lapped at least two cross wires plus 2 in. or a minimum of 8 in. at all laps.

M. Precast Beam Inspection

Beam Forms

1. Forms are clean and free of pits, bend, bowing, and uneven joints.
2. Form will provide an approved finish.
3. Form oil is applied per manufacturer's recommendations and no puddles are left in form.
4. Form is of the correct configuration and dimensions.
5. All skews and blockouts are correctly positioned.

Beam Reinforcement

1. Mill reports verifying size and strength of all reinforcement are available.
2. Proper preheat is used when welding reinforcement.
3. No unspecified tack welds are used on the reinforcement.
4. All bearing plates are in proper position and held securely so that they remain level during concrete placement.
5. Bearing plates have received the specified finish and have the correct reinforcement welded to them.
6. All reinforcement is correctly positioned.
7. All reinforcement is bent to specified tolerances.
8. Proper lap lengths are used for all lap splices.
9. All reinforcement has the specified concrete cover.
10. All reinforcement is held with nonferrous chairs.
11. All reinforcement is held securely to prevent movement during concrete placement.
12. All reinforcement is free of dirt, form oil, and hardened concrete.
13. All steel which will be exposed in the final condition has received the specified finish.
14. All inserts and plates, which are to be patched in the final condition, are recessed to provide the proper concrete cover.
15. All sleeves are lined with the specified material—PVC and steel.
16. All lifting devices are so positioned that they will not be exposed in the final position.

N. Precast Column Inspection

Column Forms

1. Forms are clean and free of pits, bends, bowing, and uneven joints.
2. Forms will provide an approved finish.
3. Form oil is applied per manufacturer's recommendations and no puddles are left in the form.

4. Form is of the correct configuration and dimensions.
5. Column corbels and beam blockouts are of correct configuration and provide for correctly sloped bearing surfaces.

Column Reinforcement

1. Mill reports are available verifying all reinforcement is of proper strength and size.
2. Proper preheat is used when welding all reinforcement.
3. No unspecified tack welds are used on the reinforcement.
4. All main reinforcement is continuous without splices.
5. All lifting devices are so positioned that they will not be exposed in the final erected position.
6. All inserts and plates, which must be patched in the final erected position, are recessed to provide proper concrete cover.
7. Column ties are properly spaced and all extra ties specified are provided.
8. All column ties are bent to specified tolerances.
9. All reinforcement has the specified concrete cover.
10. All reinforcement is held with nonferrous chairs.
11. All reinforcement is held securely to prevent movement during concrete placement.
12. All steel which will be exposed in the final condition has received the specified finish.
13. All reinforcement is free of dirt, form oil, and hardened concrete.
14. Sleeves for beam connections are lined with specified material—PVC and steel.
15. Bearing and base plates are properly positioned and held securely so that they remain level during concrete placement.
16. Provisions for future extension of columns conform to design detail.

Chapter 9
Maintenance

SAM BHUYAN

9.1 INTRODUCTION

The purpose of maintenance is to assure proper and timely preventive actions to minimize equipment failure and premature deterioration. Most of the material in this chapter has been adapted from the maintenance manual published by the National Parking Association. The reader should refer to the "Parking Garage Maintenance Manual" for further details related to the maintenance of structures.

This chapter is directed primarily at owners or operators of parking facilities. The material included will assist architects and engineers to develop an understanding of the essential elements of parking structure maintenance. An important objective in designing new and restoring existing structures is to minimize operating and maintenance costs.

Parking facilities experience unusually harsh exposure conditions compared to most buildings. Temperature extremes, dynamic loads, and deicer attack are potentially destructive to all parking facilities. Premature deterioration, such as scaling, spalling, cracking, and leaking can reduce the integrity of exposed concrete surfaces, especially the floor slab.

Deferred structural maintenance can lead to serious deficiencies. Premature deterioration of concrete floors is costly and, in extreme cases, can impair the structural system's integrity. To minimize the impact and the cost of structural deterioration, timely corrective and preventive maintenance action is needed. On the other hand, failures associated with certain operational features, such as lighting, parking equipment, or security monitoring devices are relatively easy to correct. Preventive maintenance defers major repairs and it is usually more cost effective and less disruptive to operations.

Parking facilities today are structurally more complex than ever. Challenges facing the designer include:

- Building codes and city ordinances impose design requirements not present several years ago.
- Integration of parking facility operational and structural features. Parking structures require special attention for structural maintenance in order to assure long-term structural integrity.
- Multi-use facilities or high-rise structures with integrated parking, demand life expectancy well beyond 40 yr.
- Quality control of materials, labor, and erection is more critical than in most buildings.
- The service environment is more severe and potentially more destructive than conventional structures are capable of effectively withstanding.

Parking facility operational programs also feel the demand of new challenges. These include:

- Equipment failures which occur due to high utilization rate and sensitivity of components to adverse conditions. Automation instead of manpower is required to minimize operating costs.
- Pay facilities that are quite competitive and have little money for structural maintenance or for establishing a budget to support a maintenance program.
- Extensive salt use contaminating the concrete, allowing chloride to corrode embedded reinforcing steel, causing progressive surface spalling. Other destructive processes are structural-member cracking, leaking, floor-slab spalling, and surface scaling.
- Legal liability for code or regulatory violations and negligence to enforce security.

This chapter will address general as well as specific maintenance actions required to extend the life of parking structures. General maintenance is associated with operational aspects of the facility. Specific maintenance is associated with the structural system and is required periodically during the life of the structure. It is intended to provide guidelines for maintaining parking facilities at a satisfactory level of service.

9.2   RECOMMENDED MAINTENANCE PROGRAM AND CHECKLIST

As shown in Table 9-1, maintenance needs for a parking facility can be separated into the following three broad categories:

- Structural
- Operational
- Aesthetic

Each of the above items has characteristics that are significantly different from the others, which will require each item to be treated separately. Also, maintenance must be performed at regular intervals if the full benefit of the effort is to be realized. Irregular or incomplete maintenance will provide a marginal return on the investment. To ensure that a maintenance program is functional, a schedule must be established and appropriate maintenance procedures followed.

The first step in developing and implementing a maintenance program is a walk-through review, which is a visual inspection of the entire facility. For existing or restored structures, limited nondestructive and laboratory testing may be required to qualify construction materials and as-built conditions. The walk-through review by an experienced restoration engineer assists in developing a tailored maintenance program for a specific facility based on factors such as:

- Age and geographic location of the facility
- Structural system and the design details involved

TABLE 9-1 Maintenance Category Outline

Structural System

1. Floors
2. Beams, columns, and bumper walls
3. Joint-sealant systems
4. Stair and elevator towers
5. Exposed steel

Operational

1. Cleaning
2. Snow and ice control
3. Mechanical systems
4. Electrical systems
5. Parking-control equipment
6. Security systems
7. Signs (graphics) and stripping
8. Inspection
9. Safety checks

Aesthetics

1. Landscaping
2. Painting
3. General appearance

- Quality of construction material specified
- Construction quality or deficiencies
- Existing distress in structural elements, such as spalling, cracking, scaling, or excessive deformations
- Corrosion-protection system specified or implemented
- Operational elements of the facility

Once relevant maintenance needs are identified, procedures and schedules can be set up for maintaining the structure. Regularly scheduled walk-through inspections then form the basis for implementing and monitoring the effectiveness of the maintenance programs.

The walk-through inspections can be performed by the in-house maintenance staff. The in-house maintenance crews should be on the alert to locate distress of structural elements in accordance with fixed schedules. All such areas should be noted on plans for annual examination and evaluation by an engineer. Areas of concern may then be further examined and evaluated under his guidance. If necessary, a condition appraisal of the facility and repairs should be performed as discussed in Chapter 10.

9.2.1 Structural System Maintenance

Structural elements may be divided into several distinct categories: floor slabs, beams, columns, bumper walls, stair and elevator towers, joint systems, and exposed steel. Structural maintenance consists of repairing deteriorated members, renewals of protective coatings, and replacement of sealants to extend the service life of the structure. Repairs may be cosmetic in nature or "major." Cosmetic repairs if left undone will not adversely affect the operation or integrity of structural elements. Major repairs correct distress due to spalling, scaling, and cracking, which, if left unattended, can contribute to accelerated deterioration of structural elements. A detailed explanation of deterioration and distress that are common to structural members is provided in Chapter 10. Selection of appropriate repair method and material are also discussed in that chapter.

Maintenance of the structure is considered very important since neglect of structural maintenance can lead to major problems and high repair costs. In addition to the actual repair costs, lost parking-revenue costs during facility repairs can be substantial. Table 9-2 presents the recommended program for satisfying structure maintenance. Structural system elements are listed and the appropriate action required is specified.

TABLE 9-2   Structure Maintenance Schedule

Item	Description	Annual	As Required
1.0	Concrete Slabs		
	1.1 Visual Inspection	Perform	
	A. Floor		
	B. Ceiling		
	C. Floor coatings		
	1.2 Delamination Testing	Perform	
	A. Floor		
	B. Ceiling		
	1.3 Protective Sealer	[1]	
2.0	Beams, columns, bumper walls, and connectors		
	2.1 Visual Inspection	Perform	
	A. Columns		
	B. Beams		
	C. Bumper walls		
	D. Connections		
	E. Snowchute		
	2.2 Delamination Testing	Perform	
	A. Columns		
	B. Beams		
	C. Bumper walls		
	D. Connections		
	E. Snowchute		
	2.3 Protective Sealer	[2]	
3.0	Joint-Sealant Systems	[3]	
	3.1 Visual Inspection	Perform	
	A. Expansion joints		
	B. Construction joints		
	C. Control joints		
	3.2 Crack Routing and Sealing	Inspect	Perform
4.0	Stair Towers		
	4.1 Visual Inspection	Perform	
	A. Stairs and landings		
	B. Walls		
	C. Glass		
	4.2 Apply Protective Sealer to Landings and Steps	[4]	
5.0	Exposed Steel	Inspect	

[1] Reapply every 3–5 yr. Areas which are subject to more intense and severe exposure may require retreatment on an annual basis. Testing and inspection should be performed to determine degree of exposure. A traffic coating may be more cost effective in areas of heavy leaking or floor deterioration.

[2] Apply sealer every 3 yr on those structural members subject to frequent leaking and salt-water splash.

[3] Budget for total replacement every 10 yr.

[4] Sealer application should be made every 3 yr.

Adapted from "Parking Garage Maintenance Manual," Parking Consultants' Council, National Parking Association.

9.2.1.1   Concrete Floor Slab and Surfaces. The most significant maintenance needs are associated with the floor slab and consume the largest share of the maintenance budget. Typical conditions of deterioration which influence the floor slab are delaminations, scaling, spalling, cracking, leaking, and leaching. These conditions can contribute to accelerated deterioration of the structure and adversely affect the serviceability of the floor slab.

Periodic application of surface sealer or elastomeric coatings can minimize floor slab deterioration by reducing water penetration. For structures in northern climates exposed to road salts, an unprotected concrete surface will eventually permit chloride ions to migrate in sufficient quantity to promote corrosion of embedded reinforcement. The "time-to-corrosion" and need to protect the floor surface depends upon factors such as geographic location of the structure, concrete quality (water-to-cement ratio), the clear concrete cover of embedded steel, the permeability of the concrete, and the corrosion protection system specified. Even for structures in the southern climatic region, sealer and coating will extend the service life of the structures. This is particularly true for structures that can potentially be exposed to airborne chlorides from large bodies of salt water.

The supported entrance and exit lanes, helix, turn lanes, and flat floor areas are subject to more severe exposure conditions. Special attention must also be given to turning areas at end bays, crossovers, gutterlines, and water ponding. The high-wear areas may require more frequent treatment of a sealer application than the general parking surface. All areas should be closely monitored, and if deterioration develops, heavier sealer application rates or an application of elastomeric coating may be necessary.

Elastomeric coatings installed on floors, over offices or commercial space should be examined for wear and tear. Damaged coating areas should be repaired as soon as possible to prevent leaking and contamination, since the integrity of the entire coating system and floor slab can be jeopardized if left unrepaired. These coating systems are usually proprietary and should be inspected and repaired by the system manufacturer's authorized representative.

The ability of the sealer to reduce water absorption, screen chloride ions, and resist ultraviolet exposure should be verified by laboratory tests. The test procedures for measuring sealer effectiveness are provided in a National Cooperative Highway Research Program (NCHRP) report entitled, "Concrete Sealers for Protection of Bridge Structures, NCHRP Report 244." In addition, the method of surface preparation and sealer application should be verified by trial applications in selected areas of the structure. Trial applications also help to identify sealers that can potentially "glaze" and create slip hazards. Surface sealers should be reapplied every 3–5 yr. The frequency of reapplication should be determined by annually monitoring the chloride content of the concrete at various depths from the floor surface to at least the level of the top mat of reinforcement.

Liquid-applied membrane systems (traffic toppings) provide more effective protection against moisture and chloride contamination than surface sealers. The membrane system waterproofs the surface and allows moisture penetration only at localized imperfections, such as holes and tears. The membranes are capable of bridging some active cracks and are effective for conventionally reinforced structures and floor systems with extensive through-slab cracking. Membranes are significantly more costly (four to six times) than surface sealers and are susceptible to wear, especially in the driving and turning aisles of the parking facility. Recoating of the top layer of the membrane system is often necessary in high-wear locations. However, with proper maintenance, some membranes can last well beyond 10 yr.

Preventive maintenance measures, such as applying a protective sealer and elastomeric coating, are most effective when applied to a floor slab that has not been contaminated by road salt. When sealers and elastomeric coatings are applied to older facilities with chloride-contaminated floor slab, concrete deterioration cannot be effectively controlled due to the continuing corrosion of embedded reinforcement. Therefore, the floor-slab maintenance cost for restored facilities is relatively higher than maintenance costs associated with a new parking facility; also, older facilities require more frequent repairs.

A facility should be monitored annually for concrete deterioration. Open spalls and delaminations in floor slab should be patched to reduce the impact of progressive deterioration and to maintain serviceability. Also, open spalls and exposed reinforcement are a tripping hazard for facility users. Due to time or weather constraints some spalls may have to be repaired temporarily with asphalt or prepackaged repair materials. Asphaltic repair materials generally tend to trap and retain moisture which in some instance can contribute to accelerated deterioration of the underlying and adjacent concrete. For relatively permanent repair, all unsound concrete must be removed. Corroding reinforcement should be completely excavated, cleaned, and epoxy coated. The repair area should then be patched with properly air-entrained, high-quality portland cement-based patch materials which are relatively impervious to moisture. Preventive measures and maintenance procedures for chloride-induced deterioration are fairly comprehensive, and usually help to minimize other forms of concrete deterioration.

Sealing cracks and joints in floor slabs is necessary to limit ceiling deterioration. All loose overhead concrete spalls are potential safety hazards which can damage vehicles or injure facility users. Remove loose overhead spalls as soon as possible.

Ponding is also a potential slipping and skidding hazard, particularly during the winter months, due to freezing. Also, ponded areas can contribute to more rapid floor-slab deterioration and joint leakage. Eliminate isolated ponded areas by installing supplemental drains. Relatively large areas may require resurfacing to provide adequate drainage.

9.2.1.2   Beams, Columns, and Bumper Walls. Beam and column deterioration can adversely affect the structural integrity and the load-carrying capacity of floor slabs. Extensive deterioration can sometimes result in localized wheel punch-through by vehicles. Deterioration of these underlying members can primarily be attributed to water leakage through failed joints and floor-slab cracks. Vertical surfaces of columns and bumper walls are also susceptible to damage by salt water spray from moving vehicles.

Beam and columns adjacent to and below expansion joints are most susceptible to deterioration. Beam and column deterioration can be minimized by proper maintenance of joint sealant systems of the floor surface and sealer application on the column base and bumper wall. Water leakage can contribute to the corrosion of embedded reinforcement, freeze-thaw deterioration, rust staining, and leaching.

Concrete bumper walls can potentially be subjected to vehicle impact; the bumper walls should be monitored for cracking and spalling. Structural steel connections should be monitored for corrosion and distress due to impact. Ponded areas and gutterlines adjacent to bumper walls can contribute to the corrosion of steel connections, leaching, and rust staining. These adverse conditions generally require installing new curbs, supplemental drains, or concrete wash.

9.2.1.3   Joint Sealant Systems. Expansion, construction, and control joints in parking structures accommodate movements due to the volume change of concrete. The volume-change movement can be attributed to concrete shrinkage, the seasonal temperature variations, elastic shortening in prestressed structures, and creep of concrete. The joints in the structure are sealed with flexible elastomeric sealant to minimize water leakage and accelerated deterioration of the structure. In addition to the joints, random floor-slab cracking can also contribute to leakage and to the deterioration of structural members. When and where appropriate, seal random through-slab cracks with flexible elastomeric sealant material. A discussion of concrete cracking, joint distress, and causes of sealant system failure is included in Sections 10.3.3 and 10.3.4 in Chapter 10.

Joint-sealant systems have a limited life expectancy. Depending upon the structural configuration, wear and tear, exposure conditions, the joint-sealant system can be expected to provide 8–10 yr of service prior to complete replacement. Spot-patch repair of joints is cost effective only when less than 30% of the joint-sealant system shows deterioration or leaking. Concrete surfaces adjacent to expansion joints, construction joints, control joints, and sealed random slab cracks should be treated periodically with a surface sealer. Also, the concrete surfaces adjacent to the joints and cracks must be inspected for deterioration to maintain effectiveness of the joint-sealant system.

9.2.1.4   Stair and Elevator Towers. Without regular maintenance, leaks between the floor-slab surface and stair and elevator towers can be a problem. Quite often the leakage can be attributed to poor drainage conditions around the towers. Drainage can be improved by providing curbs and washes, which will then tend to minimize deterioration of underlying elements, such as doors, light fixtures, electrical conduits, metal stairs, exposed structural steel members, connections, etc. In addition, rust staining, leaching, and paint peeling can be aesthetically unpleasing. Frequent inspections and repair of the damaged elastomeric expansion-joint seal between the tower and floor surface will also minimize distress caused by leaking.

Stairs and landings are exposed to salt contamination and concrete surfaces require periodic resealing. Masonry walls should also be maintained by sealing the surface. Stair and elevator wall cracking should be evaluated and repaired. Door and window glazing, if present, should be replaced when damaged or leaking.

9.2.1.5   Exposed Steel. Exposed steel within a concrete parking facility is generally limited to the structural steel connections, stairs, pedestrian railings, vehicular guardrails, and metal decking. Premature deterioration of metal components is caused by neglect and the chemical reaction between the metals and the corrosive environment. Check for potentially unsafe conditions due to the corrosion of connections monthly. Treatment of metals with proper surface preparation and a quality paint will minimize corrosion.

9.2.2 Operational Maintenance

Special emphasis is placed on operational maintenance because malfunctions or breakdowns can take part, or all, of the facility out of service and/or reduce user security and safety. Operational maintenance involves regular and scheduled inspection and repair of items, such as parking equipment, elevators, electrical systems, heating and ventilation systems, security monitoring, and fire-fighting equipment. Routine cleaning, including sweeping and washdown, is also a part of operational maintenance. Snow and ice control are important in climates where appropriate. Table 9-3 presents a recommended program for satisfying operational maintenance needs.

9.2.2.1 Cleaning Requirements. One of the most frequently overlooked aspects of parking-facility maintenance is proper floor cleaning. Sweeping can be accomplished by using hand brooms or mechanized sweepers designed for use in buildings. The maximum weight for

TABLE 9-3 Operational Maintenance Schedule

Item	Description	Frequency		
		Monthly	Annual	As Required
1.0	Cleaning Requirements			
	1.1 General cleaning	Perform		[1]
	1.2 Sweeping	Perform		[1]
	1.3 Remove ponded water			Perform
	1.4 Floor-surface flushdown		Perform	[1]
2.0	Snow Removal and Ice Control			Perform
3.0	Mechanical/Electrical Systems			
	3.1 Drainage system (Includes sediment trap)		Inspect	
	3.2 Elevators	Inspect		Inspect
	3.3 Ventilation equipment	Inspect		
	3.4 Fire protection	Inspect		
	3.5 General lighting	Inspect		
	3.6 Exit and emergency lighting	Inspect		[1]
	3.7 Emergency generator	Inspect		
	3.8 Parking equipment	Inspect		
	3.9 Security monitoring	Perform		[1]
	3.10 Safety checks	Perform		[1]
4.0	Graphics and Striping		Inspect	
5.0	Inspection (see structural-maintenance schedule)		Perform	

[1] More frequent performance of this task is suggested.

mechanized sweeping equipment in a parking facility is generally limited to 8000 lb gross weight or 4000 lb per axle.

All dirt and debris should be removed from the facility. Dirt and debris should be kept away from drain basins and pipes, as blockage may cause leaking and failure. Dirt and debris in expansion-joint systems can potentially damage elastomeric seals.

Grease buildup in parking spaces should be removed with appropriate degreasers. In addition, a poultice made with limestone, sodium hydroxide solution, and trisodium phosphate (TSP) is effective in cleaning oil spills. Refer to the *Concrete Construction* magazine publication, ''Removing Stains from Concrete'' for more detailed information on removing stains. Excessive grease build-up is common at the entrance gate and adjacent to the cashier's booth. Grease should be removed regularly.

Salt accumulates during winter months and should be removed each spring by flushing the surface. A flushdown with low-pressure waterhoses is advisable after the facility has been swept. Flushing can be incorporated with a check of the standpipe system, which can be coordinated with the deck washing. Flushing of critical areas such as entrance and exit lanes, turn lanes, flat areas, and main drive aisles should be performed frequently during the winter when moderate temperatures prevail. If moderate temperatures do not prevail, then in-house maintenance personnel should use sponge mops or brooms to remove accumulated salt-laden slush or water. Entrance and exit lanes would benefit most from periodic removal of snow and slush deposits.

High-pressure water jet cleaning systems should not be used on floor slabs near control joints, expansion joints, crack sealants, and deck coatings. High-pressure water jets can cause damage and leaking and can create the potential for serious deterioration. High-pressure water cleaning may be used to remove grease spots on the floor slab if care is taken to avoid damage to joint sealants and deck coating materials.

When flushing the floor surface, avoid washing sand into the drain system. Temporary burlap or straw filters are effective methods to prevent sand from plugging drains.

9.2.2.2 Snow Removal and Ice Control. It is possible to damage the joint-sealant and deck-coating systems with abnormal and/or abusive traffic. The three most common causes of damage to the systems are:

- Dropping heavy or sharp objects onto the surface
- Dragging heavy or sharp objects across the surface

- Unprotected snow removal equipment and studded tires or tire chains

To minimize damage, the following snow removal guidelines need to be implemented:

1. All expansion joint locations must be clearly marked by means visible for the equipment operator while the deck is covered with snow. It is recommended that markings such as red, yellow, or orange stripes be placed on the adjacent walls or columns at each end of the expansion joint. Where walls or columns do not exist, place safety flags, properly anchored 55-gal drums filled with sand, or other means of identification at each end of the joint.

2. Piling snow in corners or other locations within the facility is not recommended. Snow varies greatly in weight, but packed snow can be quite dense, and ice often forms at such piles during freeze/thaw cycles. While the structure can probably safely support most snow piles, the weight may be sufficient to crack the concrete. Such cracking could permit water (and dissolved salt) penetration of the concrete which could hasten the deterioration of the slab. Therefore, the accumulation of snow piles over long periods of time must be avoided.

3. Establish a snow removal pattern so that the snow removal equipment approaches the expansion joints at an angle not greater than 75 degrees and preferably parallel to the expansion joint. This will reduce the chance of catching the snowplow blade on the expansion-joint system. Snowplow damage is not usually covered by the expansion-joint warranty. Snow and ice packed in an expansion-joint system can contribute to seal failure. Follow manufacturer's guidelines and where necessary, clean snow in the expansion-joint gland to minimize failure of the seal.

4. Snow is normally plowed utilizing a vehicle with an axle weight of 4000 lb maximum. Snowplow blades and bucket loaders should be modified with a heavy rubber cutting edge attached to the bottom, or with "shoes" or other positive means designed to keep the steel blade from making contact with the concrete floor surface. It is necessary to keep the steel blade a minimum of $\frac{1}{8}$ but, preferably $\frac{1}{2}$ in. above the floor surface to avoid damage to the concrete, deck coating, and expansion-joint sealant systems. Whenever possible, use of a power brush for snow removal is suggested.

5. Snow is plowed to specific locations within the facility where a bucket loader or industrial snowblower can be used to throw the snow over the side. In congested areas snowchutes or off-peak snow removal operation may be required to effectively deal with heavy snowfall. Care must be taken with a bucket loader or snowblower to avoid damage to the concrete walls, connection hardware, deck coating, and expansion joints.

6. Schedule an inspection of the deck coating, control joints, expansion joints and concrete walls in early spring to assess the winter's wear. If damage has occurred, repairs can be scheduled in the upcoming construction season.

The slope of the floor is designed to drain surface water as quickly as possible. However, certain areas are particularly vulnerable to freezing when water drains from sun-warmed surfaces into shaded areas. This occurs on the top level and at entrance/exit lanes. In-house maintenance crews must be aware of these areas and take steps to control icy conditions as they occur.

Most common chemical de-icers can have major physical effects on concrete. Several de-icers are listed with a general description of common affects on materials typically located in and around the parking facility. The affect on any single material may progressively affect the entire concrete system. It must be noted that no de-icing compounds, including road salt, work in extremely cold temperatures.

Some of the common chemicals associated with ice control are:

1. *Urea* is the only currently available de-icer that does not damage concrete, lawns, shrubs, or metal.

2. *Calcium chloride*, a major active component of many proprietary de-icers, has little chemical effect on concrete, lawns, and shrubs, but it causes corrosion of metal, and is particularly hazardous to prestressed steel.

3. *Sodium chloride* (halite, table salt, or rock salt) has little chemical effect on concrete, but will damage lawns and shrubs and promote corrosion of metal.

4. *Ammonium nitrate or ammonium sulfate* is beneficial to most vegetation, but may lead to complete destruction of concrete because of direct chemical attack on concrete and the reinforcing steel.

5. *"CMA,"* calcium magnesium acetate, will be available once it can be processed economically. It offers the potential for being one of the few noncorrosive and nonreactive de-icers.

It is important to minimize the use or the amount of de-icing chemicals during the first 2 yr of concrete curing. It is emphasized that properly designed, air-entrained and cured concrete is required in order to have a durable concrete structural system. Also, do not use sodium chloride, calcium chloride, ammonium nitrate, or ammonium sulfate on the concrete surface.

Ice build-up can be controlled by using hot sand. Do not apply de-icing chemicals containing chloride directly to the concrete unless extremely icy conditions exist. Small amounts of salt (3–6% of weight) added to sand can be very effective at increasing traction and preventing skid problems. Apply the sand/salt mixture to ice only as needed.

Drain systems should be protected against runoff-related sand accumulation during ice-control operations. Temporary burlap or straw filters should be used to prevent drain clogging and possible damage to drain systems.

The following is a series of recommended de-icing measures in order of decreasing preference:

1. Clean, plow, and scrape off ice and snow; do not use de-icing agents.
2. Use sand to increase traction; when washing down the floor, protect the drainage system.
3. De-ice with urea.
4. Use a mixture of sand and calcium or sodium chloride, but protect the drainage system.

9.2.2.3 Mechanical Systems. All mechanical equipment should be inspected regularly and serviced as required. Mechanical systems include the drainage system, ventilation equipment, elevators and shafts, and fire-fighting equipment.

Drain basins, inlet grates, leaders, downspouts, heat tape, and all support brackets should be inspected for leaks, damage, or distress. Sediment basins should be cleaned as required to prevent clogging and water ponding. Floor sleeves should be examined and sealed against leaking also. Inspect all electrical connections and make repairs to assure a safe heat-tape system. All deficiencies noted should be recorded and appropriate action programmed.

Where water is ponding, such as near corners or at other areas of floor surface, consideration should be given to installing supplemental floor drains. A small drain and leader to the floor below can alleviate hazardous ice build-up and minimize chloride ion infiltration into the concrete. The minimum recommendation would be to broom or sponge mop the water to existing floor drains. Ponding water, as it evaporates, leaves behind a high concentration of salt, which can migrate into the concrete and contribute to corrosion of the reinforcement and progressive concrete deterioration.

Heating, ventilation fans, air-conditioning equipment ductwork, and the necessary support systems should be inspected. Service manuals provided for this equipment by the manufacturer should be checked for appropriate maintenance action. All servicing required should be performed promptly and to the specifications provided by the equipment manufacturer. Replacement belts and pulleys for fans should be kept in stock. Replacement of worn or damaged parts should be completed periodically to minimize the chances of breakdown. All questions regarding servicing should be directed to the equipment manufacturer or supplier.

The elevators, shafts, and associated hardware should be inspected and serviced in accordance with the elevator manufacturer's recommendations. A maintenance agreement with a reputable elevator service company or the manufacturer is the most effective method for servicing the elevators to minimize breakdowns. Water leaking into the elevator shaft should be corrected as soon as it is discovered. Use of an auxiliary pump system may be required if water build-up becomes excessive.

Standpipe and sprinkler systems should be periodically charged and activated to check for proper operation. Coordinate the standpipe system check with the fire department and your washdown of the floors. Portable fire extinguishers should be checked for satisfactory charge. Broken extinguishers, damaged fire hoses, and cabinets should be repaired or replaced.

Before the onset of cold weather, the parking facility must be winterized. Water risers, fire protection systems, standpipes, landscaping sprinklers, and hosebib systems must be flushed and completely drained to prevent ice build-up and bursting of pipes. Drain lines and other underground piping should be flushed and checked for blockage.

Heating systems should be started and tested. Often overlooked in winterization procedures are the heating units in the parking equipment which must be checked for proper operation. In the spring, after danger of freezing, these procedures must be reversed and plumbing systems restored to full operation. Heating systems must be shut down.

9.2.2.4 Electrical System. The effective operation of any parking facility requires adequate lighting to insure that users can move securely and easily within the facility—motorists as well as pedestrians. If the existing lighting system becomes functionally obsolete, it should be

replaced with a new system. New systems, such as high-pressure sodium vapor, can be substantially more energy efficient than an older
system. Replacement of fluorescent light fixtures should be considered
since they are inefficient in cool weather. Capital expenditure, maintenance, and energy savings must be evaluated for "cost effectiveness."

An annual detailed inspection of *all* fixtures and equipment is
required. Inspect pedestrian "EXIT" and emergency light fixtures on
a more periodic basis. Those fixtures not working properly should be
repaired or replaced; damaged lenses should be replaced; timers and
photo cells should be periodically checked and calibrated.

Electrical conduit exposed to leaking water or rusting should be
cleaned and repaired. Damaged conduit that has pulled loose from its
mount or has exposed conductors should be maintained in proper condition. Damaged or rusted electrical panels should be cleaned and
repainted or replaced.

If the parking facility is interconnected with an emergency generator, periodic checks of the system should be made per the manufacturer's recommendations to assure reliance of the generator system.

9.2.2.5   Parking-Control Equipment. All parking equipment within the
facility should be examined, and a preventive maintenance program
established to minimize breakdowns. It is prudent to maintain an inventory of critical components so that your maintenance crew can
quickly repair the equipment. Periodic servicing of the parking equipment is essential for smooth facility operation. A service agreement
should be established with the parking equipment supplier so that he
is on call to provide assistance in the event of a breakdown.

Equipment added to the facility after the original installation must
be compatible with the existing equipment. Copies of operation and
service manuals for all parking equipment should be kept nearby for
easy reference. A log of maintenance and service calls should be established for each piece of equipment.

9.2.2.6   Security. As evidenced by interest and attendance at conventions of various parking groups, security is one of the biggest problems
in the industry today. Security is an ongoing process that good design
alone will not achieve. Careful selection of equipment, training, and
management by security professionals are equally important.

Two types of security measures, "passive security" and "active
security," are employed to maximize security in a parking facility.
Passive security does not require human response. These security

measures are a physical part of the facility, such as lighting and glass-walled stair towers. The common thread among all passive features is visibility—the ability to see and be seen while in a parking facility. Lighting is universally considered to be the most important security feature in the facility. Staining concrete has also proved to be a cost-effective method of increasing brightness. White stains on ceilings, beam soffits, and walls can improve brightness; however, staining seems to encourage graffiti, which can sometimes tend to hurt the perception of security. The additional general maintenance and up-keep are also important in maintaining security.

Active security measures invoke an active human response. The active systems such as those listed below, where applicable, should be tested frequently to insure proper operation:

a. Television surveillance cameras
b. Audio-monitoring devices
c. Telephone in elevator cabs and cashier's booth
d. Panic hardware on doors
e. Security monitoring
f. Alarms
g. Other special features

Security policing of the facility, such as a scheduled drive through by trained security personnel, should be maintained to deter undesirable behavior within the facility. The importance of security monitoring cannot be overstressed; it is essential to maintain security systems in proper working order at all times. Periodic inspection of all equipment and observation by in-house personnel is required. Where deficiencies in security monitoring are found, corrective actions should be taken immediately.

9.2.2.7 Graphics and Striping. Proper graphics are essential to the smooth operation of the parking facility; they must be kept clean and visible. Graphics combining words or symbols with arrows are most effective for traffic- and pedestrian-movement control. All entrance, exit, traffic directional, and display signs should be kept clean and legible. Paint or facing material for graphics should be examined annually for deterioration and repainted or repaired as required. Also replace lights in illuminated signs.

Floor-level and stair-elevator-tower designations directing patrons to their parking locations should be kept legible and visible from all entrances and exits. Stair and elevator-tower level designations should be located on both sides of the door.

Floor striping should be inspected each spring after cleaning and should be repainted as required. In older facilities, consideration should be given to restriping so as to accommodate smaller cars. Be careful to maintain proper illumination levels when relocating designated pedestrian walkways. The restriping plan should not remove the walkway from underneath a lighted source, thus creating a tripping hazard. Restriping should be performed after resealing the concrete floor surface.

9.2.2.8 Inspection. Annual inspections are the best way to guarantee that minor conditions requiring repair are contained and corrected before they cause major problems. Inspections should be performed each spring to determine if salt and wear exposure has caused concrete deterioration. The inspection and concrete testing is recommended to be accomplished by a qualified structural engineer.

9.2.2.9 Safety Checks. It is important to minimize potentially unsafe conditions within a facility. Certain elements of operational and structural maintenance can have an impact on safety. Also, some features that enhance security tend to enhance safety.

Structural system maintenance will tend to minimize tripping hazards and potential injury to facility users due to loose overhead concrete. Concrete spalls can develop on ceilings as well as on bumper walls, precast panels, and concrete facades. Proper maintenance of the pedestrian guardrails and inspection of vehicular barriers will minimize unsafe conditions.

Ice and snow control will tend to reduce slipping and skidding hazards for pedestrians and vehicles. Proper selection and application of surface sealer is important to reduce glazing, which tends to make surfaces slippery. Repainting the face and edge of concrete curbs annually is essential to maintaining high visibility and minimize tripping hazards. Other safety considerations include maintaining proper illumination levels within the facility, lighted "EXIT" signs, emergency lights, fire-safety equipment, and active security systems. In addition, in enclosed or underground structures carbon monoxide detectors and other ventilation systems should be checked daily for proper operation.

9.2.3 Aesthetic Maintenance

In addition to operational and structural maintenance needs within the parking facility, aesthetic features also require regular maintenance. The most obvious features of the parking facility associated with aes-

TABLE 9-4 Aesthetic Maintenance Schedule

Item	Description	Frequency Annual	Frequency As Required
1.0	Landscaping	Inspect	
	1.1 Mow grass		Perform
	1.2 Prune shrubs		Perform
	1.3 Tend flower beds		Perform
2.0	Painting	Inspect	
	2.1 Clean and paint		Perform
3.0	General appearance	Inspect	
	3.1 Take corrective action		Perform

Adapted from "Parking Garage Maintenance Manual," Parking Consultants' Council, National Parking Association.

thetics are landscaping, painting, and the facility appearance in general. The parking facility deserves maintenance similar to other buildings (Table 9-4).

9.2.3.1 Landscaping. Landscaping features around the parking facility enhance its appearance. Flower beds, shrubbery, and grass plots should be well attended. Planters should be cleaned and cultivated frequently. Landscaping should be done judiciously so as not to provide hiding places and reduce security.

9.2.3.2 Painting. Painting exposed structural elements, fascia panels, stair- and elevator-tower interiors, step landings, pedestrian handrails, vehicular guardrails, and miscellaneous metals on a periodic basis is essential. Where painting is required, it should be completed as soon as time and budget constraints allow.

Painted surfaces should be inspected annually to determine their condition. Small rust spots should be cleaned and touched up each year. Repainting should be done as required by the element, type of paint, and exposure conditions. Most painted surfaces need repainting at 3- to 7-yr intervals.

Repainting the face and edge of concrete curbs should be done semiannually to minimize any potential for tripping by maintaining high visibility. Concrete surfaces are typically painted with the following types of paints:

1. Water-based Portland cement paints
2. Water-based polymer latex paints
3. Single- or two-component polymer paints
4. Oil-based paints

Refer to ACI 515 for general guidelines on the selection of paints for a particular use. Water-based portland cement paints exhibit properties similar to those of concrete. Primary uses are filling in and leveling minor imperfections of concrete surfaces. The disadvantages of portland cement-based paints are that they cannot be applied over existing paints, have a limited color selection, and the surfaces tend to show stains and dirt, and they are hard to clean.

Water-based polymer latex paints are available for interior and exterior use. Latex paints form a breathable layer and are able to resist blistering when applied to concrete with a high or varying moisture content. Latex paints are available for use in both interior and exterior formulation. Latex paints must be applied to a moistened surface and may require a primer coat of a low-viscosity-penetrant paint before application to some concrete surfaces.

Polymer paints are available in single- or two-component forms. Both form a smooth, dense, high-gloss finish that is highly resistant to humidity, stains, or dirt. Polymer paints are available in a wide range of colors. Single-component polymer paints offer flexibility and extensibility; when applied sufficiently thick single-component polymer paints are able to bridge minor cracks that may form in the concrete surface. Two-component polymer paints form a high-gloss surface that is highly resistant to chemical attack and is easy to clean. Since polymer paints form a moisture barrier, they should not be applied to moist surfaces or to a concrete substrate with a high or varying moisture content.

Oil-based paints are those that contain derivatives of fatty acids or drying oils. These paints are not resistant to the natural alkalinity of concrete. Oil-based paints should always be applied with a primer coat. The primer coat will reduce the paint's susceptibility to a reaction with the alkalinity of the concrete but will not eliminate it.

Water-based polymer latex paints should be used in most parking-facility concrete painting applications. It will provide a durable, breathable surface protection for general use on concrete and masonry walls. Contact the manufacturer and follow the recommendations for surface preparation, storage of paint, and its application.

Metal surfaces are typically painted with the following types of paints:

1. Enamel paints
2. Zinc-rich paints

Enamel paints are a general-purpose interior/exterior paint used for protection against weather. Enamel paints are readily applied to primed, previously-painted, or galvanized-metal surfaces. Surface preparation includes the removal of dirt, oil, grease, and other surface contaminants; rust and paint that is not tightly bonded should be removed and the areas spot primed. Primers should be applied as per manufacturers' recommendations. Enamel paints should not be applied to damp or wet surfaces. Enamel paint should be applied when the air, product, and surface temperatures are at least 50°F. Care should be taken not to apply the paint late in the day when dew and condensation are likely or when rain is possible.

Zinc-rich paints can be used as a one-coat maintenance coating or as a permanent primer. Zinc-rich paints can be used in areas with varying temperature conditions and high humidity. Zinc-rich paint when applied to a surface forms a coating which self heals and resumes protection when damaged. As with enamel paints application should be done when the air, paint, and surface temperatures are at least 50°F. Care should also be taken not to apply the paint late in the day when condensation is likely, or on days when rain is possible.

9.2.3.3 General Appearance. Glass in the stair towers is provided for security as well as aesthetics. Windows and floors of the stair- and elevator towers, cashiers' booth and lobbies should be swept or washed on a regular basis. Damaged window glass or deteriorated glazing should be repaired as the need dictates. Roof systems should be checked periodically for leaks.

Occupied and heated spaces beneath the parking floor, such as the office spaces, public areas, cashiers' booths, restrooms, and mechanical/electrical equipment rooms, may require special consideration. If floor-slab cracking or leaking overhead occurs, then vehicular deck coatings may be required to protect against water-related damage.

Walkways leading to and from stair-tower entries should be kept clean. The entire perimeter of the facility should be kept clean and presentable. Trash receptacles placed at convenient locations around the facility help insure proper trash disposal.

Rust stains are usually indicators of other problems, such as concrete cracking or paint or sealant failure. The cause of rust staining should be determined and corrected. Refer to the *Concrete Construction* magazine publication, "Removing Stains From Concrete."

Chapter 10
Repair

SAM BHUYAN

10.1 INTRODUCTION

Although concrete is a relatively durable construction material, preventative maintenance and necessary corrective actions are required to extend the useful life of parking structures. Parking facilities experience harsh exposure conditions which can contribute to accelerated concrete deterioration and adversely affect the life expectancy of the structure. Quite often, owners and operators have to repair deteriorated structures with only 15–20 yr of service. It is not uncommon for repair costs to exceed $10.00 per sq ft. In addition, the repairs disrupt the facility's operations and cause inconveniences for users which result in a loss of revenue.

There are several accelerated deterioration mechanisms that can reduce the life expectancy of the parking facilities. In the northern regions, corrosion of embedded reinforcement due to chloride contamination by road salt is generally the primary contributor of accelerated deterioration. Another commonly observed concrete deterioration is due to the lack of, or inadequate air entrainment, in concrete. In this instance, freezing and thawing can contribute to progressive and rapid deterioration. Some structures in the southern regions also are exposed to an equally harsh environment. Concrete structures located in the vicinity of large bodies of saltwater and in costal regions are susceptible to corrosion-induced deterioration. Airborne chlorides and marine watersprays can be just as damaging as road salts.

Other deterioration mechanisms such as cracking, leaching, carbonation, and abrasion can adversely affect structures in severe as well as moderate climates. In addition, inadequate design details, poor drainage condition, use of poor quality concrete, and joint deterioration can have a significant impact on the service life of structures.

The durability of structures is also adversely affected by a lack of corrosion-protection systems. Today, an owner can use a combination of state-of-the-art corrosion protection systems, such as concrete sealers, membrane coatings, epoxy-coated reinforcement, admixtures to reduce concrete permeability, corrosion-inhibiting admixtures, and cathodic protection to extend the life of a structure effectively. See Chapter 6 for a discussion of durability requirements for new parking structures.

Appropriate repairs must be performed to address existing deterioration due to service exposure and inherent structure deficiencies. The repair methods and materials selected must be durable and capable of extending the service life of the structure cost effectively. Also, the repairs must be timely to reduce the cost impact of accelerated deterioration. For instance, floor slab repairs performed at an early stage of distress cost less than repairs that are delayed. The increased repair cost is primarily due to the increased rate of deterioration. Much of the subject material presented in this chapter is applicable to structures exposed to severe as well as moderate climate.

10.2   APPROACH TO RESTORING A PARKING STRUCTURE

10.2.1   General

To correct problems and restore a parking structure owners must develop a comprehensive program for evaluating existing conditions, making repairs, and setting up maintenance procedures that extend the life of the facility. A comprehensive and cost-effective restoration program is built around an accurate evaluation of the condition of the structure. For a parking structure the program generally focuses on the following issues:

1. Repair deteriorated concrete floor slags to restore integrity
2. Provide durable protective wearing surfaces to minimize further deterioration
3. Perform repairs to other underlying structural members for continued safe use of the facility

A flow diagram for a systematic approach to restoring a parking structure is shown in Figure 10-1.

10.2.2   Types of Restoration Services

An owner should have a clear understanding of what his needs are in order to obtain appropriate services. Table 10-1 summarizes some of the reasons for an owner to request restoration services.

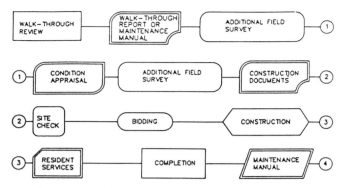

Figure 10-1. Flow diagram of a systematic approach to restoring a parking structure.

The basic types of restoration services that can be provided are as follows:

A. *Walk-Through Reports.* The walk-through service of the structure provides a general overview of the structure condition and a recommendation for additional investigations, if required. The work should be performed by an experienced restoration engineer. The walk-through report is generally framed around a one-day field observation that identifies major structural problems or emergency repair needs. The reports usually do not include testing. Observations are generally limited to visually obvious items, such as spalling, cracking, cold joints, poor formwork, debris in the concrete, and honeycombing. The report clearly indicates whether or not the owner has a problem.

B. *Condition Appraisal.* The condition appraisal provides the owner with a comprehensive examination and report. Specific recommendations for corrective action are issued, intended primarily to extend the service life of the facility. Restoration cost estimates and repair priorities are provided. Materials and nondestructive-testing-result evaluation provide a foundation for the selection of repair alternates.

C. *Construction Documents.* Construction documents implement condition appraisal findings. Construction documents should not be based only on a walk-through of the facility.

D. *Resident Services.* Field observation of repair/restoration work by project resident and contract administration. This service supplements the more traditional periodic observations.

E. *Maintenance Programs.* Maintenance program development is generally based on a brief walk-through inspection that al-

TABLE 10-1 Restoration Services

Owner's Need	Level of Service	Scope of Work
Do I have a problem?	Walk-through review	State problems, if any; recommend further work, if any.
Just how bad is my problem?	Field survey and condition appraisal	Identify problems, causes and effects; recommend repairs.
How much should I budget for repair and/or maintenance?	Field survey and condition appraisal or maintenance program	Give repair cost budget estimate. May also provide life-cycle cost analysis of various repair alternatives and may provide structural analysis.
I want to repair my deck.	Field survey and construction documents	Develop plans and specifications, final quantities, and cost estimates for bidding.
How will I know the repairs will be done correctly?	Project resident	Record of actual quantities of work done. Daily records of work progress. Implement plans and specifications.
I have a new deck, or I just had the deck restored. Now, how do I maintain it?	Field survey and maintenance program	Provide customized maintenance manual listing annual maintenance items, priorities, schedule, and estimated maintenance costs for structure's service life.
I am concerned about the water leakage through the floor slab. I am concerned about specific structural or member-durability problems.	Field survey and condition appraisal	Identify cause and effect of the specific problem. Recommend repairs and provide repair-cost estimates. Also recommend further work, if any, for areas not covered by the present scope of work.

lows the restoration engineer an opportunity to get a feel for the facility and its existing or potential problems. In addition, chloride samples are sometimes taken, and original construction documents reviewed. The maintenance inspection is intended only to qualify the conditions requiring maintenance and provide general procedures and guidelines to extend the service life of the structure.

The effort required to perform each of the tasks varies considerably; therefore, a specialist should perform a preliminary walk-through review of the structure to establish the extent of service required. For instance, a costly, time-consuming condition appraisal can be avoided if only minor repairs and implementation of a maintenance program are required. Unlike routine maintenance efforts, most repair and restoration work should be directed and implemented under the supervision of a restoration specialist.

10.3   CONCRETE DETERIORATION

This section will present the most common types of distress observed in parking facilities in need of repair or maintenance. Concrete deterioration generally falls into one of the following major categories — spalling, cracking, scaling, and leaching. Joint deterioration is also included in this discussion because it generally contributes to distress in underlying structural members. Other forms of deterioration, such as abrasion of driving surfaces, carbonation, and distress due to reactive aggregates in the concrete are generally less frequent, observed only in isolated instances. Discussion of concrete deterioration will be limited only to the major categories listed above. These deterioration mechanisms primarily contribute to the durability problems experienced by parking structures. In addition, the following conditions tend to aggravate and contribute to the deterioration process:

- Selection of a less durable structural system and inadequate design details
- Poor drainage
- Material defects
- Construction deficiencies
- Lack of an appropriate corrosion-protection system

10.3.1   Corrosion-Induced Deterioration

Spalling and delamination of concrete due to corrosion of embedded reinforcement is a common form of distress in structures located in the northern climates or other areas subjected to salt environment. The use of road salt during winter months often results in chloride contamination of the floor slabs and acceleration of reinforcement corrosion in the presence of oxygen and moisture (Figure 10-2). Many relatively older structures in the mild climatic region are also susceptible to

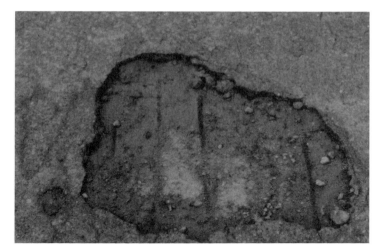

Figure 10-2. Spalling is progressive and steps must be taken to restore and protect the surface from further deterioration.

corrosion-induced deterioration to a lesser degree. The spalls in rein-forced-concrete surfaces are usually dish-shaped cavities one to several inches deep with varying surface areas. Floor slab spalling can be quite extensive, sometimes covering several hundred square feet (Figure 10-3). Corrosion-induced deterioration can also occur in structures having extremely porous concrete that is susceptible to carbonation (Figure 10-4).

Before open spalls or "potholes" can occur on the floor surface, a horizontal fracture called "delamination" will usually develop below the concrete surface. Fractures begin at the level of the corroding reinforcement or other embedded metal and migrate to the surface. These floor-slab delaminations can be detected by tapping the floor surface with a hammer or by a chain-drag survey. Freeze-thaw cycles, traffic action, and additional corrosion tend to further accelerate the rate of spall development.

10.3.1.1   Chloride Contamination. Concrete is not an impervious ma-terial. Excess water, not required for hydration, eventually dries, leaving behind an interconnected network of capillary pores. Concrete capillary pores have a relatively larger diameter, ranging from $15-1000$ Å (Figure 10-5). In comparison, the chloride ion diameter is less than 2 Å. Therefore, chloride ions can eventually penetrate concrete. The contamination process is accelerated by salt accumulation on surfaces

(a)

(b)

Figure 10-3. (a) Extensive deterioration can adversely affect load-carrying capacity of structural members. (b) Load testing of slab to determine impact of deterioration on structural integrity.

Figure 10-4.   Corrosion of reinforcement due to leaching and carbonation of poor quality concrete.

exposed to de-icing salts, shallow concrete cover, wetting and drying, and poor quality concrete.

The amount of chloride content in concrete which will contribute to corrosion is referred to as the "corrosion threshold." The National Cooperative Research Program Report #57 defines corrosion threshold as the minimum quantity of chloride required to initiate the corrosion of embedded steel in the presence of moisture and oxygen (Figure 10-6). The chloride content can be reported in various units such as:

- percentage chloride ion by weight of concrete
- ppm chloride ion by weight of concrete
- percentage chloride ion by weight of cement
- lb chloride ion per cu yd

The chloride ion content is reported as acid-soluble or water-soluble based on the analytical test procedure utilized to obtain the results. The acid-soluble test method measures the chloride which is soluble

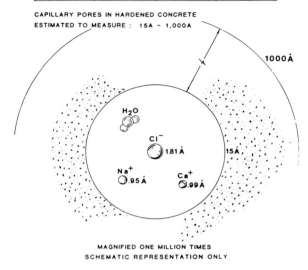

ELEMENT	ATOMIC NO.	ATOMIC RADIUS	IONIC RADIUS
Cl⁻	17	0.99 Å	1.81 Å
Ca⁺	20	1.97 Å	0.99 Å
Na⁺	11	1.86 Å	0.95 Å

CAPILLARY PORES IN HARDENED CONCRETE
ESTIMATED TO MEASURE : 15A – 1,000A

1000Å

H_2O

Cl⁻
1.81 Å

15Å

Na⁺
.95 Å

Ca⁺
.99 Å

MAGNIFIED ONE MILLION TIMES
SCHEMATIC REPRESENTATION ONLY

Figure 10-5. Concrete porosity can eventually permit migration of chloride ions to the level of reinforcement.

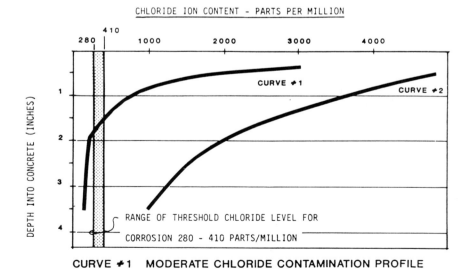

CHLORIDE ION CONTENT – PARTS PER MILLION

CURVE #1   MODERATE CHLORIDE CONTAMINATION PROFILE

CURVE #2   SEVERE CHLORIDE CONTAMINATION PROFILE

Figure 10-6. Chloride content vs. depth into concrete.

in nitric acid. Water-soluble chloride is the chloride that can be extracted by water in accordance with procedures of a specific test method.

Chlorides in concrete occur in the water-soluble form or may be chemically combined with other ingredients. Soluble chlorides initiate corrosion, while combined chloride is believed to have little effect on corrosion. When considering the probability of corrosion, it is more appropriate to consider only the water-soluble chloride ion content of the concrete since the acid-soluble chloride results include chlorides that are chemically combined.

The ACI Building Code (ACI 318) reports the chloride ion content as percentage chloride ion by weight of cement. Research done by the Federal Highway Administration indicates that corrosion threshold is 0.20% acid-soluble chloride ion by weight of cement. Therefore, for new structures, ACI 318 limits the maximum permissible water-soluble chloride ion content for mild steel reinforcement to 0.15% by weight of cement considering that 75% of acid-soluble chloride is water soluble. For prestressed-concrete structures the chloride ion content has been established at 0.06% by weight of cement. This is primarily to reflect the severity of corrosion and loss of cross-sectional area of highly-stressed steel reinforcement.

10.3.1.2 Corrosion Mechanism. Corrosion of metal in concrete is an electrochemical process, which contributes to progressive concrete deterioration. The impact of chloride contamination and corrosion can be explained by understanding the protective, as well as the corrosive, mechanism of reinforcement in concrete. The ability of metal to form a protective film greatly reduces the rate at which it corrodes. This protective film is generally an oxide and plays an important role in the corrosion resistance of metal, such as aluminum, chromium, stainless steel, lead, and other relatively noble metals. The oxide film on the steel reinforcement embedded in the concrete is relatively stable due to the high pH. A pH of 12–13 generally provides a passive environment which initially protects the embedded reinforcement.

The electrochemical reactions of the corrosion process involve the transfer of electrons and migration of ions. The essential elements of a basic corrosion cell are illustrated in Figure 10-7. The anode is the point where corrosion occurs by migration of ions into the electrolyte. The cathode is the point where electrons are consumed and no corrosion occurs. The electrolyte is usually an aqueous solution containing ions, capable of conducting current. The return circuit is a metallic path through which electrons move from the anode to the cathode, which usually consists of the metallic reinforcement itself.

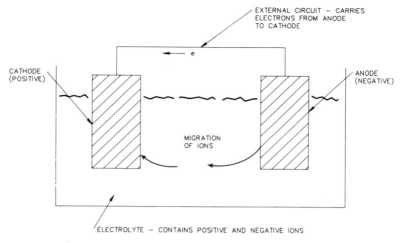

Figure 10-7. Schematic diagram of a basic corrosion cell.

The voltage between the anode and the cathode can exist due to the presence of dissimilar metals in the concrete which are electrically connected. It can also exist in a continuous metallic element, such as an embedded reinforcement, due to the difference of the environment between two areas on the same element. The difference of the environment along a reinforcement in concrete can be attributed to variations in chloride ion concentrations, variations in surface condition, extent of consolidation, availability of oxygen, pH of concrete, or moisture (Figure 10-8). This corrosion cell is also referred to as the "microcell."

In a chloride-contaminated slab the potential difference that will sustain the corrosion process can be attributed to the difference in the chloride ion concentrations along the reinforcement, as well as the

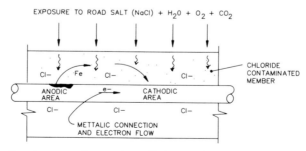

Figure 10-8. Development of corrosion cell along embedded reinforcement (micro cell).

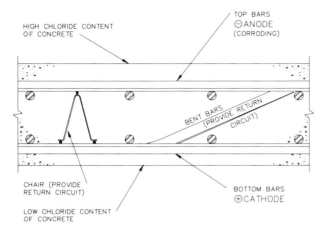

Figure 10-9. Development of corrosion cell in top and bottom layer of slab reinforcement (macro cell).

amount of chloride ions reaching the top and lower mats of the slab. The upper layer of reinforcement in the more chloride-contaminated concrete will be anodic to the bottom layer and results in development of a strong corrosion cell referred to as a "macro cell." This will result in the corrosion of the upper layer of reinforcement when the mats are electrically connected (Figure 10-9). Generally a macro-cell corrosion will contribute to a more rapid deterioration of the structure than a micro-cell corrosion: a single reinforcement electrically isolated, but embedded in chloride-contaminated concrete, is likely to corrode at a relatively slower rate than a top reinforcement also connected to the lower mat of reinforcement in uncontaminated concrete. Most reinforced concrete structures tend to deteriorate due to micro- as well as macro-cell corrosion. Another example of accelerated corrosion is a reinforcing bar exposed to both concrete and water, as at a spalled area of the floor slab. The reinforcing section in the water is likely to corrode due to the different electrolytes to which it is exposed. The rate at which corrosion occurs is also affected by the relative areas between the anode and the cathode. If the anodic area is small, relative to the area of the cathode, the anode (the upper mat of reinforcement in a slab) will tend to corrode rapidly. This is because the corrosion current is concentrated in a smaller area.

Corrosion by-products (rust) occupy a volume at least 2.5 times that of the parent metal. This expansion causes high tensile stresses which crack ("delaminate") the surrounding concrete. Concrete cracking can occur when section loss of the corroding metal is 5% or less.

Cracks first appear vertically over, and parallel to, the corroding reinforcement. These cracks permit more moisture, oxygen, and chloride
ions to the level of the reinforcement, causing accelerated corrosion
and concrete delamination. The corrosion-induced cracks running
along the length of the reinforcement are potentially more damaging
than transverse cracks running across the length of the reinforcement.
If the concrete has low permeability, relatively fine and hairline transverse cracks usually do not contribute to accelerated corrosion of
embedded reinforcement. However, wide transverse and through-slab
cracks can contribute to the corrosion of embedded reinforcement.

10.3.1.3 Corrosion-Induced Distress. Corrosion of concrete reinforcement can adversely affect structural members. The adverse impact on
serviceability and structural integrity is as follows:

- Concrete deterioration causes serviceability, maintenance, and
  operational safety problems. The corrosion mechanism is initiated when the chloride ions penetrate the level of the embedded reinforcement.
- Concrete cross section loss due to corrosion of reinforcement
  can adversely affect the load-carrying capacity of individual
  elements of the structural system, such as floors, slabs, beams,
  and columns.
- The reinforcement loses significant cross section due to corrosion, which can contribute to stress redistribution and possible overstressing of the structure.
- The reinforcement debonds from the concrete in delaminated
  areas which can result in reduced load-carrying capacities of
  members due to loss of anchorage.

Spall development in beams due to water leakage is shown in Figure
10-10.

Surface spalling near mid-span reduces the concrete section. Concrete section reduction at mid-span can significantly reduce the structural capacity of the concrete member. At the same time, severe corrosion of bottom reinforcement due to leakage can also result in overstressing and possible reinforcement yielding or failures (Figure
10-11(a)). Loose spalls and delamination as a result of corrosion-induced deterioration can be hazardous for facility users and can damage
vehicles.

Of all the structural members the floor slab is the most susceptible
to corrosion-induced deterioration, since the floor surface is directly
exposed to the elements. In addition, concrete permeability (quality),

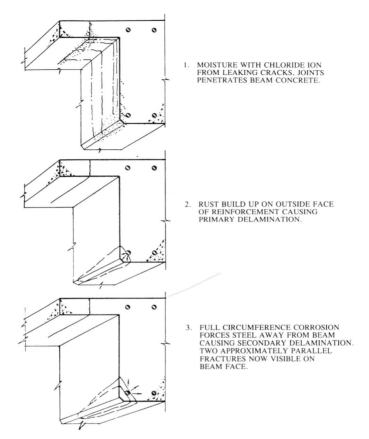

1. MOISTURE WITH CHLORIDE ION
    FROM LEAKING CRACKS, JOINTS
    PENETRATES BEAM CONCRETE.

2. RUST BUILD UP ON OUTSIDE FACE
    OF REINFORCEMENT CAUSING
    PRIMARY DELAMINATION.

3. FULL CIRCUMFERENCE CORROSION
    FORCES STEEL AWAY FROM BEAM
    CAUSING SECONDARY DELAMINATION.
    TWO APPROXIMATELY PARALLEL
    FRACTURES NOW VISIBLE ON
    BEAM FACE.

Figure 10-10.   Beam deterioration mechanism.

and cover over reinforcement, has a direct impact on the "time-to-corrosion" and the extent of the floor-slab deterioration. As stated in ACI 318, the maximum water-to-cement ratio should be limited to 0.40 for concrete with $1\frac{1}{2}$ in. of cover. The concrete cover and quality are the primary elements of a corrosion-protection system. Flexural cracks and joints in structural members, especially conventionally reinforced-concrete slabs and beams, provide extensions of these avenues for moisture and chloride ion penetration to embedded reinforcement. The corrosion-induced deterioration of the floors is also adversely affected by cracking, joint leakage and poor drainage conditions (Figure 10-11(b)–10-11(h)). For long-term durability, it is beneficial to design precast or cast-in-place, post-tensioned concrete structures which limit

(a)

(b)

Figure 10-11. (a) Deterioration of underside of slab due to water leakage through slab. (b) Beam-deterioration.

(c)

(d)

Figure 10-11. (*Continued*) (c) Deterioration of underside of slab. (d) Joist deterioration at leaking construction joint.

(e)

(f)

Figure 10-11. (*Continued*) (e) Beam deterioration at leaking expansion joint. (f) Corrosion of metal deck.

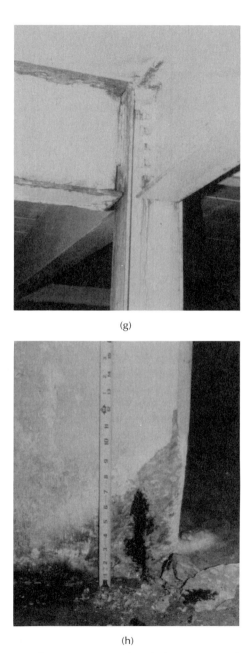

(g)

(h)

Figure 10-11. (*Continued*) (g) Corrosion of underlying structural steel beam and column. (h) Deterioration of column base.

the exposure of mild-steel reinforcement and reduce the cracking of floor slabs.

Most parking facilities built prior to 1977 generally did not utilize an effective corrosion-protection system, such as surface sealer, epoxy-coated reinforcement, or corrosion inhibitor, except that some structures were occasionally sealed with boiled linseed oil. Concrete surface treatment with boiled linseed oil is considered to be relatively ineffective in screening chloride ions. Also, structures built prior to 1977 utilized concrete with a water-cement ratio of 0.53, to meet requirements specified by the then-current ACI Building Code. These structures have a significantly higher potential for corrosion-induced deterioration in comparison to structures built to meet requirements of the 1977 ACI Building Code. The repair cost for these less-durable structures is usually high due to the progressive nature of chloride-induced deterioration.

The above discussion is intended to provide an overview of the corrosion process and familiarize the reader with those conditions which impact upon this distress mechanism and its influence on structural members. A more detailed discussion is provided in the American Concrete Institute's (ACI) committee reports, entitled "Guide to Durable Concrete," ACI 201.2R, and "Corrosion of Metals in Concrete," ACI 222R.

10.3.2   Freeze-Thaw-Induced Deterioration

Concrete floor surfaces of parking facilities are susceptible to freeze-thaw deterioration, especially if the concrete is not adequately air-entrained and critically saturated due to ponding or poor drainage. The most common form of surface deterioration is scaling. Scaling is characterized by the progressive deterioration of the concrete surface through paste (sand/cement) failure. It results from the disruptive forces generated in the paste when the concrete is saturated and freezes. Scaling is common in those areas of the continent subject to freeze-thaw cycling.

Scaling begins with a slight surface flaking or internal horizontal delamination close to the surface, which becomes deeper with continuing exposure. Initially, only the surface texture and a small amount of paste are eroded. Surface flaking and scaling create depressions which can retain water and contribute to progressively deeper and more extensive deterioration. Eventually, however, coarse aggregate is exposed, and larger surface areas are affected (Figure 10-12).

Scaling can significantly impair the serviceability of concrete in-

1. CONCRETE BECOMES SATURATED BY WATER PENETRATING THROUGH PORES AND CAPILLARIES.

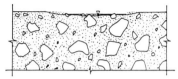

2. CONCRETE IS FROZEN IN A SATURATED STATE CAUSING HIGH STRESS. LOOSE FLAKES APPEAR ON SURFACE AS THE MORTAR BREAKS AWAY.

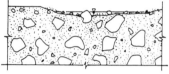

3. AS FLAKING PROGRESSES, AGGREGATE IS EXPOSED AND EVENTUALLY BREAKS AWAY, THEREBY EXPOSING MORE PASTE TO FREEZE-THAW DAMAGE. IN EXTREME CASES, APPARENTLY SOUND CONCRETE CAN BE REDUCED TO A GRAVEL-LIKE STATE IN A SHORT PERIOD OF TIME.

Figure 10-12. Concrete surface scaling mechanism. Source: "Parking Garage Maintenance Manual," Parking Consultants Council, National Parking Association.

tended as driving or walking surfaces. Flat portions of floor slabs, gutter lines, areas near drains, and ponded areas are more susceptible to scaling due to greater potential for saturation of the surface and the presence of de-icing chemicals (Figures 10-13 and 10-14). Also, exposed surfaces, such as the upper-most floors, are subjected to more freeze-thaw cycling and therefore are more susceptible to scaling.

A concrete mixture with a proper entrained air-void system is required to protect concrete against freeze-thaw deterioration (Figures 10-15, 10-16, and 10-17). Air-entrained concrete is generally produced by using an air-entraining admixture during mixing. Unlike entrapped air, the entrained air voids in air-entrained concrete are microscopic in size and uniformly distributed to accommodate the expansive forces generated by frozen moisture in the saturated concrete.

10.3.2.1   Deterioration Mechanism. Concrete is naturally porous. Excess water not required for hydration (hardening), but needed for workability during mixing, placement, consolidation, and finishing even-

Figure 10-13. Scaling adjacent to drain can contribute to ponding and progressive deterioration of concrete.

Figure 10-14. Scaling and joint deterioration.

Figure 10-15. Deterioration of non-air entrained concrete accelerated by poor drainage.

tually dries, leaving behind a continuous network of pores and capillaries. This network gives concrete its porosity. Porosity, or "permeability," is generally high for concrete mixes with a high water/cement ratio and low for mixes with a low water/cement ratio. High

Figure 10-16. Undermining of control joint by deterioration of concrete adjacent to the sealant.

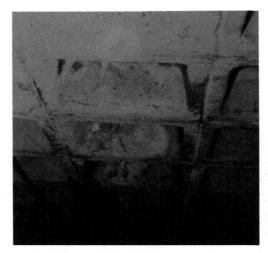

Figure 10-17. Scaling on underside of slab.

porosity allows the concrete to absorb significant free water during exposure to rain or snow. If concrete cannot dry and becomes saturated during a freeze cycle, ice accumulates in the pore structure.

The destructive mechanism is not ice accumulation itself, but rather water pressure generated during ice development. Water migration through the pore network exerts significant pressures during freezing. It has been substantiated that water pressures cause the paste failure.

10.3.2.2 Influencing Factors. There are a number of factors that influence the nature and extent of scaling on concrete surfaces. The following discussion is not intended to convey any particular order of importance for the factors reviewed, which are divided into two categories.

- The first category defines and describes those factors related to the service environment. Factors associated with the environment are number and intensity of freeze-thaw cycles, presence of de-icer chemicals, and degree of saturation.
- The second category of influencing factors is that associated with the particular concrete and its design features. Material properties which greatly influence the susceptibility of concrete to scaling and freeze-thaw deterioration are air entrainment, strength, water/cement ratio, and the mix design.

As previously discussed, freezing is the principal cause of scaling. If there were no freeze-thaw cycles, scaling could not occur. It has been established that the number of freeze-thaw cycles directly influences the deterioration rate. For similar concretes subjected to equivalent degrees of saturation, concrete exposed to the higher number of freeze-thaw cycles will disintegrate earlier and more severely than concrete subjected to fewer freeze-thaw cycles.

In addition to the number of cycles, the rate or cycle intensity is also significant. Rapid freeze-thaw cycling is far more destructive to concrete than slow freeze-thaw cycling due to the redistribution of the pressures in the concrete matrix. Concrete surfaces exposed to direct sunlight during winter periods are subject to more frequent and rapid cycling than concrete which is exposed to ambient temperatures, but shaded from direct sunlight.

The impact which de-icer chemicals (salt) have on scaling is both mechanical and chemical. High concentrations of salt depress the pore-water freezing point and increase the osmotic pressures which cause paste failure. Also, high salt concentrations can set up a counter system of pressures caused by the alkaline/acid relationships between the concrete and pore water, respectively.

As previously discussed, excess water is required within the pore network during freezing to induce disruptive pressures. Concrete that is relatively dry and subject to freeze-thaw cycling experiences minimal disruption. Continually moist concrete will disintegrate rapidly during freeze-thaw cycling because the water cannot escape without generating disruptive pressures.

Air entrainment has been used successfully for the past 40 yr to protect concrete against scaling. Air entrainment consists of a uniform dispersion of small bubbles in the paste matrix. These bubbles compete with the pore network for water during freezing and thus relieve the destructive pressures. Research has shown that the bubbles must have a particular size and spacing to be effective at protecting concrete.

In addition to air entrainment, the development of minimum strength prior to the first frost exposure is needed to insure adequate resistance against freeze-thaw damage. Concrete strength must be at least 3500 psi prior to exposure to the freezing cycle if it is to remain durable in service. Properly air-entrained concrete that has not gained sufficient strength before freezing, will be subject to premature freeze-thaw deterioration.

The water/cement ratio directly influences concrete porosity (permeability). Highly permeable concretes are more susceptible to rapid saturation than are those of lower permeability. Concrete has a certain tolerance for moisture. Moisture diffusion within a relatively dry matrix can influence the concentrations of water and can minimize saturation, thus preventing premature deterioration.

Design of the concrete mix, especially the cement factor, water/cement ratio, and use of the maximum-size coarse-aggregate fraction can also enhance long-term durability. The mix design should be tested prior to concrete placement in order to insure that the air system specified is achieved during construction. It is common to find differences between the specified and measured air entrainment in the plastic concrete and in the air content of the finished hardened slab.

Concrete design details and concepts also influence susceptibility to scaling. Concrete floor surfaces subject to frequent freezing and de-icer chemical application can be designed to drain rapidly, minimizing critical saturation potential. Parking-facility floor slabs with a minimum of $1\frac{1}{2}$–2% grade will drain rapidly and will be inherently less susceptible to scaling due to the limited potential for saturation. Well-designed gradients for drainage and an adequate number of surface drains will reduce excess water and keep the floor surface fairly dry. Floor slabs that are unusually flat or have few drains can experience rapid deterioration due to their high potential for saturation.

The above discussion is intended to provide an overview of the scaling process and familiarize the reader with those conditions which impact upon this distress mechanism and its influence on structural members. A more detailed discussion is provided in the American Concrete Institute's (ACI) committee report, entitled "Guide to Durable Concrete," ACI 201.2R.

10.3.3   Concrete Cracking

Concrete is strong in compression but relatively weak in tension. Therefore, concrete cracking is caused by development of tensile stress in concrete members. Concrete cracking can occur in plastic as well as in hardened concrete. Plastic concrete cracking can be attributed to improper concrete placement, consolidation, and/or curing or plastic shrinkage of the concrete. Cracking in hardened concrete is usually due to the internal stresses induced by the normal response of structural members to applied loads, temperature changes, support settlement, or drying shrinkage (Figures 10-18, 10-19(a), (b), and 10-20). Concrete cracking can occur in plastic as well as in hardened concrete. In some instances cracking in slabs, beams, columns, walls, and load-bearing areas can be attributed to restraints to volume change of con-

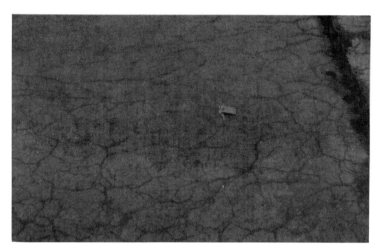

Figure 10-18.   Pattern cracks on floor surface due to drying shrinkage.

crete due to design or construction deficiencies. Cracking is also an indication of concealed problems in the concrete floors or supporting members, such as the initial stages of corrosion of embedded reinforcement (Figure 10-21). Localized loss of prestressing forces due to prestressing strand or tendon deterioration, or embedded anchorage failure, can also result in cracking.

Concrete cracking is expected to occur when the concrete member is subjected to tensile stresses and reinforcement is provided to transfer stress across the cracks. Properly designed and positioned reinforcement help to distribute and control crack widths. For floor slab exposed to de-icing chemicals ACI 224R suggests limiting crack widths on the tension face of members to 0.007 in. Cracking can be detrimental when it will permit water leakage or contribute to concrete deterioration. These cracks should be sealed to minimize the effect of cracking on the long-term durability of the structure.

Concrete cracking can be minimized by proper selection of structural systems at the time of design. Conventional reinforced concrete floors are highly susceptible to cracking due to shrinkage, thermal, and flexural stresses. Precast pre-tensioned or cast-in-place post-tensioned concrete floor systems are much less susceptible to cracking.

Proper design and installation of control and expansion-joint systems will also limit concrete cracking. Floor-slab joints allow for shrinkage and temperature-related volume changes which limit the tensile stresses in concrete. When volume change of concrete is restrained, random concrete cracking can occur. The restraint to volume

(a)

(b)

Figure 10-19.   (a) Radial cracks over column due to flexural stresses. (b) Floor slab cracks around column.

Figure 10-20. Through-slab crack in pan joist system contributing to water leakage.

change is provided by various parts of the structure, such as shear walls, stair towers, and rigid columns. The likelihood of crack formation from volume change is increased by the presence of geometrical discontinuities, construction discontinuities (joints), and tension al-

Figure 10-21. Beam cracks along embedded reinforcement. Note crack along bottom reinforcement has been injected.

Figure 10-22. Water leakage through control joints in precast double-tee floor.

ready existing from applied loads. Concrete spalling and cracking can occur due to inadequate design details that can result in "binding" of expansion joints and slip-bearing joints. All floor-slab joints should be detailed and installed to effectively seal and minimize leakage and potential deterioration of underlying beams, columns and connections (Figure 10-22).

10.3.4   Joint Distress and Leakage

Construction joints are installed at preselected locations to limit the size of concrete placements. These joints are tooled and filled with a flexible sealant to prevent leakage. Control joints provide for concrete volume change movements by creating a series of weakened planes for cracking at predetermined points in the floors and walls. Control joints are usually tooled or formed in the plastic concrete and then filled with a flexible sealant to prevent water leakage.

Flexible-joint sealant material installed in joints in the structure will deteriorate due to exposure to ultraviolet rays, abrasion, and age (Figures 10-23 and 10-24). Joints on supported floor slabs must be maintained by sealing against water leakage and intrusion of debris. To maintain the effectiveness of floor-slab joints, most sealant material needs to be removed and replaced at 8- to 10-yr intervals. Often, joint-sealant deterioration is not the only cause for failed and leaking con-

Figure 10-23.   Damaged wideseal expansion joint.

Figure 10-24.   Control joint failure.

struction and control-joint systems. A joint sealant for a particular application should be selected based on the required degree of flexibility, hardness, bond, strength, or durability. Deterioration of the concrete joint edge also effects joint-system performance. Edge deterioration due to wheel or snow plow impact are typical observations. Freeze-thaw related distress from entrapped moisture between a failed sealant material and the adjacent concrete can contribute to edge deterioration (Figure 10-16). Overfilled sealant material is typically ripped from joints or cracks by vehicular traffic. Inappropriate application of the flexible sealant material, either in too deep a groove or too wide and shallow a joint, can result in premature material failure.

Expansion joints provide a practical limit on structure dimensions to accommodate movements associated with temperature changes, concrete creep, and long-term concrete shrinkage. Expansion joint openings are also sealed with a flexible material. The expansion-joint system's effectiveness to seal the opening varies considerably. Some systems that are properly designed, installed, and maintained can be effective; however, design and installation of expansion-joint systems are often inadequate. Some joint systems are susceptible to damage due to a lack of cleaning of dirt and debris and the potential corrosion of exposed metal components. Also, expansion joints are susceptible to vandalism.

Premolded flexible urethane expansion joints are quite often specified for parking structures (Figures 5-1 through 5-3). When properly designed and installed these joints can be effective; however, improper design considerations, such as undersized sealant width, lack of provisions for shear transfer across the joint, positioning of joints in or adjacent to turn aisles, and excessive vehicular and snow plow abuse usually result in premolded joint failure. Failures can also be attributed to improper preparation of the concrete joint edge and variations in thickness of the installed joint sealant.

Metal-edged expansion-joint systems with flexible glands are capable of withstanding greater abuse then premolded joint systems (Figures 5-5 and 5-6). It is preferable to specify a metal-edged system on exposed levels of a structure. However, these systems are susceptible to dirt and debris accumulation, and require periodic cleaning. The metal-edged joint systems are more costly to install and repair.

Leakage through a failed and unmaintained joint system in a parking facility can create concerns for corrosion-induced and freeze-thaw related distress on interior levels. Serious concerns for patron safety, damage to vehicles, and general aesthetics are also affected by failed joint systems.

Figure 10-25. Leaching and rust staining.

10.3.5   Leaching

Leaching is caused by frequent water leakage through cracks. The water leaking through the crack carries along part of the hydrated lime and other water-soluble products and deposits them as a white film, stain, or in extreme cases, stalactites on the ceiling below (Figure 10-25). Continued leakage will weaken the concrete over a period of years and the deterioration rate is affected by the concrete quality. Leaching is generally more noticeable in cracks along gutterlines and areas that are susceptible to ponding. Water that leaks through cracks can leave deposits that will damage automobile paints.

10.4   CONDITION APPRAISAL

10.4.1   General

Parking-structure restoration and maintenance programs begin with a thorough facility appraisal. The condition appraisal assists in locating existing distress, qualifying materials, and quantifying the extent of deterioration. Therefore, the condition appraisal provides a foundation for selecting repair materials, repair methods, and evaluating specific repair alternatives to restore the facility. This section contains details which can assist practicing professionals to develop a restoration program and enables owners to evaluate proposed programs. Again, a proper condition appraisal is necessary to assure success in restoring a structure.

An appraisal requires an indepth review of existing conditions by observing individual elements and performing selective material and nondestructive testing. Evaluation of laboratory and field survey data collected requires review both by a materials specialist and a structural engineer. The primary objectives of the condition appraisal are to:

1. Define and describe conditions which exist within the facility, focusing mainly on the deterioration of floor slabs and supporting structural elements.
2. Identify cause(s) of the observed deterioration.
3. Describe the extent of the observed deterioration.
4. Evaluate the impact of the observed deterioration on the serviceability, durability, and structural integrity of the facility. (Detailed survey must often be supported by materials and nondestructive testing to identify causes of deterioration.)
5. Develop repair alternatives for the facility. Repair alternatives should include discussion of only feasible repairs which can effectively extend the useful life of the facility.
6. Recommend repairs for the facility based on technical interpretation of data and establish repair priorities to stage construction over several years.
7. Identify items which can adversely affect safety.
   a. Accelerated deterioration of structural members
   b. Code items, such as barrier heights, barrier wall lateral-load-carrying capacity, and mechanical/electrical systems
   c. Structural distress of as-built members due to material quality, design deficiencies, or construction practices

Evaluation of structural distress requires a more extensive testing program to verify results of the analytical work performed to determine the cause(s) of the problem. In some cases, full-scale load testing of the structural element is required to verify results of field, laboratory, and analytical work.

10.4.2 Field Survey

Prior to preparing a comprehensive condition appraisal of a structure, it is essential to perform an accurate and complete field survey. Three key words characterize the field survey: locate, qualify, and quantify. Deterioration found in the survey is recorded on field survey plan sheets which may become part of the condition appraisal report. Materials testing also constitutes an essential part of the survey.

The field survey consists of six phases: preparation, initial walk-through, visual examination documentation, photographic recording, materials testing, and preliminary evaluation. Each of these phases will be discussed in detail. Emphasis placed upon any one phase is a function of the condition of the structure being surveyed. Usually those structures in a more advanced state of deterioration require more effort than structures showing minimal deterioration. Effective preparation is essential to insure efficient use of field survey time.

10.4.3   Preparation for Field Survey

Effective preparation is essential to insure efficient use of field survey time and systematic accumulation of relevant information. Readiness for the survey consists of reviewing the document, preparing field survey sheets, and obtaining appropriate equipment.

10.4.3.1   Document Review. The first step in preparing for the field survey is to attempt to obtain and review plans and specifications used in construction of the structure. It is also helpful to review other available construction records, such as shop drawing submittals, mix-design data, testing and inspection reports, and any maintenance or restoration work that may have been performed. This review is necessary to:

- Become aware of problems characteristic of the type of construction.
- Identify potential problem areas that may require more extensive investigation and preliminary selections of test methods and locations.
- Reduce amount of start-up orientation period (time in the field getting your bearings, etc.)
- Obtain details on materials (i.e., type of steel, concrete, and admixtures) and cover requirements.
- Determine type, size, and orientation of embedded reinforcing.
- Determine strength of concrete and steel.
- Determine design loads for supported tiers and roof.
- Examine drainage characteristics and locate gutterlines and drains.
- Locate stair towers, elevators, and other features.

Prepare a checklist of items to be covered during the survey, including quantity and types of testing to be performed. Also review ACI 201.1R,

"Guide for Making a Condition Survey of Concrete in Service," which contains a checklist for making a condition survey of structures. Contact the appropriate testing agencies to establish requirements and lead time required to perform the evaluation of samples collected during the field survey. Contact testing laboratory or coring company to take cores at the job site and arrange for shipment of the cores back to the office.

When construction or previous repair documents are unavailable, try to retrieve from other sources. Potential sources for obtaining construction records and documents include:

- Municipal building and zoning Departments or Inspectorial services departments
- Original architects (or later-generation firms)
- Historical society archives
- Contractor's or fabricator's records (or later-generation firms)
- Physical plant or engineering departments of major corporations, educational, or health-care institutions

Have as much information as possible on-hand when making inquiries. The following will prove useful:

- Building location
- Original name or owner of facility (if different)
- Date of construction
- Architect, structural engineer, fabricator, and/or contractor

Any of the above information may provide a clue which could lead to tracking down a set of drawings. Especially with older facilities, even as obscure a clue as a rebar identification tag, can provide essential information that may result in obtaining a source of drawings.

Always try to supplement paper review with verbal discussions with the owner (or other staff member). From your conversations, try to get a feel for the past history of problems and obtain information regarding specific problems that are most troublesome to the owner.

Reviewing contract documents and construction records can be invaluable in determining the cause(s) of problem(s). Results of any analysis performed should accurately predict measurement of crack widths, deflections, relative movement, etc. made during the field survey. Analysis is based on actual existing loads, geometry, and material properties, rather than design values.

When design and construction documents are not available for review, in addition to use of a pachometer (refer to Section 10.4.8), other nondestructive test methods that use X-ray, radar, and infrared thermography are also required to determine reinforcement location and pattern. In most instances, nondestructive test results are visually verified by limited exploratory excavations or test wells. Developing design information by these methods is time consuming. Therefore, the need for and extent of information required should be established based on results of an initial walk-through review of the structure.

10.4.3.2   Field Survey Sheet. Prior to the field survey determine the scale to be used for the condition survey sheets. Points to consider are: 1) minimize number of sheets required for report; 2) keep scale large enough to accurately record data, typically $\frac{1}{16}$ or $\frac{3}{32}$ in. scale; 3) choose a sheet size that can be easily handled during the field survey; and 4) where possible use the same scale that will be used for the contract documents. Typically photocopies of original design drawings can be utilized for the condition survey. (Architectural plan sheets typically are the "cleanest" drawings available and usually indicate all "landmarks" within the facility.) Otherwise, draft a grid and produce sufficient blank sheets for the survey. Two sample field survey blank sheets are illustrated in Figures A-1 and A-2 included in the Appendix.

10.4.3.3   Equipment. Get the appropriate equipment ready. Decide which members of the field survey team are to bring which equipment. Avoid field delays by making sure that all equipment is clean, charged, complete, and ready to use.

10.4.4   Initial Walk-Through

Conduct an initial walk-through of the structure during the initial stages of the site visit. Items or issues of particular concern can be identified, and the relative condition of the various structural elements determined. Note special conditions and established code requirements. It is advisable to first perform the tasks that require the largest amounts of time (chain drag and visual examination) so one can re-evaluate the time schedule. Also, it may be advisable to perform reflected ceiling surveys prior to floor-slab surveys in certain types of facilities (thin one-way slabs, pan joist systems or waffle slabs, or slabs with a waterproofing membrane system) where either 1) the condition of the ceiling holds the key to gaining a quicker appreciation of the facility conditions, and/or 2) the floor-slab deterioration is extensive, and the ceiling survey may provide the insight required to expedite the floor survey.

10.4.5   Visual Examination and Documentation

During this phase of the survey, distress in members is identified as potential work items. Perform a visual examination and a general review of the structural and operational elements summarized in Figure A-3 in the Appendix. Also note all work item locations on the field sheets and any other forms of distress or features in the various structural members that may help in the evaluation of the structure's performance, but not necessarily identified as potential work items. (Refer to Section 10.3 for the cause and significance of typical forms of deterioration.) Also review ACI 201.1R, which contains definitions of terms associated with concrete durability and forms of commonly observed distress.

In most instances, the upper surface of floor slabs receive the first and greatest attention. Due to the presence of embedded reinforcement near the top surface, there is a greater potential for concrete deterioration. Floor surveys consist of two phases: 1) locate deteriorations, and 2) record them.

The person recording data accurately locates and scales the spalled/deteriorated areas onto field survey sheets using the appropriate coding. The square footage of the patch required to repair the deteriorated area should be estimated as carefully as possible. (Refer to Figures A-4 through A-8 in the Appendix for a field survey legend and illustration of procedures to systematically record survey data.) Document size, location, and depth of scaling and spalling. Use a tape measure if necessary; it is of course better to estimate higher than lower.

Cracks may also be an early indication of corrosion of embedded reinforcement. Refer to ACI 224.1R for determining other causes of cracking. Floor-slab areas with cracks oriented parallel to embedded reinforcement should be examined with a pachometer to locate the proximity of the embedded reinforcement to the crack. Also, record crack patterns (if present), types, widths, and lengths. It is important to note whether the cracks are through-slab. Through-slab cracks should be documented during the visual survey of the underside of the slab.

As suggested in the National Cooperative Highway Research Program Report #57, cracks can be classified with respect to width, orientation, and, where possible, cause. Precise measurement of crack widths is neither feasible nor desirable, though a description in the following terms is useful:

Fine(F)   less than 0.01 in.
Medium(M)   0.01–0.03 in.
Wide(W)   greater than 0.03 in.

The crack width may be indicated by the abbreviation F, M, or W beside each crack. If the cracks are described, their orientation is usually classified according to one of the following terms: transverse, longitudinal, diagonal, radial, pattern, or random. Care is required in ascertaining crack widths; cracks usually look wider at the surface because of their broken edges. Consequently, the width ($\frac{1}{4}$ in. below the surface) is often reported. The crack widths reported by the petrographer during the microscopic examination of core samples are generally more accurate than those obtained in the field survey. The depth of cracks can be verified by taking core samples. The amount of moisture on the deck has a dramatic effect on the degree of cracking that is visible. Fine cracks are difficult to identify on wet or dry decks. Conversely, if a deck is examined under drying conditions, when moisture is associated with each crack, all the cracks are visible and appear to be wider than they are. The inexperienced observer may produce an exaggerated report.

Active (moving) cracks of any width are much more troublesome than inactive (dormant) cracks because they tend to enlarge and also limit the options when selecting the method of repair. It is often difficult to determine whether or not a crack is active, though a crack that is visible on both the top and bottom surfaces of the deck will tend to be active. In some instances, it may be advisable to use a crack comparator to more accurately record crack width. A crack comparator is a small hand-held microscope with a scale on the lens closest to the surface being viewed. Cracks should be recorded simultaneously with floor delaminations, indicating lineal footage and width.

Previously repaired spalls and cracks should also be noted, as well as signs or remnants of previous floor-surface sealer or coating applications. It is critical to determine variations in extent of deterioration for floor areas that are protected compared to those areas that are not. It is also important to note variations in the extent of deterioration for different levels of structures. The selection of the repair method and materials is influenced by the extent of the floor-slab deterioration. For instance, the lower (first-supported) level tends to have more deterioration than the roof (exposed) level. This can be attributed to natural wash-down off the roof by rain and periodic removal of the accumulated road salt. In contrast, the first (supported) level surface tends to have a relatively greater accumulation of road salt which is tracked in and deposited by vehicles entering the facility from the street.

Freeze-thaw-related deterioration, such as scaling and shallow sub-parallel delaminations, should be also noted. Items to consider when quantifying freeze-thaw-related deterioration are as follows:

1. Exposure conditions: top levels, entrance areas, and perimeter parking bays typically exhibit greater levels of freeze-thaw deterioration.
2. Drainage: ponded areas, flat turn aisles or end bays, and gutterlines typically exhibit greater levels of freeze-thaw deterioration.
3. Location of construction joints: resistance to freeze-thaw deterioration can change for different concrete placements.
4. Location and condition of drains: ponding will contribute to increased freeze-thaw deterioration.

As described in ACI 201.1, surface scaling may be qualified as follows:

- *Scaling—light:* loss of surface mortar without exposure of coarse aggregate. (Figure 10-26(a)).
- *Scaling—medium:* loss of surface mortar up to 5–10 mm in depth and exposure of coarse aggregate (Figure 10-26(b)).
- *Scaling—severe:* loss of surface mortar 5–10 mm in depth with some loss of mortar surrounding aggregate particles 10–20 mm

Figure 10-26. (a) Light scaling.

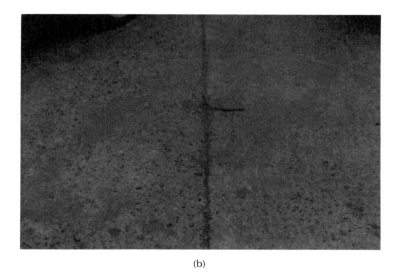

(b)

(c)

Figure 10-26.   (*Continued*) (b) Moderate scaling. (c) Severe (heavy) scaling.

in depth, so that aggregate is clearly exposed and stands out from concrete (Figure 10-26(c)).

Scaling should not be confused with abrasion. Abrasion will typically be concentrated in drive and turn aisles, and will generally appear as a more ''smooth'' and ''polished'' surface than that which is evident

Figure 10-27. Surface abrasion and cracking.

in scaled floor areas (Figure 10-27). Concrete popouts are also an early indication of freeze-thaw deterioration (Figure 10-28).

Systematic passes should be made through the structure, concentrating on negative-moment regions, gutterlines, and turns. The slab-on-grade area of the structure should also be inspected for potential problems, such as differential settlement, drainage and scaling concerns, and trip hazards.

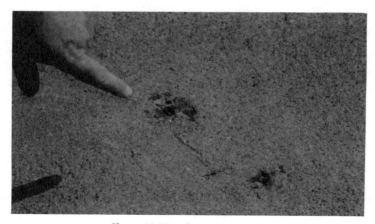

Figure 10-28. Concrete popout.

The inspection of ceilings is typically the next step in the survey. Beam distress is also recorded in this step. Beam, column, and ceiling delaminations are located by sounding the member with a hammer or tap rod. Longitudinal cracks near beam and column corners are usually reliable indicators of delaminated concrete. Ceilings showing evidence of cracking or salt and water staining should also be sounded. Ceiling deterioration is recorded in plan view on the reflected ceiling sheets.

The next step is to survey the columns to determine the extent of deterioration present on columns, connections, bumper walls and re-taining walls. Since beam, column, and bumper-wall distress cannot be shown in plan view, the coding system shown in Figure A-4 in the Appendix is very helpful in documenting deterioration. Refer to Figure A-5 in the Appendix, where a column spall for example, is denoted as "3C" with an arrow drawn to the appropriate face of the column.

Stair towers are systematically surveyed, followed by electrical and drain system inspection. Finally, a perimeter survey is made to observe conditions not visible from within the structure.

Concerning the recording of information during a survey, it may be practical to separate top-of-slab data and ceiling/column data on different sheets, especially where extensive deterioration of slabs is encountered. Color coding can be helpful when separate sheets are not used.

10.4.6 Photographic Recording

It is extremely helpful to record deterioration forms by means of pho-tographs or video cameras. First, the photos and video tapes serve to further qualify distress found. Second, forms of distress which cannot be explained immediately in the field can be presented to other engi-neers to obtain their opinions. Remember, film is relatively cheap, and a photo may prevent mistakes or an unnecessary return to the job site.

ASA 200V or 400 are good all-around film speeds when variable lighting conditions are encountered. Color film should always be used. Location of photographs should be recorded on field-survey sheets or on separate log sheets with an explanation of the subject being pho-tographed.

10.4.7 Delamination Survey

When the steel begins to corrode and before spalls are visible on the deck surface, horizontal cracks, or delaminations, occur at or above the level of the top reinforcing steel. Delaminations need to be detected

Figure 10-29. Chain drag delamination survey.

because they indicate a high level of corrosion activity and represent areas of unsound concrete that must be removed and repaired. It is not uncommon for more than one delamination to occur on different horizontal planes above the reinforcing steel.

Floor-slab delaminations are located by sounding the surface with a hammer, iron rod, or chain. When the delaminated area is struck, a distinct hollow sound is heard. Further sounding defines the limits of an area, and the boundaries are then marked with a spray paint or chalk (Figures 10-29 and 10-30).

Research is in progress to detect delaminated areas by remote sensing through the application of thermograph and radar systems. These methods are not currently in routine use for delamination survey of parking structures.

10.4.8 Pachometer Survey

The pachometer magnetically locates the embedded steel and measures the intensity of the magnetic field produced by the embedded steel, which can be correlated to a specific depth from the concrete surface, provided the size of the embedded steel is known (Figure 10-31). It cannot detect the presence of aluminum or PVC conduit. However, it will respond to electric current in its vicinity. One should stay clear of electrical equipment to obtain accurate readings.

Always review available structural drawings prior to performing the survey. The pachometer will give accurate results if the structural

Figure 10-30. Boundaries of delaminated areas marked by spray chalk.

Figure 10-31. A pachometer is used to locate and measure concrete cover over embedded reinforcement.

member is lightly reinforced. In heavily reinforced members, the effect of secondary reinforcement does not allow accurate measurement of cover. Also, reinforcing bars that run parallel to the bar to be measured will influence the reading if the distance between the bars is less than two or three times the cover distance. The pachometer cannot be accurately used at temperatures below 40°F due to the batteries used in these devices.

When the diameter of the embedded steel is known, the depth of concrete cover can be read from the needle position on the dial of the pachometer. As nearby steel can have an additive influence on needle deflection, the cover reading should be taken at a few locations along a bar. This will help to determine the influence of other elements. Large needle deflection of the meter indicates shallow reinforcement.

Rebar cover measurements should be taken at ten to fifteen random points on each supported tier and recorded on field-survey sheets. Extra readings may be required over beams on prestressed decks to locate tendons and measure cover. Pachometer readings should also be taken in the area of coring and in the vicinity of test locations.

It is necessary to locate and determine the depth of concrete cover over the reinforcement for the following four reasons:

- Establish the chloride ion content of the concrete at the level of the embedded reinforcement.
- Correlate extent of observed deterioration.
- Locate areas with shallow cover that would restrict concrete removal by rotary or scarifying equipment.
- Relate position of reinforcement to concrete removal limits in repair details.

10.4.9 Materials Testing

In conjunction with the field survey, it is necessary to perform materials and nondestructive testing to supplement the results of visual observations. Testing should be performed by experienced personnel and, where necessary, the results evaluated by a materials consultant. The testing assists in verifying the extent of deterioration and in evaluating the condition of the existing concrete. This information is useful in selecting appropriate repair methods and materials. In some instances, test wells are opened to visually verify the extent of the problem. Some of the commonly-used laboratory and nondestructive tests for a condition survey of a concrete parking structure along with their applications are listed in Table 10-2. Where applicable, standard test

TABLE 10-2   Commonly Used Laboratory and Nondestructive Tests

Tests	Standard Designation	Application
Materials Testing		
1. Chloride-ion content	FHWA-RD77	Determining the chloride content of concrete to establish potential for corrosion of reinforcement and extent of chloride ion penetration.
2. Compressive strength	ASTM C42	Obtaining and testing drilled core samples to establish quality of concrete.
3. Petrographic examination	ASTM C856	Microscopic examination of concrete core samples to evaluate quality and durability of concrete.
4. Shear bond strength or pull-out test by applying direct tension		Determining bond strength of core sample to evaluate integrity of concrete topping, overlays, and patches.
5. Air-void system of the concrete	ASTM C457	Evaluate surface scaling or freeze-thaw resistance of concrete.
6. Metallurgical examination	Scanning microscopy/X-ray analysis/hardness.	Examination of prestressing tendons to detect corrosion or embrittlement.
Nondestructive Testing		
7. Delamination survey	Chain drag	Determining extent of concrete deterioration in structural members including floor slabs (chain drags often used).
8. Pachometer survey	—	Measuring concrete cover over reinforcement and size.
9. Half-cell testing	ASTM C876	Detection of corrosion activity.
10. Pulse velocity	ASTM C597	Locating internal discontinuities, such as voids, cracking, etc.
11. Radar	—	Locating internal discontinuities and measure size and location of reinforcement.

methods are referenced. The testing program for a structure is generally established by considering the following items:

1. The type of structural system involved
2. The types and forms of deterioration observed, such as corrosion-induced spalling and scaling cracking. The testing should be representative of the type and extent of deterioration noted.

3. The need to qualify previous repairs
4. Probable repair solutions and repair costs

10.4.9.1 Chloride Ion Content of Concrete. Concrete powder samples are taken at various locations throughout a facility by means of a rotary hammer, and the pulverized samples are taken in equal increments of depth at each location to establish chloride content as a function of depth. This method aids in determining the corrosion potential of the steel at a certain depth within the concrete. The use of the rotary hammer to obtain concrete powder samples has the advantage of portability, light weight, speed, and economy (Figure 10-32). An alternate method to the use of a rotary hammer is to obtain pulverized concrete samples in a laboratory from individual core samples. The use of core samples permits the preparation of test samples under controlled conditions and provides better accuracy.

It is generally desirable to sample for chloride content at two to three locations from each level of the structure. Sites should include drive lanes, parking stalls, turns, and speed ramps on all supported levels. In addition, a "baseline" sample from an uncontaminated area should be obtained. Also, determine the minimum depth of the reinforcement in the vicinity of the sample location.

The significance of chloride ion testing is to determine the chloride ion concentration at the level of the reinforcing steel. High chloride ion concentration at the level of reinforcing steel correlates well

Figure 10-32. Obtaining concrete powder sample in field.

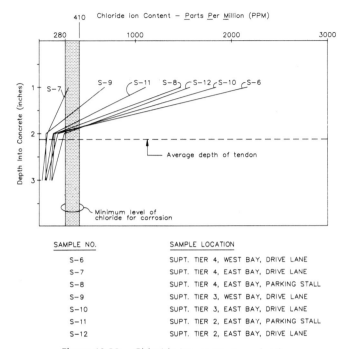

Figure 10-33.   Chloride ion content vs. depth.

SAMPLE NO.	SAMPLE LOCATION
S-6	SUPT. TIER 4, WEST BAY, DRIVE LANE
S-7	SUPT. TIER 4, EAST BAY, DRIVE LANE
S-8	SUPT. TIER 4, EAST BAY, PARKING STALL
S-9	SUPT. TIER 3, WEST BAY, DRIVE LANE
S-10	SUPT. TIER 3, EAST BAY, DRIVE LANE
S-11	SUPT. TIER 2, EAST BAY, PARKING STALL
S-12	SUPT. TIER 2, EAST BAY, DRIVE LANE

with the presence of active corrosion. Research by the Federal Highway Administration (FHWA) has established that a acid soluble chloride ion content of 280 to 410 ppm at the reinforcing bars results in accelerated corrosion. This information is valuable for investigation of rehabilitation alternatives. A graph is used to interpret chloride ion test results. Parts per million (ppm) are plotted on the horizontal axis, and depth into the concrete is plotted on the vertical axis. A typical plot is shown in Figure 10-33.

Performance of traffic-bearing membranes or coatings can also be established by chloride ion penetration testing. Samples taken from concrete before membrane application and after several years of service provide insight into system effectiveness. Surface sealers can be evaluated for effectiveness in a like manner. Performance of a floor-slab protection system can be monitored simply by repeating the sampling, analysis, and evaluation of results on an annual basis.

10.4.9.2   Coring and Testing. Drilled concrete cores provide valuable information into the types of deterioration encountered within the structure and to the types of repairs required to return the structure to a serviceable condition. Note the location of all cores removed from the structure on the field-survey sheets for future reference. Do not core prestressed structures prior to confirming location of tendons by X-ray examination.

Remove core samples from selected locations after performing the visual observations, delaminations survey, and other nondestructive testing. This assists in correlating, confirming, or resolving conflicting data obtained during the field survey. The majority of the cores should be taken in areas of deterioration such as, delaminations, scaling, and cracking, (Figures 10-34 to 10-36). For instance, core through cracks aligned directly over embedded mild steel reinforcement to obtain the depth of concrete cover and the extent of the corrosion of the steel. However, do not attempt to core through prestressing tendons. Always obtain a few core samples from apparently good areas of the structure to establish a "reference" for comparison of the extent of deterioration in other samples.

Figure 10-34. Coring for concrete samples.

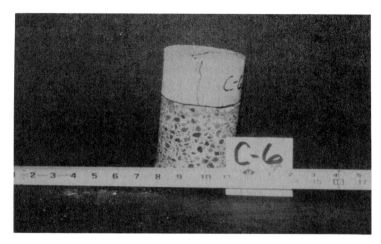

Figure 10-35. Core sample over crack in overlay to observe depth of cracking.

Figure 10-36. Core sample over previously patched areas with a waterproofing membrane and asphalt wearing surface.

Locations for coring must be marked by the survey team. Use the pachometer to locate reinforcement in the area. Mark coring location so that the chance of reinforcement damage is minimized, especially when coring near post-tensioning tendons. Cores for testing concrete compressive strength should be specifically sited to avoid reinforcing steel. However, it is desirable sometimes to obtain samples of reinforcing steel in cores selected for petrographic examination. Try to obtain core samples by cutting steel near the ends of the rebar (to compromise only the development of the rebar) or in structurally redundant locations of the facility.

The minimum number of samples obtained depends on the type and size of the facility being examined. Visual examination of cores considered useful can generally be gained from three to four samples for a one-level facility, four to five for two levels, five to six for three levels, etc.

Upon return to the office, photograph all the cores taken with a scale reference object. At this time select cores for compressive or petrographic examination, and if necessary, bond and/or air content testing depending upon the type of the deterioration observed (Figure 10-37).

10.4.9.2.1 Concrete Compressive Strength. Concrete core samples are usually obtained from selected areas of the structure to determine com-

Figure 10-37. Several concrete core samples are obtained from selected locations for laboratory examination and testing.

pressive strength. Core samples are obtained and tested in accordance with the Standard Test Method ASTM C42. A minimum of three core samples should initially be tested from a facility to obtain an average concrete compressive strength. Additional cores should be taken when noticeable variations in extent of concrete deterioration is related to different concrete placements or different levels of the structure.

Compressive strengths quantify the relative quality of concrete in the floor slab. They provide confirmation of results obtained by visual observation, materials testing, and nondestructive examination. For instance, extensive floor-slab spalling due to corrosion of reinforcement in a structure with adequate concrete cover may be attributed to poor quality concrete. This can be confirmed qualitatively by petrographic examination and measured by core compressive strength results. Concrete compressive strength results assist in estimating concrete removal methods and costs. Unit price for concrete removal and preparation is higher for concrete with higher compressive strengths. Also, the selection of concrete removal methods is affected by the concrete compressive strength.

10.4.9.2.2 Petrographic Examination. Microscopic examination, sometimes in combination with other techniques, is used to examine samples of concrete. Features that can be evaluated include denseness of cement paste, depth of carbonation, occurrence of bleeding, presence of contaminating substances, air content, and other properties. Standard recommended practice is given in ASTM C 856.

Perform an examination of core samples both from deteriorated areas as well as potentially sound areas. Microscopic examination can provide valuable information on the extent of concrete removal during repairs. For example, the examination can determine the extent of freeze-thaw damaged concrete removal that will be required to provide a sound substrate for subsequent placement of an air-entrained overlay to protect the surface. When indicated by petrographic examination and visual observations, determine the characteristics of the entrained air-void system to evaluate the durability of the concrete. The air-void characteristics are determined in accordance with Standard Test Method, ASTM C 457. In accordance with ACI 345, air-void characteristics of an adequate system are: 1) calculated spacing factor less than 0.008 in.; 2) a surface area of the air voids greater than about 600 sq in. per cu in. of air void volume; and, 3) a number of air voids per linear inch of traverse significantly greater than the numerical value of the percentage of air in the concrete.

10.4.9.2.3 Shear Bond Test. Shear bond testing is performed to evaluate the bond strength of concrete overlays, precast topping, and patches to the substrate. The National Cooperative Highway Research Program Report #99 indicates that a value of 200 psi is a desirable bond strength, which has generally been accepted and used as a guide in designing bonding mediums. In areas of questionable bond integrity, microscopic examination of the bond interface should be performed to determine the potential cause(s) of debonding. Debonding of topping in some instances can contribute to a reduction in the load-carrying capacity of structural precast double-tee floor systems.

10.4.10 Nondestructive Testing

The delamination and pachometer surveys described earlier in this section are also a form of nondestructive examination. There are several other nondestructive test methods which can be utilized to evaluate the condition of the structure.

10.4.10.1 Half-Cell Corrosion-Potential Testing. The half-cell potential method consists of estimating the electrical half-cell potential of reinforcing steel in concrete for the purpose of determining the potential of corrosion activity of the reinforcing steel. The method is limited by the presence of electrical continuity. A concrete surface that has dried to the extent that it is a dielectric, and surfaces that are coated with a dielectric material (epoxy, for example) will not provide an acceptable electrical circuit. The testing apparatus consists of a copper-copper sulfate half cell, which is effectively a rigid tube or container composed of a dielectric material that is nonreactive with copper or copper sulfate, a porous wooden or plastic plug that remains wet by capillary action, and a copper rod that is immersed within the tube in a saturated solution of copper sulfate. A detailed description of the apparatus and parameters for use of the apparatus are included in ASTM C876. Means and methods for electrical connections to embedded steel, the half cell and prewetting of the concrete surface are also described in the test method (Figure 10-38).

Except in very specialized cases, half-cell potential testing should be used to evaluate corrosion potential of floor-slab surfaces only. A spacing of 4 ft has been found satisfactory for evaluation of bridge decks, and is appropriate in most instances in conventionally reinforced-parking structures. A wider spacing is not recommended.

Record the electrical half-cell potentials to the nearest 0.01V. By convention, a negative (−) sign is used for all readings. Report all half-

Figure 10-38.   Half-cell potential measurement in field.

cell potentials in volts and correct the temperature if half-cell temperature is outside the range of $72 \pm 10°F$. The temperature coefficient
for correction is given in ASTM C876.

According to ASTM C876 the significance of the numerical value
of the potentials measured is as shown below. Voltages listed are referenced to the copper-copper sulfate (CSE) half cell.

a. If potentials over an area are less negative than $-0.20V$ CSE,
there is a greater than 90% probability that no reinforcing corrosion is occurring at that area at the time of measurement.

b. If potentials over an area are in the range of $-0.20--0.35V$
CSE, corrosion activity of the reinforcing steel in that area is
uncertain.

c. If potentials over an area are more negative than $-0.35V$ CSE,
there is a greater than 90% probability that reinforcing-steel
corrosion is occurring in that area at the time of measurement.

Results of half-cell potential testing only indicate probability of
corrosion occurring in areas with a potential reading greater than
0.20V. The numerical value of the results is not intended to indicate
the relative rate of corrosion. Measurement of corrosion rates requires
measuring the corrosion current and the resistance to flow of current.

10.4.10.2   Pulse-Velocity Test.   The ultrasonic-pulse-velocity method
consists of measuring the time of travel of an ultrasonic pulse passing

through the concrete to be tested. The pulse generator circuit consists of electronic circuitry for producing pulses of voltage, and a transducer for converting these pulses into a form of mechanical energy having vibration frequencies of 15–50 kHz. Contact with the concrete is made through a suitable connection; another transducer is connected to the concrete at a measured distance from the first. The time of travel of the pulses between the two transducers is measured electronically. The apparatus requirements are defined in ASTM C597.

There are three ways of measuring pulse velocity through concrete. The best and most accurate method is by direct transmission through concrete where the transducers are held on opposite faces of the concrete specimen. Only personnel experienced in the use of this equipment should be utilized to operate and interpret results of testing.

Some applications of pulse-velocity testing for use in the evaluation of concrete structures are listed below.

- Establishing the uniformity of concrete
- Establishing acceptance criteria for concrete; generally, high pulse velocity readings indicate good-quality concrete.
- Estimating the strength of concrete through correlation of known strengths of core samples
- Measuring and detecting cracks
- Inspecting reinforced concrete members

10.4.10.3 X-Ray Examination. X-ray examination can be helpful to identify the location of embedded reinforcement in very thick structural members where pachometer survey is likely to be ineffective. However, x-ray examination is costly, and should only be used if effective data are anticipated. Some of the concerns regarding x-ray examination are as follows:

1. Access to both surfaces of the member is required—one surface for the x-ray source transducer, the opposite surface for the recording film.
2. The x-ray does not distinguish reinforcement placement relative to depth. Images from rebar on the near face of a member may "shadow" an image from rebar deeper in the member.
3. The x-ray signal as it moves through a member is conical; i.e., the range of area the x-ray examines increases with depth of penetration. Careful interpretation of results is required.

X-ray examinations are recommended prior to obtaining core samples from post-tensioned decks by locating post-tensioning tendons. Other potential uses of x-ray examination are to locate end zone reinforcement in a post-tensioned beam or beam-column connections, or to locate end zone stem reinforcement in a precast tee beam.

In all instances, the best method to improve the quality of x-ray examination is to locate the film which records the examination on the side of the member nearest the major reinforcement which is desired to be located. By positioning the film accordingly, the location of the reinforcement in the member is nearer the image on the film, and the interpretation of x-ray examination results should be much easier and more accurate.

10.4.10.4 Radar Examination. Radar examination is especially useful in locating large areas of embedded reinforcement within a floor slab or structural member. A graphic representation of the reinforcement location is continuously developed during the examination, which can be easily calibrated by coring at predetermined locations to visually record the depth and size of reinforcement.

Welded wire fabric reinforcement can be easily located by radar examination where pachometer examination is generally ineffective. Agencies utilizing radar examination present the opinions that the equipment is capable of detecting voids beneath concrete pavements on grades, determining concrete thickness when only one side is accessible, determining the asphalt overlay thickness, and surveying slab delamination.

In radar profiling, echoes from a pulsed electromagnetic wave are received by an antenna. The penetration and resolution of the signal are a function of the frequency of the electromagnetic pulse. A high-frequency signal provides high resolution but has shallow penetration. Lower-frequency signals have greater penetration but poorer resolution. Depending on the thicknesses of individual layers and the desired penetration, two or possibly three different radar antennas ranging from a few hundred to a thousand or more megahertz may be utilized. The data-recording rate is chosen to yield high-resolution continuous profiles over multiple traverses. Field evaluation and preliminary interpretation of data are generally made from a group of parallel profiles.

Radar profiling detects the presence of delaminated conditions and may indicate deterioration caused by cracking. Chemical deterioration of concrete and consequent changes in the dielectric constant may be detected if radar is used in conjunction with other methods, particularly electrical resistivity. Although radar examination is expensive, it proves to be cost effective when large areas of floor slabs are required to be surveyed for reinforcement placement, particularly when graphic representation of the survey information is desired.

10.4.11   Exploratory Excavation

Exploratory excavations are very useful to verify conditions of the underlying concrete or embedded reinforcements. In some instances, exploratory excavations may simply consist of coring through the slab surface. In other instances, one may need to perform partial or full-depth excavation of the floor slab to verify conditions. Examples of test wells are shown in Figures 10-39 and 10-40.

10.4.11.1   Waterproofing Membranes. Quite often waterproofing membrane systems have to be removed to physically observe the condition of the underlying slab and to obtain core samples in order to verify the effectiveness of the system. Select test locations in areas with poor drainage conditions and over areas with noticeable water leakage, wet spots, cracks, or efflorescence on the underside of the slab.

Whenever an existing waterproofing membrane is encountered, it is advisable to bring back samples to the office for analysis. The membrane can be observed for weathering, embrittlement, tensile

Figure 10-39.   Test well to observe condition of waterproofing system and underlying concrete.

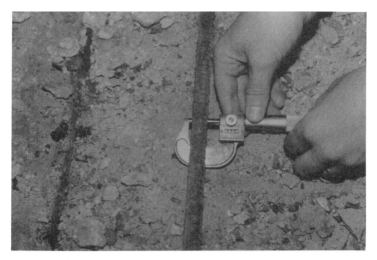

Figure 10-40.   Measuring extent of rebar corrosion.

properties, and thickness. In the field the membrane should be observed for cracking, adhesion, and excessive wear. Also, obtain information as to the membrane type and the date of its installation.

10.4.11.2   Post-Tensioning Tendons. Post-tensioning tendon conditions, especially in a multi-wire buttonhead system, are visually examined for distress by opening test wells. One has to be extremely careful when attempting to open test wells around post-tensioning systems. Also, it is extremely unsafe to stand directly over tendons that are being excavated. These excavations should be directed by an engineer familiar with the design of the structural system and the restoration of prestressed structures.

Locations typically selected for excavations are tendon high points over beams, intermediate anchorages at construction joints, and at end anchorages (Figures 10-41, 10-42, and 10-43). For end anchorages do not remove the concrete behind the bearing plate. Since the concrete behind the embedded-plate assembly is precompressed, concrete removals may result in anchorage failure due to movement or rotation of the anchorage assembly. At each excavated location, note concrete cover of tendons, and condition of tendons. If corrosion damage is not apparent and a tendon is broken, a sample of the tendon should be obtained at the fracture location for further metallurgical examination to determine the cause of the failure.

(a)

(b)

Figure 10-41. (a) Excavation over intermediate anchorage of a button-head post-tensioning system to determine condition of tendon and anchors. (b) Excavated intermediate anchorage of tendon being prepared for patching.

10.4.12 Data Evaluation

The data evaluation consists of: 1) Analyzing and correlating the results of the field observations, measurements, and testing, and 2) re-

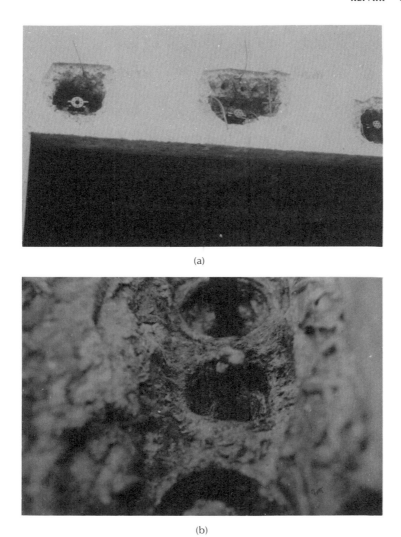

(a)

(b)

Figure 10-42. (a) Excavation of tendon end anchorages. (b) Excavated end anchorages of beam tendons.

viewing laboratory test results. The objective is to determine the primary cause(s) of observed deterioration and its impact on structural members. Analysis of the field, laboratory, and testing data require the application of knowledge and judgment by a restoration specialist in determining the cause(s) of distress.

Figure 10-43.   Excavation over through-slab crack to observe condition of tendon.

10.4.12.1   Cause(s) of Deterioration and Contributing Factors. The primary cause of deterioration must be supported by field observations, measurements, and laboratory testing. For example, corrosion-induced deterioration must be verified by results obtained from delamination, pachometer, and half-cell testing survey and the chloride ion content of the concrete at the level of the reinforcement. Coring and test well can further provide information regarding the extent of reinforcement corrosion. Another observation may be floor-slab cracking. Cracking is a symptom of distress which may have a variety of causes and contributing factors. (Refer to Section 10.3.3.) The selection of the correct repair method will depend on determining the primary cause of cracking and the contributing factor(s), if any. The crack pattern, location, and size observed in the structure must be supported by analytical and/or laboratory test results.

Very often there are other factor(s) that also contribute to the observed deterioration; try to determine them. For instance, a primary cause of deterioration is reinforcement corrosion, but contributing factors may also include lack of air entrainment and freeze-thaw deterioration, shallow concrete cover, poor-quality concrete, and inadequate drainage. Understanding the contributing factor will help to recommend better repairs by specifying appropriate materials and methods.

Review of original drawings, specifications, inspection reports, and shop drawings can also be useful in determining the cause(s) of a problem. For instance, it is not unusual to find a variation in as-built condition from that specified in structural drawings, which can contribute to distress, such as concrete cracking and spalling in structural elements. For such deficiencies, supplemental reinforcement or strengthening may be required to restore the load-carrying capacity or integrity of the member.

10.4.12.2 Impact of Deterioration. The next step is to determine the impact of the observed deterioration. This is done by tallying quantities of the various forms of deterioration, such as spalling, delaminations, cracking, and scaling from the field-survey sheet on a tier-by-tier basis. For example, determine the extent of floor slab spalling and delaminations. Consider the impact from the standpoint of the extent of present deterioration, as well as the potential for future deterioration. Present floor slab spalling and delaminations may only be 5% of the floor area; but, half-cell potential survey may indicate a strong probability for continuing corrosion over most of the remaining floor area. The repairs recommended must correct the present deterioration and address the potential for continuing deterioration of the floor slab due to corrosion of the embedded reinforcement.

The impact of the deterioration observed on the individual elements of the structure can be categorized based on the priority of repairs as follows: 1) Structural/safety related, 2) serviceability/durability related, or 3) aesthetic/cosmetic in nature.

Deterioration can adversely affect structural integrity, safety, and serviceability of members. As stated in the ACI Committee 362 report, estimating the loss of structural integrity due to deterioration is one of the most difficult aspects of an investigation. Structural engineers should consider the effect of cracking, partial loss of the reinforcement bond due to delaminations, loss of the cross-sectional area of primary reinforcement, and loss of strength and toughness. In some cases load testing is the only alternative to determine structural capacity of a deteriorated structure. This will determine the load-carrying capacity of the member at the time of the test. The influence of other large forces, such as those due to restrained volume change or lateral load forces, can be evaluated analytically, according to Chapter 20 of ACI 318, or by modifying the load test to account for such factors (Figures 10-44 and 10-45.)

Repairs must be performed when the load-carrying capacity of members is reduced or serviceability of the facility is affected by deterioration. Repairs correct the effects of deterioration and attempt to return structural elements close to their original condition and ser-

Figure 10-44. Setup to monitor displacements of precast members during a load test.

viceability. Measures taken primarily to minimize deterioration and prolong the life of the structure are considered to be maintenance *actions* as opposed to repairs. For example, application of a membrane traffic topping or surface sealer is considered to be maintenance action, since this is a preventive action to protect the structure from chloride ion penetration.

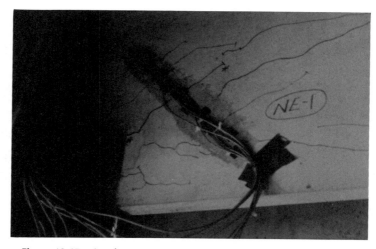

Figure 10-45. Load test instrumentation to measure concrete strains.

10.5   REPAIR METHODS AND MATERIALS

10.5.1   General

There is a significant amount of literature on concrete repair methods and materials that has been published during the past several years. Many committees of the ACI have now published reports that directly or indirectly relate to the repair and restoration of concrete structures. The results of research and the application of repair methods and materials have been reported in publications presented by the Federal Highway Administration research programs, the Portland Cement Association, the Corps of Engineers, and in articles presented in various trade journals. Therefore, this section of the chapter will only attempt to summarize the basic requirements for durable repair techniques that are more commonly used to restore a parking structure. Other less frequently used techniques have not been included. However, do not be limited by the basic material presented here, and where necessary, attempt to incorporate special repair techniques.

10.5.2   Repair Methods

Most of the commonly used repair methods can be categorized into the following five repair approaches:

- Patching
- Protective coating
- Replacement
- Cathodic protection
- Sealing

10.5.2.1   Patching. Patching replaces deteriorated concrete on the surface of horizontal and vertical members. When properly implemented, patching will restore structural integrity as well as improve serviceability or correct cosmetic damage. However, in some instances, this may not be the correct repair approach. Evaluation of appropriate repair approach is discussed in Section 10.6.

Patching can be referred to as "partial-depth" or "full-depth" based on the extent of concrete removed. Quite often, for thin slab sections (less than 5 in. thick) it is difficult to perform shallow concrete removals and usually results in full-depth concrete removal. As a general rule of thumb, a full-depth patch is specified when concrete removal equals or exceeds half the slab's thickness.

As shown in Figures 10-46, 10-47, 10-48, and 10-49, patching

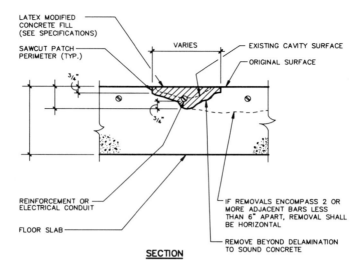

LATEX MODIFIED
CONCRETE FILL
(SEE SPECIFICATIONS)

VARIES

EXISTING CAVITY SURFACE

SAWCUT PATCH
PERIMETER (TYP.)

ORIGINAL SURFACE

3/4"

3/4"

REINFORCEMENT OR
ELECTRICAL CONDUIT

FLOOR SLAB

IF REMOVALS ENCOMPASS 2 OR
MORE ADJACENT BARS LESS
THAN 6" APART, REMOVAL SHALL
BE HORIZONTAL

REMOVE BEYOND DELAMINATION
TO SOUND CONCRETE

SECTION

NOTES:

1. REMOVE AND REPLACE ALL SOUND AND UNSOUND CONCRETE WITHIN
    SECTION SHOWN CROSS—HATCHED.

2. SHALLOW REMOVAL SHALL CONSIST OF CONCRETE REMOVALS THAT EXTEND
    BELOW ONE LAYER OF REINFORCEMENT.

3. PAY UNIT = S.F.

4. DETAIL NOT TO SCALE.

Figure 10-46.   Patch detail with shallow concrete removals.

consists of removing the unsound concrete, cleaning reinforcing steel
exposed by removals, preparing the exposed surface, and installing a
specialty concrete patching material. Patch edges for partial depth removals are often chipped or saw-cut to near vertical to a depth of at
least $\frac{3}{4}$ in., as opposed to leaving a "feather-edge." Refer to the sample
specification included in Appendix A for surface preparation and other
requirements.

Although patches of high-quality (low permeability) material are
installed, the adjacent surface tends to have lower durability. In chloride-contaminated slabs, the durability of this repair system is adversely affected by delamination and spalling of floor-slab areas due
to continuing corrosion of reinforcement beyond patch limits. Patching
can, however, rapidly restore the structural integrity of the member
and limit further damage to embedded reinforcement. The emphasis
is on repairs that address only existing damage.

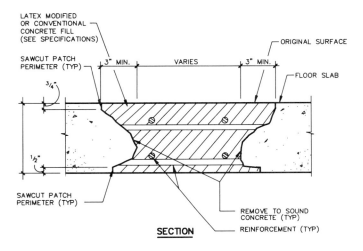

LATEX MODIFIED
OR CONVENTIONAL
CONCRETE FILL
(SEE SPECIFICATIONS)

ORIGINAL SURFACE

SAWCUT PATCH
PERIMETER (TYP)

3" MIN.   VARIES   3" MIN.

FLOOR SLAB

3/4"

1/2"

SAWCUT PATCH
PERIMETER (TYP)

REMOVE TO SOUND
CONCRETE (TYP)

SECTION

REINFORCEMENT (TYP)

NOTES:

1. REMOVE AND REPLACE ALL SOUND AND UNSOUND CONCRETE WITHIN SECTION
   SHOWN CROSS-HATCHED.

2. FORMS SHALL REMAIN IN PLACE UNTIL CONCRETE HAS ATTAINED 3000 PSI
   MINIMUM COMPRESSIVE STRENGTH.

3. PAY UNIT = S.F.

4. DETAIL NOT TO SCALE.

Figure 10-47. Full-depth patch detail.

10.5.2.2 Coating. Coating consists of applying surface sealers, elastomeric traffic-bearing membrane systems, rigid concrete, or specialty concrete overlays. Refer to Chapter 6 for a description of these repair materials. Although a sealer can be applied to vertical and overhead surfaces, membranes and overlays are usually applied only on floor slabs. Some of the characteristics of protective coatings are discussed here, relative to their use as a repair material.

10.5.2.2.1 Sealer. A repaired slab surface is usually protected by sealers to minimize moisture and chloride ion penetration by application of a concrete sealer. The original concrete surface requires cleaning by special means, such as high-pressure water, sand or shotblasting to remove all old coatings or sealer, and/or surface laitance. Regular reapplication of the sealer every 3–5 yr is to be anticipated depending on the severity of exposure conditions and sealer selection. A sealer is ineffective when applied to a cracked concrete surface, primarily due

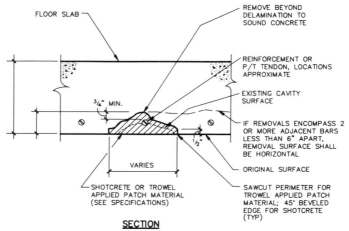

FLOOR SLAB

REMOVE BEYOND
DELAMINATION TO
SOUND CONCRETE

REINFORCEMENT OR
P/T TENDON, LOCATIONS
APPROXIMATE

EXISTING CAVITY
SURFACE

3/4" MIN.

IF REMOVALS ENCOMPASS 2
OR MORE ADJACENT BARS
LESS THAN 6" APART,
REMOVAL SURFACE SHALL
BE HORIZONTAL

1/2"

VARIES

ORIGINAL SURFACE

SHOTCRETE OR TROWEL
APPLIED PATCH MATERIAL
(SEE SPECIFICATIONS)

SAWCUT PERIMETER FOR
TROWEL APPLIED PATCH
MATERIAL; 45° BEVELED
EDGE FOR SHOTCRETE
(TYP)

SECTION

NOTES:

1. REMOVE AND REPLACE ALL SOUND AND UNSOUND CONCRETE WITHIN
   SECTION SHOWN CROSS-HATCHED.

2. SHALLOW REMOVAL SHALL CONSIST OF CONCRETE REMOVALS THAT
   EXTEND BEYOND ONE LAYER OF REINFORCEMENT.

3. PAY UNIT = S.F.

4. DETAIL NOT TO SCALE.

Figure 10-48. Ceiling repair detail.

to its inability to bridge cracks. Sealers are also relatively ineffective when applied to surfaces of poorly air-entrained concrete or concrete that is chloride contaminated. Sealers are usually most effective when applied to a relatively new structure, primarily as a preventive measure for corrosion-induced deterioration. However, a sealer will eventually permit chloride ions to penetrate to the level of the embedded reinforcement. Regular sealer application only helps to extend the "time-to-corrosion," but is not capable by itself of preventing corrosion-induced deterioration and ensuring long-term durability.

10.5.2.2.2 Waterproofing Membranes (Traffic Topping). A traffic-bearing membrane waterproofs the slab and minimizes chloride ion penetration more effectively than a sealer. It also reduces the oxygen supply and moisture that supports corrosion and assists in extending the service life of slabs. A membrane can effectively bridge active cracks and is suitable for floor systems with extensive through-slab cracking.

Where membranes have been applied to existing decks, surveys indicate the continuation of corrosion activity if all chloride-contaminated concrete is not removed. The long-term effect of membranes on

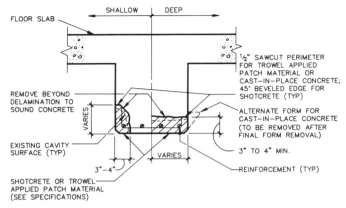

Figure 10-49. Shallow and deep beam repair detail.

continuing corrosion of embedded reinforcement is not well established. The performance of these systems on chloride-contaminated decks is highly variable due to membrane effectiveness, existing deck conditions, and the extent of removal of contaminated concrete. Some future concrete spalling beneath the membrane is to be expected due to corrosion-induced concrete spalling.

In heavy-traffic areas, such as drive aisles, frequent maintenance will be required to extend the life of the membrane system. Use of a "wear-balanced" membrane system consisting of different grades of the membranes will reduce maintenance frequency. Membranes are also susceptible to snow-plow damage, but plows can be raised or tipped with rubber guards and follow guidelines provided by the manufacturer for snow removal to minimize damage.

10.5.2.2.3 Bonded-Concrete Overlay. A bonded-concrete overlay provides a more durable repair of deteriorated concrete floors. A bonded-concrete overlay requires the removal and preparation of all deteriorated exposed surfaces prior to system installation. The surface is scarified to remove all contaminants, such as oil spots, from the surface. Exposed reinforcing is either cleaned or removed and replaced with one that is epoxy coated. After surface preparations are complete, a bonded overlay is installed to restore slab integrity while providing a new wearing surface (Figures 10-50 and 10-51, and refer to sample specification for overlay in the Appendix).

With overlays, the new surface profile can be adjusted for improved drainage and reduced ponding. Concrete overlays are not appropriate for floor slabs with active cracking. Reflective cracking in overlay is likely to reduce the service life of restored slabs. Where applicable, the additional superimposed load due to the overlay must be evaluated. The reader should refer to the National Cooperative Highway Research Program's Report #57 for further details regarding surface preparation, placement, and curing of concrete overlays. A more-detailed discussion on advantages and disadvantages of overlays is also provided in the report.

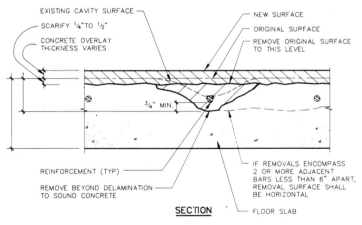

NOTES:

1. REMOVE AND REPLACE ALL SOUND AND UNSOUND CONCRETE WITHIN SECTION SHOWN CROSS-HATCHED.

2. SHALLOW FLOOR SURFACE REMOVAL SHALL CONSIST OF CONCRETE REMOVALS THAT EXTEND BELOW ONE LAYER OF REINFORCEMENT.

3. FILL IS NOT REQUIRED PRIOR TO OVERLAY PLACEMENT.

4. PAY UNIT = S.F.

5. DETAIL NOT TO SCALE.

Figure 10-50. Concrete overlay detail.

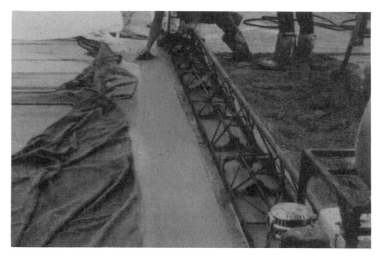

Figure 10-51. Concrete overlay placement with vibratory screed.

10.5.2.3 Replacement. When a floor slab is extensively deteriorated, removal and replacement of the slab may be a viable repair alternative, provided the underlying members are in relatively good condition. This is referred to as "partial-depth" replacement. Floor slabs which are less than 5 in. thick are difficult to repair. Concrete removals on pan-joist, waffle-slab, and one-way slab systems usually result in complete removal of the slab (Figures 10-52 and 10-53). The existing underlying beams and waffle or pan-joist ribs are used to support the new slab, provided adequate measures are taken to ensure composite behavior of the rebuilt floor system. The new slab can be reconstructed with durable concrete and epoxy-coated reinforcement to extend the service life of the facility. However, the new slab is susceptible to cracking due to volume-change restraint offered by the existing underlying members. In extreme cases, "full-depth" replacement of the floor system (slab and underlying elements) or demolition of the entire structure may be necessary.

10.5.2.4 Cathodic Protection. In concept, the only method which will effectively stop the corrosion of embedded reinforcement in chloride-contaminated slabs is cathodic protection. For a description of the system refer to Chapter 6. Industry experience in the application of cathodic protection to parking structure floor systems is limited. However, there are several reported applications of cathodic protection to

Figure 10-52. Full-depth slab removal over beam.

bridge decks. The results to date, in general, have been encouraging. Field applications of cathodic-protection systems to bridge decks have been in operation since 1974.

Application of cathodic protection can only mitigate corrosion, and repairs to restore structural integrity and serviceability must still

Figure 10-53. Full-depth slab placement.

be performed. Therefore, the protection system is most cost-effective when it is applied to structures having floor slabs in the initial stages of deterioration. Also, a cathodic protection system is not economical when applied to structures with less than 10 yr of planned or anticipated life expectancy. Presently, only conventionally reinforced concrete structures have been cathodically protected. Cathodic protection of prestressing steel is still in an experimental stage.

At present, several types of cathodic-protection systems are being marketed. Some of these proprietary cathodic-protection systems have been developed within the last 2 to 3 years, and it is difficult to evaluate the life expectancy and effectiveness of the installed systems over an extended period. Presently cathodic-protection-system cost can range from $4 to $6 per square foot. This cost is in addition to that required for concrete repairs. Also, if the floor slab is extensively cracked, the structure will require protective measures to minimize leakage and continuing deterioration of underlying members; therefore, the initial cost to repair the structure can sometimes be very high.

10.5.2.5   Sealing. Sealing consists of performing repairs that will minimize water leakage through floor-slab cracks and joints. Since sealing by itself cannot be considered to be a repair approach, it must always be performed in conjunction with the other approach described earlier. Potential sources of water leakage are: 1) expansion joints, 2) construction joints, 3) control or isolation joints and, 4) construction- or service-related cracks.

Refer to Sections 10.3.3 and 10.3.4 for a discussion of the distress associated with cracking and failure of joints. Also, refer to Figures 10-54, 10-55, and 10-56 for typical details related to sealing cracks and joints. Design, detailing, material, and installation methods for joint seals are also discussed in detail in ACI 504R and ACI 224.1R.

Under certain circumstances, cracks can be repaired by epoxy injection (Figure 10-57). The material and its application are described in ACI 224.1R. The procedure for epoxy injection is discussed in ACI 503R and 503.1. Epoxy injection of service-related (active) cracks usually results in cracking adjacent to previously injected cracks. For these instances, cracks should be treated with a flexible-joint sealant material (Figure 10-58(a), and (b)).

10.5.3   Basic Requirements

A repair is generally successful if the repair material is compatible with the original substrate and has the required strength and durabil-

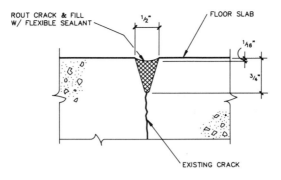

SECTION

NOTES:

1. CENTER ROUTED GROOVE ON CRACK.

2. PREPARE AND ALLOW FOR PRIMER TO CURE PROPERLY PRIOR TO INSTALLING SEALANT.

3. SEE SPECIFICATIONS FOR APPROVED MATERIALS.

4. INSTALL SEALANT EVENLY AND RECESS $1/16$" BELOW SURFACE. INSTALL SEALANT FLUSH UNDER TRAFFIC TOPPING. DO NOT OVERFILL JOINT.

5. PAY UNIT = S.F.

6. DETAIL NOT TO SCALE.

Figure 10-54. Detail for sealing random floor slab cracks.

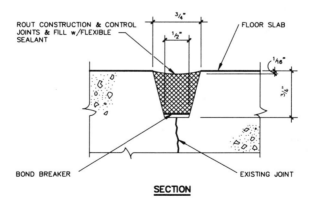

SECTION

NOTES:

1. CENTER ROUTED GROOVE ON CONSTRUCTION/CONTROL JOINTS.

2. PREPARE AND ALLOW FOR PRIMER TO CURE PROPERLY PRIOR TO INSTALLING SEALANT.

3. SEE SPECIFICATIONS FOR APPROVED MATERIALS.

4. INSTALL SEALANT EVENLY AND RECESS $1/16$" BELOW SURFACE. INSTALL SEALANT FLUSH UNDER TRAFFIC TOPPING. DO NOT OVERFILL JOINT.

5. PAY UNIT = L.F.

6. DETAIL NOT TO SCALE.

Figure 10-55. Detail for sealing construction and control joints.

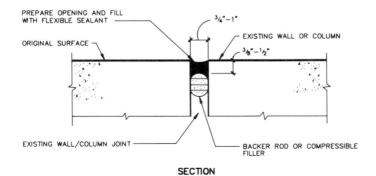

PREPARE OPENING AND FILL
WITH FLEXIBLE SEALANT

$\frac{3}{4}" - 1"$

ORIGINAL SURFACE

EXISTING WALL OR COLUMN

$\frac{3}{8}" - \frac{1}{2}"$

EXISTING WALL/COLUMN JOINT

BACKER ROD OR COMPRESSIBLE
FILLER

SECTION

NOTES

1. REMOVE EXISTING JOINT SEALANT MATERIAL.

2. GRIND JOINT EDGES.

3. INSTALL BACKER ROD.

4. PREPARE AND ALLOW FOR PRIMER TO CURE PROPERLY PRIOR TO
    INSTALLING SEALANT.

5. SEE SPECIFICATIONS FOR APPROVED MATERIALS.

6. INSTALL SEALANT EVENLY AND RECESS $\frac{1}{16}"$ BELOW SURFACE
    DO NOT OVERFILL JOINT.

7. PAY UNIT = L.F.

8. DETAIL NOT TO SCALE.

Figure 10-56. Detail for sealing existing control joints.

Figure 10-57. Injected slab crack.

(a)

(b)

Figure 10-58. (a) Crack routing. (b) Routed and sealed random crack.

ity. Other considerations are appearance and economy. The four basic requirements for a satisfactory repair are:

• Concrete removal and surface preparation
• Application of bonding medium
• Proper selection of repair material
• Proper material application

10.5.3.1 Concrete Removal. For all concrete repair situations, regardless of the type of structural member, a basic requirement is to remove all the deteriorated, delaminated, and unsound concrete prior to placing any new patch material. When complete removal of the deteriorated concrete is not accomplished, there is a good probability of patch failure at the bond interface between the existing concrete and the new patch material.

Concrete removal in parking structures is generally performed by light (15 lbs maximum) chipping hammers (Figure 10-59). These light chipping hammers are very convenient for concrete removal around and below the existing reinforcement. The size of the chipping hammer is limited to minimize damage to the surrounding area. Under certain circumstances, such as a relatively thick (8–12 in.) slab, the use of heavier hammers can be permitted by only preapproved operators. Other factors which can influence the selection of hammer size are overall thickness of the member, relative depth of removal to the

Figure 10-59. Concrete removal prior to patching floor spall.

overall member-thickness (partial or full depth) removals, inclination of members (vertical or horizontal), and location (vertical or confined area) of removal. Special care should be applied in removing unsound concrete from around reinforcing steel and embedded anchorages to prevent a loss of bond in the remaining sound concrete.

Large areas requiring the removal of relatively thin layers of concrete, such as concrete removal in preparation for placement of overlay, may be more effectively done with a scabbler, scarifier or planner, than with chipping hammers (Figures 10-60(a),(b)). These machines are particularly effective in cleaning the surface by removing the top surface contamination of traffic oils and greases. In addition, high-pressure sand and water blasters are capable of removing deteriorated concrete and many surface contaminants. Scarification of concrete surfaces using an abraded metal-shot-rebound method has also been used in the preparation of surfaces for installing a membrane or overlay.

The reader is referred to the Corps of Engineers' Manual, EM

(a)   (b)

Figure 10-60.   (a) Scabbler for concrete removal. (b) Scarifier for concrete removal.

1110-2-2002 and the National Cooperative Highway Research Program's Report #99 for a more detailed treatment of concrete removal techniques and advantages and disadvantages of the various removal methods.

10.5.3.2 Surface Preparation. Another important step in the repair of concrete structures is the preparation of the surface to be repaired. The repair is only as good as the surface preparation, regardless of the repair method or materials selected. For reinforced concrete structures, repairs must include proper preparation of the reinforcing steel in order to develop a bond with the replacement concrete.

Removal of concrete using mechanical cleaning devices, such as scabblers, grinders, and impact hammers may cause "bruising" of the concrete surface remaining in place. Be particularly careful of bruising of lightweight concrete surfaces due to mechanical scarification. The bruising can contribute to the debonding of the patch overlay material. Also, cleaning the surface by sand blasting or high-pressure water is required as a final step to remove any damaged surface material that can potentially contribute to the debonding of the repaired section (Figure 10-61).

A sample specification for surface preparation is included in the Appendix. This specification covers requirements for locating and marking work areas on a structure, inspection of the surface and reinforcement after concrete removal, replacement of deteriorated or

Figure 10-61. Sandblasting in preparation for patch placement.

damaged reinforcement, placement of supplemental reinforcement, cleaning of reinforcement, and preparation of surface for patch/overlay placement.

10.5.3.3 Bonding Medium. Bonding of the new patch or overlay to the concrete substrate is essential for a durable repair. An adequate bond between the patch or overlay material is required to resist stresses due to differential volume change between the patching material and the substrate. The failure can occur either at the bond interface or adjacent to the interface within the section of the lower strength material.

Once debonding is initiated, the effects of freeze-thaw cycling and dynamic impact of vehicle wheel loads can contribute to the progressive deterioration of the repaired area. Debonded areas are generally prone to cracking. The cracking is usually through the entire thickness of the patch or overlay material, which can permit water leakage to the interface and the underlying substrate.

As indicated in the National Cooperative Highway's Research Report #99, various bonding media have been studied, including sand-cement and water-cement grouts, neat cement, epoxies, and latex. From these studies and field experiences, the sand-cement and water-cement grout are considered to be the most practical and commonly used bonding mediums. The sand-cement bonding grout consists of one part cement to one part sand by volume, and sufficient water to achieve the consistency of "pancake batter." Limit the water-to-cement ratio of the grout to be at least the same as or better than the concrete repair material. The grout should be applied at a uniform thickness of $\frac{1}{16}$ in. and should not exceed $\frac{1}{8}$ in. (Figure 10-62). The grout should be applied to a damp (but not saturated) concrete surface. The surface is dampened to assist in preventing rapid drying of the grout. The bond mechanism consists of the grout penetrating the surface pores of the existing substrate. When the substrate is saturated the pores are filled with water, which can adversely affect the penetration of the bonding grout into the pores of the substrate. See a sample specification for latex-modified concrete or low-slump concrete in Appendix A for additional requirements of bonding mediums.

10.5.3.4 Selection of Repair Materials. The selection of the repair material is based on the consideration of the following five characteristics, as they relate to the member being repaired:

- Thermal compatibility or incompatibility
- Shrinkage
- Strength of repair material and the substrate

Figure 10-62.   Application of bonding grout precedes concrete overlay placement.

- Durability of the repair material and the substrate
- Ability to permit vapor transmission

The compatibility of the repair materials with the existing concrete is an important concern in the selection of appropriate repair materials. Since parking structures are exposed to temperature extremes, a difference in thermal properties of the repair material and the existing concrete will contribute to the debonding and failure of repaired areas. For instance, failure of epoxy-mortar patches can often be attributed to the thermal incompatibly. Even for sand-filled epoxy mortars the thermal coefficient of expansion may be four to five times that of the underlying concrete. The high stress developed between the epoxy patch and the underlying concrete is less forgiving of deficiencies in workmanship related to patch preparation and placement resulting in more frequent failures.

For parking structures, Portland cement-based patching and overlay materials generally perform better than any other material. Even in moderate climatic regions, the use of Portland cement-based material should be preferred over other polymer concrete materials. Portland cement-based materials also minimize failures associated with a difference in the modulus of elasticity of the repaired material and existing concrete.

The differential shrinkage between the original concrete and the repair material can also contribute to debonding and cracking due to development of shear stresses along the interface. Minimize differential shrinkage of concrete repair materials by using a low water-cement ratio concrete, the maximum permissible size of coarse aggregate, and the lowest slump that will permit proper consolidation and finishing of the concrete. Minimizing the shrinkage potential of the concrete repair material is particularly important for full-depth patching and floor-slab replacements. Cracking of full-depth repair patches is a common occurrence. The use of additional reinforcement and control joints is another effective way to control cracking. Also refer to ACI 224R.

Always specify a strength of the repair material that is equal to or exceeds the strength of the underlying concrete. Also, patches placed over poor-quality concrete with less than 3000 to 4000 psi concrete-compressive strength is susceptible to debonding. In such instances, the failure may occur in the underlying concrete. The extent of the failure will be affected by the concrete removal method, bonding medium, and patch material used. For suspect or questionable quality of the underlying concrete, verify performance of the repair technique by field testing of various concrete removal methods, surface preparation, and placement methods based on shear-bond test results.

A substrate that is marginally or inadequately air entrained is susceptible to progressive deterioration due to freeze-thaw damage if enough moisture is available or trapped at the interface to critically saturate the underlying concrete. Under these conditions the repair material selected should not behave as a vapor barrier. Also, cracking or defects in the repair material that permit water to reach the interface will contribute to progressive deterioration of the repair. For instance, an elastomeric-waterproofing membrane system (traffic topping) placed over a non-air-entrained concrete surface is likely to be subjected to progressive deterioration at pinholes and defects due to scaling of the underlying concrete.

The repair material itself must be durable enough to resist freeze-thaw cycling. The patch or overlay material should be properly air entrained. Performance of rapid-setting prepackaged patching materials has generally been poor due to its inability to control air entrainment. However, prepackaged Portland cement and polymer-concrete-based patching materials have been used to maintain serviceability for a limited time as a temporary measure to reduce safety hazards. In some instances, asphalt concrete has also been used. Since asphalt tends to trap moisture and chlorides, patching with asphalt material may accelerate damage to concrete due to corrosion and freeze-thaw.

The patching material for overhead and vertical surfaces is less susceptible to freeze-thaw deterioration than the floor slab or horizontal surface. For areas that are protected from direct exposure to moisture, such as the ceiling, rapid-setting prepackaged Portland-cement-based repair materials have been used successfully. Other successfully but not widely used patching materials are various epoxy and polymer concretes. Polymer concretes are classed as thermosetting and hydrating. Examples of thermosetting polymer concretes are those containing epoxy and those containing methyl methacrylate. Examples of hydrating polymer concretes are those containing styrene-butadene ("latex") additives which enhance the bond and reduce permeability. In all cases installation of patching should be performed by a contractor experienced in such work. Limit the use of these materials only to address cosmetic or aesthetic repairs. For large shallow areas, pneumatically applied (shotcrete) concrete has also been used effectively. Specify the wet process with air entrainment when there is a potential for saturation of the surface by moisture.

Surface sealer selection should be based on the results of tests performed in accordance with NCHRP 244 and ASTM C 672. Also, identify sealers that can potentially "glaze" through trial applications.

The present ASTM test methods for testing properties of traffic-bearing membranes with a wearing course are not adequate for the evaluation of abrasion resistance and service life of systems. Some other basic characteristics which need to be evaluated are:

- Impermeability—Should be impermeable to water under normal use.
- Tear Resistance—Membrane should be capable of bridging cracks under normal as well as cold-weather conditions.
- Adhesion—Intercoat as well as adhesion to the substrate.
- Moisture Vapor Transmission—The membrane should be capable of breathing.
- Material Stability—Stability under service-exposure conditions to perform over extended time period.
- Chemical Resistance—Should be resistant to gasoline, oil, and antifreeze spills.
- Ease of Installation—The waterproofing material and installation procedures must be tolerant of site conditions, as opposed to ideal laboratory conditions.

Select membrane systems based on performance history, compatibility with other sealant systems, cost and the manufacturer's reputation to properly install and service the topping. Polyurethane- and neoprene-based membranes have performed well, but all membrane systems do not perform equally. The above systems are susceptible to abrasion. Also, improper application of polyurethane membranes can result in localized imperfections, such as blistering and pinholes.

The most widely used specialty concrete overlay systems that have demonstrated a satisfactory long-term performance history are latex-modified concrete (LMC) and low-slump high-density concrete (LSDC). LMC is more effective at preventing additional water and salt penetrations into the base slab than LSDC; however, the long-term durability of both systems appears to be equivalent. Polymer-concrete overlays have been used only on a limited scale and have not been fully evaluated. Such systems, whether referred to as polymer or epoxy concrete can offer solutions to surface deterioration problems and should not be excluded from consideration. Another specialty concrete overlay utilizing silica-fume-modified, high-density concrete is currently available. Its performance history is presently limited; however, the potential for the long-term durability of this system is excellent. Also, the installation cost of the silica-fume-modified overlay is lower than that of the LMC system.

10.5.3.5 Material Application. Concrete repair materials must be properly placed, consolidated, and cured. Follow the appropriate placement and curing procedures of repair materials specified in ACI Recommended Practices or in accordance with each manufacturer's instructions. Also, refer to sample specifications in Appendix A for placement, consolidation, and curing requirements for concrete patches and overlay.

10.5.3.6 Summary. The intent of these sections was to discuss the commonly used repair approaches, methods, and materials-selection guidelines considered appropriate for repairing parking structures. Published literature on the repair of concrete structures can be used by the reader to specify and implement repairs effectively.

Construction and design deficiencies, along with errors in design or construction, may require strengthening or stiffening of the structural element. In these instances the primary cause of the distress must be determined, and the appropriate corrective actions recommended, by following the approach discussed in Section 10.4. These situations are generally uncommon. If construction and design deficiencies are present, then repair methods must address the specific conditions encountered.

More recently, an increase in the number of cast-in-place post-tensioned structures requiring repair has been noted. Most of these structures were built in the late 60s and early 70s, and are now showing evidence of tendon corrosion. Repairing these structures requires a proper understanding of each structural system in order to provide appropriate bracing and shoring during construction. Removal of large floor areas may also require schemes to counteract existing prestressing forces. Moreover, it is quite difficult to estimate the remaining service life of the repaired structures, due to the potential for continuing corrosion of tendons in chloride-contaminated concrete. These repairs should only be attempted by experienced specialists.

10.6   SELECTION OF REPAIR APPROACH AND METHOD

10.6.1   General

Floor slab surfaces experience more deterioration due to direct exposure to the environment and road salt in the northern climates. Design deficiencies or construction practices that contribute to poor-quality concrete, shallow cover over embedded reinforcement, inadequately air-entrained concrete, cracking, and ponding, can have a significant impact on the extent of floor-slab deterioration. In most instances, the repair selected is only intended to extend the life of the structure cost-effectively and may not be able to correct all the existing deficiencies.

It is not unusual for floor-slab repair items to consume as much as 50–80% of the total restoration cost. Therefore, floor-slab repairs generally offer the greatest potential for cost savings. In most instances repairs to underlying members, such as columns, haunches, bearing ledges, and walls represent a much smaller portion of the total restoration cost. Repair of the structural members, other than the floor slab, are performed by using well-established and proven procedures discussed in Section 10.5. The selection of a repair scheme to restore the floor slab of a parking structure is related to the following six basic issues:

- Nature of distress
- Extent of deterioration
- Type of structure
- Repair strategies
- Life expectancy of the repaired structure
- Economics

10.6.2   Repair Approach

The same repair approach cannot be used for all structures. The repair scheme selected to restore a structure damaged by corrosion of embedded reinforcement will be different from that selected for a slab damaged by freezing and thawing. In addition, the repair scheme selected must address the adverse effect of other contributing factors, such as the quality of the concrete, poor drainage, floor-slab cracking, shallow concrete cover over reinforcement, and lack of adequate air entrainment.

The extent of the deterioration and type of structural system will also influence the selection of the repair scheme. For instance, if a 4-in. thick slab of a pan-joist system is extensively damaged due to corrosion, then patching, sealing, or cathodic protection may not be an acceptable solution. The appropriate repair scheme in this instance is probably going to be the replacement of the slab of the floor system. Slab replacement will be required, since it is difficult to perform partial-depth repair of slabs that are less than 5 in. thick. On the other hand, if the 4-in. slab is damaged by surface scaling, an elastomeric-waterproofing membrane or an overlay may be acceptable solutions. However, if the extent of the freeze-thaw damage extends 1–2 in. below the surface, replacement may be a more appropriate repair method.

In summary, from a technical standpoint, consider the nature and extent of the deterioration, the pros and cons of the repair methods that are technically acceptable, and the impact of the repair on factors contributing to the deterioration. Also, make certain that the structure can be repaired (as opposed to replaced), and that all elements of the structure will support additional loads imposed by the repair work.

The advantages and disadvantages of the various schemes were discussed in Section 10.5.2. This information can be conveniently qualified to assist in the selection of technically appropriate repair schemes by using a decision matrix as shown in Table 10-3. The concept of the decision matrix was developed by the Ontario Ministry of Transportation, Research and Development Branch, primarily for selection of bridge-deck rehabilitation methods. The decision matrix presented in Table 10-3 has been adapted from the material published in the Ministry of Transportation's manual. The table can assist in the selection of repair schemes for parking structures. Note that patching by itself is seldom an appropriate repair approach for parking structures. Also, sealing will invariably consist of patching unsound and delaminated areas prior to applying a surface sealer, traffic topping, or an overlay.

The decision matrix leads, by elimination, to the selection of repair schemes with the least disadvantages. In some cases, all schemes may be inappropriate. For instance, a structure that is extensively cracked, consisting primarily of active cracks and delaminated over

TABLE 10-3 Selection of a Floor-Slab Repair Approach

Criterion	Patching (Partial or Full-Depth)	Protective Coatings			Replacement (Partial or Full-Depth)	Cathodic Protection[1,2]
		Sealer	Traffic Topping	Overlay		
1. Corrosion-induced deterioration—10% of the floor area	No	No				
2. Corrosion-induced deterioration—30% of the floor area	No	No	No	No		No
3. Moderate scaling—10% of the floor area	No	No				Yes/No
4. Non air-entrained concrete	No	No	No			Yes/No
5. High concrete permeability	No	No				
6. Need to improve drainage	No	No	No			Yes/No
7. Shallow concrete cover	No	No				
8. Limited structural capacity				No		Yes/No
9. Limited floor clearance				No		Yes/No
10. Remaining life less than 10 yr				No	No	No
11. Active cracks	No	No		No	No	

[1]Items 3, 4, and 7 are appropriate if the C.P. system selected consists of anode embedded in a concrete overlay.
[2]Items 8 and 9 will be appropriate if the C.P. system consists of anode embedded in slots cut in the structure.
Adapted from: ''Bridge Deck Rehabilitation Manual,'' Part Two: Contract Preparation, Ontario Ministry of Transportation.

30% of the floor area, will necessitate working through the selection process again and examining the implication of violating each criterion in turn. If the structure is considered to be important, then the scheme may consist of slab replacement with a liquid-applied membrane to minimize leakage through active floor-slab cracks. The criteria contained in Table 10-3 are not rigid, but serve only as a useful starting point from a technical standpoint. As previously mentioned, repair strategies, life expectancy, and economic issues usually influence the selection of the final repair scheme.

10.6.3   Repair Strategies

It is not uncommon to develop several technically acceptable repair alternatives based on the following overall repair strategies:

1. Do nothing and use up the remaining useful life of the structure.
2. Perform repairs to address only potentially unsafe conditions that presently exist. This amounts to performing only "band-aid-type" repairs prior to either implementing a comprehensive restoration program or demolishing the structure.
3. Perform necessary repairs to extend the life of the structure 5–10 yr.
4. Perform necessary repairs to extend the life of the structure 10–20 yr.
5. Perform repairs to extend the life of the structure 20 yr or more.

The selection and evaluation of repair alternatives are important elements of the restoration process. Select repair alternatives based on the overall strategies. This assists in selecting schemes that will address future plans for use of the structure based on funds that are presently available or obtainable. However, do not consider technically unacceptable alternatives primarily to limit restoration costs.

The nature and the extent of the deterioration will also limit the selection of repair alternatives. For instance, it may not be appropriate to extend the life of a structure 5–10 yr simply by patching, if the slab is likely to undergo progressive damage due to freezing and thawing. Also, it may not be possible to assure safe operating conditions by performing only limited repairs to a structure that is extensively damaged.

10.6.4   Continuing Corrosion

Specifying the extent of sound concrete removal, in addition to the removal of unsound concrete, can have a significant impact on the life expectancy of the repairs and future maintenance efforts. This is particularly true for chloride-contaminated slabs, where the life expectancy of the repairs is adversely affected by continuing corrosion of the embedded reinforcement. At present, the only way to be assured of a "permanent" repair requiring minimal maintenance, is to remove all concrete that contains chlorides in excess of the corrosion threshold. However, in most instances, this can amount to complete removal of the floor slab. Therefore, the emphasis should be on selective, but cost-effective, removals of chloride-contaminated concrete, based on consideration of overall repair strategies and the desired life expectancy of the repairs. Refer to Section 10.3.1. for an explanation of the corrosion-mechanism process, chloride contamination, and continuing corrosion.

10.6.5   Life Expectancy of Repairs

The life expectancy of repair methods is at best an estimate. Also, estimating the service of repaired structures is only an educated opinion, based on experience gained from conditions observed in structures with a similar framing system. Therefore, difficulty in estimating the service life of repaired structures complicates the selection of a cost-effective repair method. However, removal of sound but contaminated concrete has a significant impact on extending the service life. The impact of concrete removal on the various repair methods can be best illustrated by considering the life expectancy of structures repaired by patching.

A distinction can sometimes be made between temporary and permanent repair patches. However, because of the progressive nature of corrosive processes, the service life of even a "permanent" patch is limited. As shown in Figure 10-63, in a temporary patch the concrete is removed only to the level of reinforcement. This situation contributes to progressive deterioration within and adjacent to the patch boundary as illustrated in Figure 10-64. Also, the life expectancy of the patch may be limited to only 1 or 2 yr. This method of patch repair may be appropriate for structures when serviceability is to be maintained for a limited time, or when constraints are imposed due to available funds or weather conditions.

In the instance of a relatively permanent patch, the concrete is removed below the existing reinforcement to minimize potential for corrosion within the patch boundary (Figure 10-65). Also, to control accelerated corrosion adjacent to the patch boundary, the existing reinforcement may be epoxy coated. The entire floor surface is then sealed to reduce the deterioration rate of areas beyond the patch boundary.

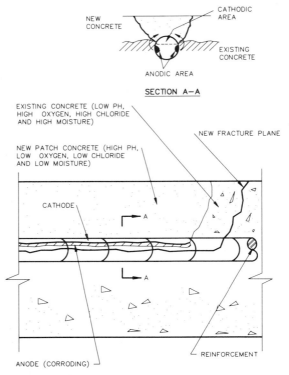

Figure 10-63. Development of corrosion cell in repair with shallow concrete removal.

Figure 10-64. Delamination due to continuing corrosion within and beyond patch boundary.

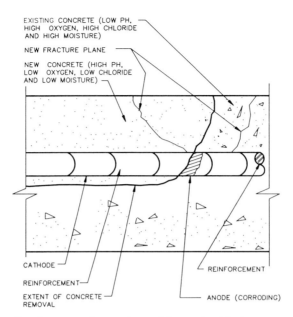

EXISTING CONCRETE (LOW PH,
HIGH OXYGEN, HIGH CHLORIDE
AND HIGH MOISTURE)

NEW FRACTURE PLANE

NEW CONCRETE (HIGH PH,
LOW OXYGEN, LOW CHLORIDE
AND LOW MOISTURE)

CATHODE

REINFORCEMENT

EXTENT OF CONCRETE
REMOVAL

REINFORCEMENT

ANODE (CORRODING)

Figure 10-65. Development of corrosion cell in repair with deep concrete removal.

Under these conditions, patching can extend the service life of the structure 3–5 yr.

In certain instances, where longer life expectancy is desirable, concrete removal along the entire length of reinforcement is specified. For conditions shown in Figure 10-66, the life expectancy of the strip-patch repair may be estimated at 10–20 yr, limited primarily by other contributing factors, such as cracking, lack of air entrainment and poor drainage that may adversely affect the service life of the structure. However, it is not feasible to implement the strip-patch-repair approach in structures with relatively thin slabs. Therefore, considerations, such as the structural system involved and the existing reinforcement pattern, will limit the practical service life of structures that can be realized.

Based on the extent of concrete removals, the structural system involved, and the concrete cover over existing reinforcements, patching and then coating the floor slab with a waterproofing membrane is likely to extend the service life of the structure 5–10 yr. An overlay can extend the service life of structures 10–20 yr. In concept the only method that will mitigate corrosion of the embedded reinforcement without removal of sound concrete is cathodic protection. Application

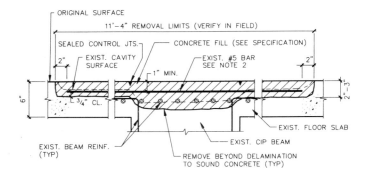

ORIGINAL SURFACE

11'-4" REMOVAL LIMITS (VERIFY IN FIELD)

SEALED CONTROL JTS.   CONCRETE FILL (SEE SPECIFICATION)

2" EXIST. CAVITY
SURFACE

EXIST. #5 BAR
SEE NOTE 2

2"

1" MIN.

6"

3/4" CL.

2-3"

EXIST. FLOOR SLAB

EXIST. BEAM REINF.
(TYP)

EXIST. CIP BEAM

REMOVE BEYOND DELAMINATION
TO SOUND CONCRETE (TYP)

SECTION

NOTE:

1. REMOVE AND REPLACE ALL SOUND AND UNSOUND CONCRETE WITHIN
   SECTIONS SHOWN CROSS-HATCHED.

2. REMOVE EXISTING TOP REINFORCEMENT AND REPLACE WITH NEW
   EPOXY COATED REINFORCEMENT OF SAME SIZE.

Figure 10-66. Strip patch repair detail.

of cathodic protection is estimated to extend the service life of structures beyond 20 yr. Full-depth slab and floor removal can be designed to be rebuilt with a life expectancy of 20–40 yr.

10.6.6 Economics

The selection of a cost-effective repair method consists of: 1) Preparing cost estimates of technically acceptable repair alternatives, and 2) estimating the service life of the repaired structure. Repair costs can vary significantly even for the same method of repair. Factors contributing to cost variations are geographic location of the structure, scope of the overall contract, size and volume of the repair work, and availability of materials and qualified contractors. Constraints associated with maintaining traffic during construction and the overall volume of construction work at the time of bidding can also vary the overall repair costs. Realistic estimates are obtained by using costs from an historical record and assigning appropriate contingency factors to the total cost of the work.

In some instances life-cycle cost analysis of repair methods is also performed to select an economical repair method. Once again, the economics are difficult to estimate due to the possible inaccuracy in assessed costs and assumed service life of the repaired structure.

10.7 REPAIR DOCUMENTS

Repair documents implement the findings of the condition appraisal. The four tasks associated with preparation and implementation are as follows:

- Development of plans, details, and specifications
- Priority assignments and preparation of cost estimates
- Contractor selection and bidding or negotiation
- Construction observation

The two most common contract forms are the negotiated contract and the lump-sum contract, both based on unit prices and quantities. The lump-sum contract is preferred for restoration work. Using the unit-price approach, repair procedures are identified as separate work items (Table 10-4). They are repair procedures which will correct the distress in structural members. The plan sheets are usually condition appraisal field sheets converted to work items based on the field-survey observations. The specifications contain the description of the scope of individual work items, material specifications, and repair procedures. Details describe concrete removal limits, and provide dimensions and material requirements.

For owners facing budget constraints, assign priority to repair items to allocate available funds. Also, when repair costs are high, restoration work may sometimes be planned over several years. The first priority is to repair structural defects to assure a safe and serviceable condition. Except for emergency repairs, priorities are assigned on a tier basis, not a work-item basis. A tier (or area) of a parking deck should be closed only once for repairs. Contract priorities assigned by work items, instead of areas or tiers, can be more annoying to the facility users. Also, confining the work to a designated area minimizes the disruption of traffic circulation during construction.

TABLE 10-4   Repair Work Items

Work Item	Description
1.0	Repair damaged floor slab
2.0	Repair damaged beams and columns
3.0	Install new electrical conduit
4.0	Paint exposed structural steel
5.0	Install new control joints
6.0	Repair bumper guards
7.0	Apply concrete surface sealer
8.0	Install new drain system

Cost estimates are prepared using the work-item listing developed earlier, and estimated quantities for each item. Quantities should be estimated on a tier-by-tier basis to aid the contractor in planning and scheduling the work with minimum disruption to the operation of the facility (Table 10-5). Always provide funding for contingencies, otherwise, latent conditions found during the repairs may contribute to cost overruns.

When restoration work is planned over more than 1 yr, repair costs will be higher due to the following reasons.

1. Contractor must move on and off site more than once.
2. Smaller repair quantities each year will have higher unit costs.
3. Items not repaired in the first stage will continue to deteriorate, so volume of repairs will increase.
4. Inflation normally results in an annual repair-cost increase.

Where possible, only experienced general and specialty contractors should be permitted to bid on restoration projects. Qualifications should be closely examined to determine if the contractor is able to perform the work according to specifications. Such an experienced contractor is one who has performed the type of restoration work being bid on, or has personnel who have completed similar work in the past. Evaluate contractors based on the number of similar projects completed, size of projects completed, experience and skill of personnel, bonding capacity, and reputation.

TABLE 10-5   Repair Cost Estimate

Work Item	Description	Estimated Cost
1.0	Repair damaged floor slab	$325,000
2.0	Repair damaged beams and columns	75,000
3.0	Install new electrical conduit	25,000
4.0	Paint exposed structural steel	30,000
5.0	Install new control joints	24,000
6.0	Repair bumper guards	28,000
7.0	Apply concrete surface sealer	42,000
8.0	Install new drain system	30,000
	Subtotal	$579,000
	Plans and Specifications	30,000
	Resident Services	30,000
	Materials Testing	18,000
	Contingency	100,000
	Total	$757,000

A qualified project resident should observe repair work, verify repair areas, and document actual work-item quantities. The resident can also monitor the compliance to material requirements, repair procedures, equipment, and construction load restrictions specified by the contract documents. Field problems can be resolved promptly with the assistance of a project resident maintaining the construction schedule and minimizing any inconvenience to facility users.

10.8   MAINTENANCE PROGRAM

Effective maintenance can prolong the life of a restored facility by minimizing the deterioration of the structure, thus protecting the owner's investment. Refer to Chapter 9 for information on a maintenance program.

Appendixes to Chapter 10

APPENDIX 10-1
Sample Specification for Surface Preparation

PART 1   GENERAL

1.01   WORK INCLUDED
This work consists of furnishing all labor, materials, equipment, supervision, and incidentals necessary to locate and remove all delaminated and unsound concrete, and preparing the cavities created by said removal to receive patching or overlay material. Also included in this work is the preparation of existing surface spalls and potholes to receive patching or overlay material.

1.02   RELATED WORK
The following is directly related to, and shall be coordinated with, surface preparation for patching:
1. Concrete formwork
2. Concrete shores and reshores
3. Concrete reinforcement
4. Cast-in-place concrete
5. Preplaced aggregate concrete
6. Low-slump dense concrete

7. Latex-modified concrete and mortar
8. Trowel-applied mortar

1.03 REFERENCES

A. "Specifications for Structural Concrete for Buildings," (ACI 301), American Concrete Institute.

B. Comply with the provisions of the following codes, specifications, and standards, except where more stringent requirements are shown on the drawings or specified herein:
   1. "Guide for Repair of Concrete Bridge Superstructures" (ACI 546.1R), American Concrete Institute.

PART 2   PRODUCTS—NOT USED

PART 3   EXECUTION

3.01 LOCATION AND MARKING OF WORK AREA

A. Floor Slabs
   1. Floor-slab delaminations shall be located by sounding the surface with a hammer, rod, or chain drag.
   2. When a delaminated area is struck, a distinct hollow sound is heard.
   3. The contractor shall sound all designated floors for delaminations.

B. Vertical and Overhead Surfaces
   1. Vertical and overhead surface delaminations shall be located by sounding the appropriate member with a hammer or rod.
   2. Cracks, usually horizontal in orientation along beam faces, and vertical in orientation near column corners are indicators of delaminated concrete.
   3. The contractor shall only sound vertical and overhead surfaces that show evidence of cracking and/or salt and water staining.

C. Delaminated areas, once located by the contractor, will be further sounded to define their limits. These limits or "boundaries" once defined shall be marked with chalk or paint.

D. The contractor shall locate spalls by visual inspection and mark their boundaries with chalk or paint after sounding the surface.

E. The engineer may define and mark additional unsound concrete areas for removal, if required.

F. Areas to be removed shall be as straight and rectangular as practical to encompass the repair and provide a neat patch.

G. The contractor shall locate all embedded post-tensioning tendons in the repair area and mark these locations for reference during concrete removal.

3.02 CONCRETE REMOVAL AND CAVITY PREPARATION

A. Temporary shoring may be required at concrete floor repair areas exceeding 5 sq ft in area and at any beam, joist, or column repair. Contractor shall review all marked removal and preparation areas and request clarification by the engineer of shoring requirements in questionable areas. Shores must be in place prior to start of concrete removal and cavity preparation in any area requiring shores.

Note 1: *(Five-square-feet area of concrete removal can be increased based on evaluation of the structural framing system.)*

B. Delaminated, spalled, and unsound concrete floor areas shall have their marked boundaries sawcut to a depth of $\frac{3}{4}$ in. into the floor slab, unless otherwise noted. For vertical and overhead surfaces the marked boundary may be sawcut, ground or chipped to a depth of $\frac{1}{2}$–$\frac{5}{8}$ in. into the existing concrete, measured from the original surface. All edges shall be straight and patch areas square or rectangular shaped. A diamond blade saw or grinder with abrasive disk suitable for cutting concrete is acceptable for performing this work. The edge cut at the delamination boundary shall be dressed perpendicular to the member face. It shall also be of uniform depth, for the entire length of the cut. Extra caution shall be exercised during sawcutting operations to avoid damaging existing reinforcement (*especially post-tensioning tendons and sheaths*) near the surface of concrete. Any damage to existing reinforcement, post-tensioning tendons, or sheaths during removals shall be repaired by the contractor with engineer approved methods at no cost to the owner.

C. All concrete shall be removed from within the marked boundary to a minimum depth of $\frac{3}{4}$ in. using 15- to 30-lb chipping hammers equipped with chisel point bits. When directed by the engineer, chipping hammers less than 15 lb shall be used to minimize damage to sound concrete. If delaminations exist beyond the minimum removal depth, then chipping shall continue until all unsound and delaminated concrete has been removed from the cavity.

D. Where embedded reinforcement is exposed by concrete removal, extra caution shall be exercised to avoid damaging it during removal of additional unsound concrete. If bond between exposed embedded reinforcement and adjacent concrete is impaired by the contractor's removal operations, then the contractor shall perform additional removal around and beyond the perimeter of the reinforcement for a minimum of $\frac{3}{4}$ in. along the entire length affected at no cost to the owner.

E. If rust is present on embedded reinforcement where it enters sound concrete, then additional removal of concrete along and beneath the reinforcement will be required. Such additional removal shall continue until nonrusted reinforcement is exposed, or removal may be terminated as the engineer directs.

3.03   INSPECTION OF THE CAVITY SURFACES AND EXPOSED
         REINFORCING

A. After removals are complete, but prior to final cleaning, the cavity and exposed reinforcement shall be inspected by the contractor and verified by the engineer for compliance with the requirements of Section 3.02. Where the engineer can detect unsatisfactory cavity preparation, the engineer may direct the contractor to perform additional removals. The engineer shall reverify that the additional removals have been performed as directed.

B. The contractor shall inspect embedded reinforcement exposed within the cavity for defects due to corrosion or damage resulting from removal operations. Replacement of damaged or defective reinforcement shall be performed according to this section and as directed by the engineer.

3.04 REINFORCEMENT IN REPAIR AREA

A. All embedded reinforcement exposed during surface preparation that has lost more than 15% of the original cross-sectional area due to corrosion shall be considered *defective*. All nondefective exposed reinforcement that has lost section (to the extent specified above) as a direct result of contractor's removal operations, shall be considered *damaged*.

Note 2: (*Loss percentages may be filled in only after analysis of the structural capacity of the as-built structure to see if any loss of reinforcing at all is permissible.*)

B. Supplement defective or damaged embedded reinforcement with a reinforcement of equal diameter having a Class "C" minimum splice (ACI 318) beyond the damaged portion of the reinforcement. Secure the new reinforcement to the existing reinforcement with wire ties and/or approved anchors. Supplemental reinforcing bars shall be epoxy coated, ASTM A615 grade 60 steel. Tendon supplement or repair materials, when applicable, shall be as required by the specifications.

C. Loose reinforcement exposed during surface preparation shall be securely anchored to the original floor prior to patch placement. Loose reinforcement shall be adequately secured by wire ties to bonded reinforcement or shall have drilled-in anchors installed to the original deck. The engineer shall determine adequacy of wire ties and approve other anchoring devices prior to their use. Tying loose reinforcement to bonded reinforcement is incidental to surface preparation and no extras will be allowed for this work. Securing loose reinforcement with drilled anchors shall be paid for at the unit-price bid for that work item.

D. Concrete shall be removed to provide a minimum of $\frac{3}{4}$-in. clearance on all sides of defective or damaged exposed embedded reinforcement that is left in place. A minimum of $1\frac{1}{2}$-in. concrete cover shall be provided over all new and existing reinforcement. Concrete cover over reinforcement may be reduced to 1 in. with engineer's approval if coated with an approved epoxy resin.

E. Supplemental reinforcement and concrete removals required for repairs of defective or damaged reinforcement shall be paid for as follows:

1. Concrete removals and supplemental reinforcement required for repairs of *defective* reinforcement shall be paid for by the owner at the unit-price bid.

2. Concrete removals and supplemental reinforcement required for repairs of *damaged* reinforcement shall be paid for by the contractor.

3.05 CLEANING OF REINFORCING WITHIN DELAMINATION AND
SPALL CAVITIES
A. All exposed steel shall be cleaned of rust to bare metal by sandblasting.
Cleaning shall be completed immediately before patch placement to en-
sure that the base metal is not exposed to the elements and further rusting
for extended periods of time.
B. Paint exposed steel with an approved epoxy resin and protect from damage prior to and during concrete placement.
3.06 PREPARATION OF CAVITY FOR PATCH PLACEMENT
A. Cavities will be examined prior to commencement of patching operations. Sounding the surface shall be part of the examination. Any delamination noted during the sounding shall be removed as specified in this
Section.
B. Cavities shall be sandblasted. Airblasting is required as a final step to
remove sand. All debris shall be removed from the site prior to the start
of patching or placement of overlay.

APPENDIX 10-2
Sample Specification for Latex-Modified Concrete and
Mortar for Patching and Overlay

PART 1   GENERAL

1.01 WORK INCLUDED
This work consists of furnishing all labor, materials, and equipment
necessary for the production and installation of latex-modified concrete
or mortar for patching floor spalls and delamination voids.
1.02 RELATED WORK
The following work is directly related to and shall be coordinated with
the placement of the latex-modified concrete and mortar:
1. Surface preparation for patching and overlay
2. Concrete shores and reshores
3. Concrete reinforcement
4. Concrete accessories
5. Concrete curing
1.03 REFERENCES
A. "Specifications for Structural Concrete for Buildings," American Concrete Institute, herein referred to as ACI 301, is included unabridged as
specification for this structure except as otherwise specified herein.
B. Comply with the provisions of the following codes, specifications, and
standards except where more stringent requirements are shown on the
drawings or specified herein:
1. "Building Code Requirements for Reinforced Concrete," American
Concrete Institute, herein referred to as ACI 318.

    2. "Hot Weather Concreting," reported by ACI Committee 305 (ACI 305R).

    3. "Cold Weather Concreting," reported by ACI Committee 301 (ACI 306R).

    4. "Guide for Concrete Floor and Slab Construction," reported by ACI Committee 302 (ACI 302.1R).

C. The contractor shall have the following ACI publications at the project construction site at all times:

    1. "Specifications for Structural Concrete for Buildings (ACI 301) with selected ACI and ASTM References," *ACI Field Reference Manual*, SP15.

    2. "Hot Weather Concreting," reported by ACI Committee 301 (ACI 305R).

    3. "Cold Weather Concreting," reported by ACI Committee 306 (ACI 306R).

    4. "Guide for Concrete Floor and Slab Construction," reported by ACI Committee 302 (ACI 302.1R).

1.04   QUALITY ASSURANCE

A. The Testing Agency will be an independent testing laboratory employed by the owner and approved by the engineer.

B. The testing agency is responsible for conducting, monitoring, and reporting to the owner the results of all tests required under this section with copies of all reports sent to the engineer and contractor. Testing agency has authority to reject concrete not meeting specifications.

C. Submit the following information for field testing of concrete unless modified in writing by the engineer:

    1. Project name and location

    2. Contractor's name

    3. Testing agency's name, address, and phone number

    4. Concrete supplier

    5. Date of report

    6. Testing agency technician's name (sampling and testing)

    7. Placement location within the structure

    8. Elapsed time from batching at plant to discharge from truck at site

    9. Concrete mix data (quantity and type)

        a. Cement

        b. Fine aggregates

        c. Coarse aggregates

        d. Water

        e. Water/cement ratio

        f. Latex emulsion

        g. Latex emulsion/cubic yard of concrete

        h. Other admixtures

   10. Weather data

        a. Air temperatures

     b. Weather
     c. Wind speed
  11. Field test data
     a. Date, time, and place of test
     b. Slump
     c. Air content
     d. Unit weight
     e. Concrete temperature
  12. Compressive test data
     a. Cylinder number
     b. Age of concrete when tested
     c. Date and time of cylinder test
     d. Curing time (field and laboratory)
     e. Compressive strength
     f. Type of break
1.05 SUBMITTALS
  A. Submittals shall be in accordance with requirements of the specifications.
  B. Before beginning patching or overlay operations, the contractor shall submit to the engineer a mix design reviewed by a representative of the latex manufacturer. Include the following information for each concrete mix design:
     1. Method used to determine the proposed mix design (ACI 301, Article 3.9)
     2. Gradation of fine and coarse aggregates—ASTM C33.
     3. Proportions of all ingredients, including all admixtures added either at the time of batching or at the job site
     4. Water-cement ratio
     5. Slump—ASTM C143
     6. Certification of the chloride content of admixtures
     7. Air content: Of freshly mixed concrete by pressure method, ASTM C173.
     8. Unit weight of concrete—ASTM C138.
     9. Strength at 3, 7, and 28 days—ASTM C39.
     10. Chloride ion content of the concrete.

                    PART 2   PRODUCTS

2.01 MATERIALS
  A. Portland cement shall be Type 1, ASTM C150 non-air entraining.
  B. Latex emulsion shall be:
     1. "Dow Modifier A," Dow Chemical Company of Midland, Michigan.
     2. Dylex 1186 as manufactured by Polysar, Incorporated, Chattanooga, Tennessee.

3. Approved equivalent used in accordance with the manufacturer's recommendations.

C. Concrete aggregates for all concrete work exposed to the weather shall be gravel or crushed limestone, containing a maximum of 2% clay lumps; friable particles and chert shall not exceed 3% in accordance with ASTM C33, Class Designation 5S. All other concrete aggregates shall be in accordance with ASTM C33. Aggregates not meeting the requirements of ASTM C33 may be used only upon written approval from the engineer who will require additional information and data.

1. Concrete fine aggregate shall be natural sand conforming to ASTM C33 and having the preferred grading shown for normal weight aggregate in ACI 302.1R, Table 4.2.1.

2. Concrete coarse aggregate shall be as specified by ASTM C33, Table 2 for aggregate size number 7 ($\frac{1}{2}$ in. to #4).

3. Concrete shall be produced using an approximate 1:1 ratio by volume of fine and coarse aggregates listed above.

2.02 MIX DESIGN

A. The selection of concrete proportions shall be in accordance with ACI 301, Article 3.8. Before any concrete is placed for the project, the contractor shall submit to the engineer data showing the method used for determining the proposed concrete mix design, including fine and coarse aggregate gradations, proportions of all ingredients, water/cement ratio, slump, air content, 28-day cylinder breaks and other required data for each different concrete type specified. The mix design shall meet the following minimum requirements:

Mix Design Requirements

	Mortar	Concrete
Compressive strengths:	4500 psi at 28 days	4500 psi at 28 days
	2500 psi at 3 days	2500 psi at 3 days
Water/cement ratio:	0.25–0.40	0.25–0.40
Latex content per sack of cement:	$3\frac{1}{2}$ gal	$3\frac{1}{2}$ gal
Slump:	4 in. +/− 2 in.	4 in. +/− 2 in.
Minimum cement content:	658 lb/cu yd	658 lb/cu yd
Air content:	4–10%	4–8%
Maximum coarse aggregate size:	$\frac{3}{8}$ in.	$\frac{1}{2}$ in.

B. For concrete placed and finished by vibrating screeds the slump shall be limited to 4 in.

C. The water-soluble chloride ion content of the concrete shall not exceed 0.15% by weight of cement in the mix; perform test to determine chloride content of concrete in accordance with FHWA-RD-85.

D. Bonding Grout: The bonding grout shall consist of a sand, cement, and latex emulsion in proportions similar to the mortar in the concrete with sufficient water to form a stiff slurry to achieve the consistency of "pancake batter."

PART 3   EXECUTION

3.01   PRODUCTION OF MORTAR OR CONCRETE

A. Production of latex-modified mortar or concrete shall be in accordance with requirements of ACI 301, Chapter 7, except as otherwise specified herein.

B. Concrete shall be produced by on-site volumetric batching done in accordance with requirements of ASTM C685.

C. On-site mortar or concrete batching in a mixer of at least $\frac{1}{3}$-cu-yd capacity shall be permitted only with approval of the engineer. On-site concrete batching and mixing shall comply with requirements of ACI 301, Section 7.2.

D. Latex-modified mortar shall be used to patch floor cavities $\frac{3}{4}$–$1\frac{1}{4}$ in. deep. Latex-modified concrete shall be used to patch cavities greater than $1\frac{1}{4}$ in. deep and for overlays.

3.02   PREPARATION (PATCH AND OVERLAY INSTALLATION—FLATWORK)

A. Cavity surfaces shall be clean and dry prior to beginning of patch or overlay installation. Preparation of surfaces to receive new concrete shall be in accordance with Sections 02030 and/or 02530.

B. Bonding grout shall be applied to a damp (but not saturated) concrete surface in uniform thickness of $\frac{1}{16}$–$\frac{1}{8}$ in. over all surfaces to receive patching or overlay. Grout shall not be allowed to dry or dust prior to placement of patch or overlay material. If concrete placement is delayed and the coating dries, the cavity or surface shall not be patched or overlaid until it has been recleaned and prepared as specified in Section 02530. Grout shall not be applied to more area than can be patched or overlaid within $\frac{1}{2}$ hr by available manpower.

3.03   PLACING AND FINISHING

A. Do not place concrete when the temperature of the surrounding patch area is less than 45°F or when mix temperature is above 85°F unless the following conditions are met:

1. Place concrete only when the temperature of the surrounding air is expected to be above 40°F and rising and expected to be above 45° F for at least 24 hr.

2. When the above conditions are not met, concrete may be placed only if insulation or heating enclosures are provided in accordance with ACI 306, "Recommended Practice for Cold Weather Concreting" and are approved by the engineer.

3. When air temperature exceeds 80°F or any combination of high temperature, low humidity, and high-wind velocity, which causes rate of evaporation in excess of 0.10 lb/p/sq ft p/hr as determined by Figure 2.1.5. in ACI 305R, "Hot Weather Concreting," requirements shall be met.

B. The concrete shall be manipulated and struck off slightly above final grade. The concrete shall then be consolidated and finished to final grade with surface vibration devices. Consolidation equipment used shall be approved by the engineer.

C. Fresh concrete 3 in. or more in thickness shall be vibrated internally in addition to surface vibration.

D. For overlays the vibrating device shall consist of vibrating screeds meeting the following requirements:

1. Placing and finishing equipment shall not exceed a maximum weight of 6000 or 3,000 lb per axle.

2. The screed shall be designed to consolidate the concrete to 98% of the unit weight determined in accordance with ASTM C138. A sufficient number of identical vibrators is provided for each 5 ft of the screed length.

3. The bottom face of screeds shall not be less than 4 in. wide and shall be metal covered with turned up or rounded leading edge to minimize tearing of the surface of the plastic concrete.

4. The screed shall be capable of forward and reverse movement under positive control. The screed shall be provided with positive control of vertical position and angle of tilt.

5. The screed shall be capable of vibrating at controlled rate, adjustable to between 3000 and 6,000 vpm.

E. Concrete shall be deposited as close to its final position as possible. All concrete shall be placed in continuous operation and terminated only at bulkheads or a designated control joint.

F. On ramps with greater than 5% slope, all concreting shall begin at the low point and end at the high point. The contractor shall make any necessary adjustment to slump or equipment to provide a wearing surface without any irregularities or roughness.

G. When a tight uniform concrete surface has been achieved by the screeding and finishing operation, the surface shall be textured using a coarse broom, as approved by the engineer from sample panels.

H. Brooming shall not tear out or loosen particles of coarse aggregate. Finishing tolerance: ACI 301, Paragraph 11.9; Class B tolerance.

I. Finish all concrete slabs to proper elevations to ensure that all surface moisture will drain freely to floor drains, and that no puddle areas exist. The contractor shall bear the cost of any corrections to provide for positive drainage.

3.03 CURING

Latex-modified mortar and concrete shall be cured according to the latex manufacturer's recommendations and according to the following minimum requirements:

1. The surface shall be covered with a single layer of clean, wet burlap as soon as the surface will support it without deformation. Cover the burlap with a continuous single thickness of polyethylene film for 24 hr.

2. After 24 hr remove the polyethylene film and allow the burlap to dry slowly for an additional 24–48 hr.
3. Remove the burlap and allow the concrete to air dry for an additional 48 hr.
4. Curing time shall be extended, as the engineer directs, when the curing temperature falls below 50°F.
5. If shrinkage cracks appear in the overlay when the initial 24-hr curing period is completed, the overlay shall be considered defective, and it shall be removed and replaced by the contractor at no extra cost.

3.04   FIELD-QUALITY CONTROL
   A. Mortar and concrete cylinders 3 in. in diameter by 6 in. long will be fabricated, cured, and tested in accordance with ACI 301 except as noted in this specification. Six cylinders will be made for each 10 cu yd of mortar prepared or each day's concrete pour, whichever is less. Cylinders shall be field cured for 3 days then transported to the testing laboratory where they shall be cured in air at 73°F, 50% relative humidity. Two cylinders will be tested at 3 days, two at 7 days and two at 28 days. Compressive strength at 3 days shall be 2000 psi minimum. Two additional cylinders shall be fabricated for each 20 cu yd of concrete produced. These shall be field cured until needed.
   B. The patch and overlay areas shall be sounded by the contractor with a chain drag after curing for 7 days. The contractor shall repair all hollowness detected by removing and replacing the patch or affected area at no extra cost to the owner.

APPENDIX 10-3
Sample Specification for Low-Slump Dense Concrete for Patching or Overlay

PART 1   GENERAL

1.01   WORK TO INCLUDE
   This work consists of providing all labor, materials, and equipment necessary for the production and installation of low-slump dense concrete for patching and overlays.

1.02   RELATED WORK
   The following work is directly related to, and shall be coordinated with, the placement of the low-slump dense concrete:
   1. Surface preparation for patching
   2. Surface preparation for overlay
   3. Concrete shores and reshores

    4. Concrete reinforcement

    5. Concrete accessories

    6. Concrete curing

1.03   REFERENCES

A. "Specifications for Structural Concrete for Buildings," (ACI 301-84), American Concrete Institute, herein referred to as ACI 301, is included unabridged as specification for this structure except as otherwise specified herein.

B. Comply with the provisions of the following codes, specifications, and standards except where more assignment requirements are shown on the drawings or specified herein:

    1. "Building Code Requirements for Reinforced Concrete," American Concrete Institute, herein referred to as ACI 318.

    2. "Specifications for Structural Concrete for Buildings" (ACI 301).

    3. "Hot Weather Concreting," reported by ACI Committee 305 (ACI 305R).

    4. "Cold Weather Concreting," reported by ACI Committee 301 (ACI 306R).

    5. "Guide for Concrete Floor and Slab Construction," reported by ACI Committee 302 (ACI 302.1R).

C. The contractor shall have the following ACI publications at the project construction site at all times:

    1. "Specifications for Structural Concrete for Buildings (ACI 301) with Selected ACI and ASTM References," *ACI Field Reference Manual,* SP15.

    2. "Hot Weather Concreting," reported by ACI Committee 305 (ACI 305R).

    3. "Cold Weather Concreting," reported by ACI Committee 306 (ACI 306R).

    4. "Guide for Concrete Floor and Slab Construction," reported by ACI Committee 302 (302.1R).

1.04   QUALITY ASSURANCE

A. The testing agency will be an independent testing laboratory employed by the owner and approved by the engineer.

B. The testing agency is responsible for conducting, monitoring, and reporting to the owner the results of all tests required under this section with copies of all reports to the engineer and contractor. Testing agency has authority to reject concrete not meeting specifications.

C. Submit the following information for field testing of concrete unless modified in writing by the engineer:

    1. Project name and location

    2. Contractor's name

    3. Testing agency's name, address, and phone number

    4. Concrete supplier

    5. Date of report

6. Testing agency technician's name (sampling and testing)
7. Placement location within the structure
8. Elapsed time from batching at plant to discharge from truck at site, where applicable
9. Concrete mix data (quantity and type):
   a. Cement
   b. Fine aggregates
   c. Coarse aggregates
   d. Water
   e. Water/cement ratio
   f. Air-entraining admixtures
   g. Water-reducing admixture and/or high-range water-reducing admixture
   h. Other mixtures
10. Weather data:
    a. Air temperatures
    b. Weather
    c. Wind speed
11. Field test data:
    a. Date, time, and place of test
    b. Slump
    c. Air content
    d. Unit weight
    e. Concrete temperature
12. Compressive test data:
    a. Cylinder number
    b. Age of concrete when tested
    c. Date and time of cylinder test
    d. Curing time (field and laboratory)
    e. Compressive strength
    f. Type of break

1.05 SUBMITTALS
   A. Make submittals in accordance with the requirements of the specifications.
   B. Contractor shall submit concrete mix design to the engineer 2 wk prior to placing concrete. Proportion mix designs as defined in ACI 301 Article 3.9. Include the following information for each concrete mix design:
      1. Method used to determine the proposed mix design (ACI 301, Article 3.9)
      2. Gradation of fine and coarse aggregates—ASTM 033
      3. Proportions of all ingredients, including all admixtures added either at the time of batching or at the job site
      4. Water/cement ratio
      5. Slump—ASTM C143
      6. Certification of the chloride content of admixtures

7. Air content:
   a. Of freshly mixed concrete by the pressure method, ASTM C231, or the volumetric method, ASTM C173
   b. Of hardened concrete by microscopical determination, including parameters of the air-void system (ASTM C457)
8. Unit weight of concrete—ASTM C138
9. Strength at 7 and 28 days—ASTM C39
10. Chloride ion content of the concrete

C. Testing agency shall submit to the engineer, concrete producer, contractor, and owner during construction the results of field-quality-control concrete tests. Include the following information:
   1. Weight of concrete—ASTM C138
   2. Slump—ASTM 143
   3. Air content of freshly mixed concrete by the pressure method, ASTM C231 or the volumetric method, ASTM C173
   4. Air content and parameters of the air-void system by microscopical determination—ASTM C457, as when directed by the engineer
   5. Concrete temperature (at placement time)
   6. Air temperature (at placement time)
   7. Strength determined in accordance with ASTM C39

PART 2   PRODUCTS

2.01 AGGREGATES
A. General:
   1. Coarse aggregates for all concrete work exposed to the weather shall be gravel or crushed limestone containing a minimum 2% clay lumps; friable particles and the sum of clay lumps, friable particles and chert shall not exceed 3% in accordance with ASTM C33, Class Designation 5S.
   2. All other concrete aggregates shall meet requirements of ASTM C33 and additional requirements as specified in this section.
   3. Aggregates not meeting the requirements of ASTM C33 may be used only upon written approval from the engineer who will require additional information and data.
B. Coarse Aggregates:
   1. Concrete coarse aggregates shall meet gradation requirements as specified by ASTM C33, Table 2.
   2. Coarse aggregate shall be nominal maximum size as indicated below:
      a. $\frac{3}{8}$ in. for patch cavities $\frac{3}{4}$–$1\frac{1}{2}$ in. deep
      b. $\frac{1}{2}$ in. for patch cavities greater than $1\frac{1}{2}$ in. deep and for overlay work

      Maximum size of aggregate shall be limited to one-third the thickness of the patch or overlay. Also use the largest practical size aggregate to minimize shrink-related concrete cracking.
C. Fine Aggregates: Fine aggregates shall be natural sand conforming to

ASTM C33 with respect to soundness and control of deleterious materials. Gradation shall be as shown for normal weight aggregates in ACI 302.1R, Table 4.2.1.

2.02   CEMENT (ACI 301, ARTICLE 2.1)
   A. Portland cement, Type I, ASTM C150. Use one brand throughout the project.

2.03   WATER (ACI 301, ARTICLE 2.3)
   A. Mixing water for concrete shall meet requirements of ASTM C94.

2.04   ADMIXTURES (ACI 301, ARTICLE 2.2)
   A. Use of any admixtures must be approved in writing by the engineer prior to its use.
   B. Use approved admixtures in strict accordance with manufacturer's recommendations.
   C. Air-entraining admixture—ASTM C260
   D. Prohibited admixtures: Calcium chloride, thiocyanates, or admixtures containing more than 1% chloride ions, by weight of the admixture, are not permitted. Additionally, each admixture shall not contribute more than 5 ppm, by weight, of chloride ions to the total concrete constituents.
   E. High-range water-reducing admixture (superplasticizer):
      1. ASTM C494, Type F or G
      2. Use shall not change the requirements for:
         a. Maximum water/cement ratios
         b. Concrete strength
         c. Air content of the placed concrete
         d. Minimum cement content

2.05   STORAGE OF MATERIALS (ACI 301, ARTICLE 2.5)

2.06   MIX DESIGN
   A. Approved mix design is as follows:
      Low-slump dense concrete mix proportion requirements
      1. Compressive strength (28 days)   5000 psi
      2. Cement content   799 lb/cu yd
      3. Water/cement ratio (by weight)   0.32 maximum
      4. Air content   7 ± 1%
      5. Slump*   1 in. maximum

      *Concrete slump with addition of high-range water reducer (superplasticizer) shall not exceed 6 in.

   B. All concrete shall contain the specified high-range water-reducing admixture (superplasticizer).
   C. The water-soluble chloride ion content of the concrete shall not exceed 0.15% by weight of cement. Perform test for determination of chloride content of concrete in accordance with FHWA-RD-77-85.
   D. The contractor shall submit the mix design data in accordance with 1.05 of this section.
   E. Bonding grout shall be equal parts by weight of cement and sand mixed with sufficient water to form a stiff slurry. Consistency shall be such that it can be applied to the old concrete in a thin, even coat using a stiff

broom or brush without running or puddling. Grout shall be thinned to paint consistency for application to vertical surfaces.

F. For concrete placed and finished by vibratory screeds the slump shall be limited to 4 in.

PART 3   EXECUTION

3.01   PRODUCTION OF CONCRETE

A. Production of concrete shall be in accordance with requirements of ACI 301, Chapter 7, except as otherwise specified herein.

B. Concrete shall be produced by on-site volumetric batching or ready-mixed concrete batched in accordance with requirements of ASTM C685 and ASTM C94.

C. On-site concrete batching in a mixer of at least $\frac{1}{8}$-cu-yd capacity shall be permitted only with approval of the engineer. On-site concrete batching and mixing shall comply with requirements of ACI 301, Section 7.2.

3.02   TRANSPORTATION AND DISCHARGE

A. Concrete transported by truck mixer or agitator shall be completely discharged within 1 hr after water has been added to the cement or cement has been added to the aggregates. During hot weather concreting, the discharge time shall be limited to 45 min.

3.03   PREPARATION (PATCH AND OVERLAY INSTALLATION— FLATWORK)

A. Cavity surfaces shall be clean and dry prior to start of patch or overlay installation. Preparation of surfaces to receive new concrete shall be in accordance with requirements of the project specification.

B. Bonding grout shall be applied to a damp (but not saturated) concrete surface in uniform thickness of $\frac{1}{16}-\frac{1}{8}$ in. over all surfaces to receive patching or overlay. Grout shall not be allowed to dry or dust prior to placement of patch or overlay material. If concrete placement is delayed and the coating dries, the cavity or surface shall not be patched or overlaid until it has been recleaned and prepared as specified in Section 02530. Grout shall not be applied to more area than can be patched or overlaid within $\frac{1}{2}$ hr by available manpower.

3.04   PLACING AND FINISHING

A. Do not place concrete when the temperature of the surrounding patch area is less than 45°F or when mix temperature is above 90°F unless the following conditions are met:

1. Place concrete only when the temperature of the surrounding patch is expected to be above 40°F and rising and expected to be above 45°F for at least 24 hr.

2. When the above conditions are not met, concrete may be placed only if insulation or heating enclosures are provided in accordance with ACI 306, "Recommended Practice for Cold Weather Concreting" and have been approved by the engineer.

B. Temperature of concrete as placed shall not exceed 90°F. Cost for precautionary measures required by engineer shall be borne by contractor.

C. The concrete shall be manipulated and struck off slightly above final grade. The concrete shall then be consolidated and finished to final grade with surface-vibration devices. Consolidation equipment used shall be approved by the engineer.

D. Fresh concrete 3 in. or more in thickness shall be vibrated internally in addition to surface vibration.

E. For overlays the vibrating device shall consist of vibrating screeds meeting the following requirements:

1. Placing and finishing equipment shall not exceed a maximum weight of 6000 lb or 3000 lb per axle.

2. The screed shall be designed to consolidate the concrete to 98% of the unit weight determined in accordance with ASTM C138. A sufficient number of identical vibrators shall be effectively installed such that at least one vibrator is provided for every 5 ft of screed length.

3. The bottom face of screeds shall not be less than 4 in. wide and shall be metal-covered with turned-up or rounded leading edge to minimize tearing of the surface of the plastic concrete.

4. The screed shall be capable of forward and reverse movement under positive control. The screed shall be provided with positive control of vertical position and angle of tilt.

5. The screed shall be capable of vibrating at controlled rate, adjustable to between 3000 and 11,000 vpm.

F. Concrete shall be deposited as close to its final position as possible. All concrete shall be placed in continuous operation and terminated only at bulkheads or a designated control joint.

G. On ramps with greater than 5% slope, all concreting shall begin at the low point and end at the high point. The contractor shall make any necessary adjustment to the slump or equipment to provide a wearing surface without any irregularities or roughness.

H. When a tight uniform concrete surface has been achieved by the screeding and finishing operation, give the surface coarse transverse-scored texture using a broom. Texture shall be as accepted by the engineer from sample panels. Brooming shall not tear out or loosen particles of coarse aggregate.

I. Finishing tolerance—ACI 301, paragraph 11.9: Class B tolerance.

J. Finish all concrete slabs to proper elevations to ensure that all surface moisture will drain freely to floor drains, and that no puddle areas exist. The contractor shall bear the cost of any corrections to provide for positive drainage.

3.05 CURING

Concrete shall be cured according to the following minimum requirements:

1. The surface shall be covered with a single layer of clean, wet burlap and a layer of plastic as soon as the surface will support it without deformation. Keep the wet burlap and plastic in place for 72 hr.

2. The plastic may then be removed and the burlap allowed to dry slowly for an additional 48 hr.

3. Remove the burlap and allow the concrete to air dry.

4. Curing time shall be extended, as the engineer directs, when the curing temperature falls below 50°F.

3.06   FIELD QUALITY CONTROL

A. Three-inch by six-inch cylinders will be fabricated, cured, and tested in accordance with Chapter 16 of ACI 301, except as noted in this specification. Six cylinders will be made for every 10 cu yd of concrete batched. Two cylinders shall be tested at 7 days, two at 28 days, and two held in reserve. Two additional cylinders shall be fabricated for every 20 cu yd of prepared concrete. These shall be field cured in accordance with requirements of ASTM C31 and tested as directed by the engineer.

B. For each set of cylinders made, a slump and air-content test shall also be performed. Slump tests shall be in accordance with ASTM C143. Air-content tests shall be in accordance with ASTM C231. Temperature of the concrete shall also be taken at this time.

C. Patched and overlaid areas shall be sounded by the contractor with a chain drag after 7 days of cure, and any hollowness shall be corrected by removing and replacing unsound areas at no extra cost to the owner.

Appendix A

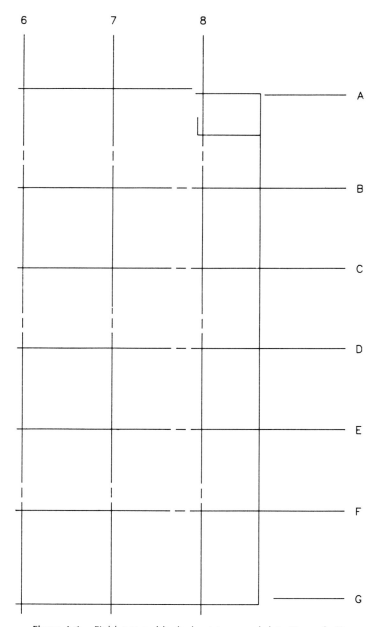

Figure A-1. Field survey blank sheet to record data (Example 1).

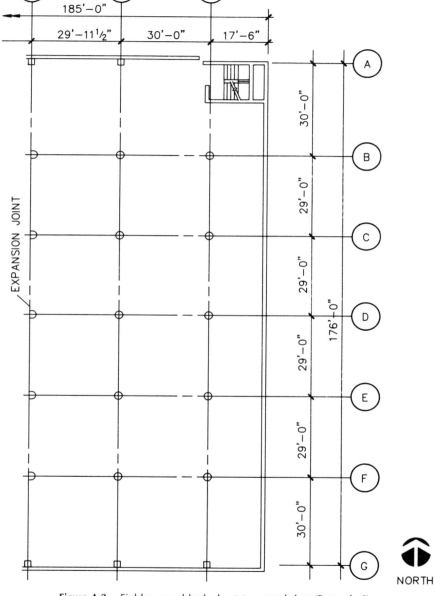

Figure A-2.   Field survey blank sheet to record data (Example 2).

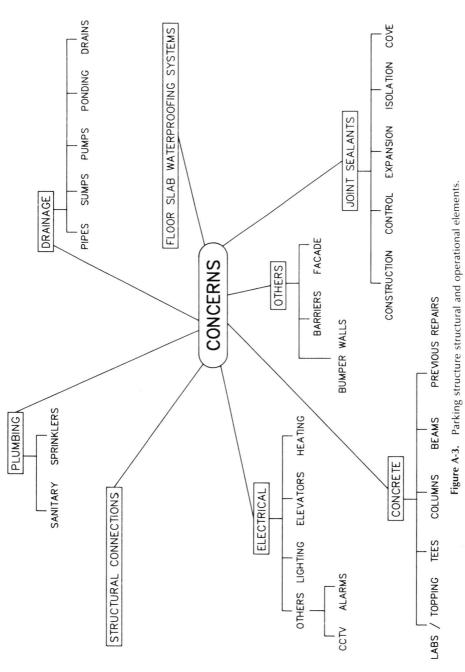

Figure A-3. Parking structure structural and operational elements.

369

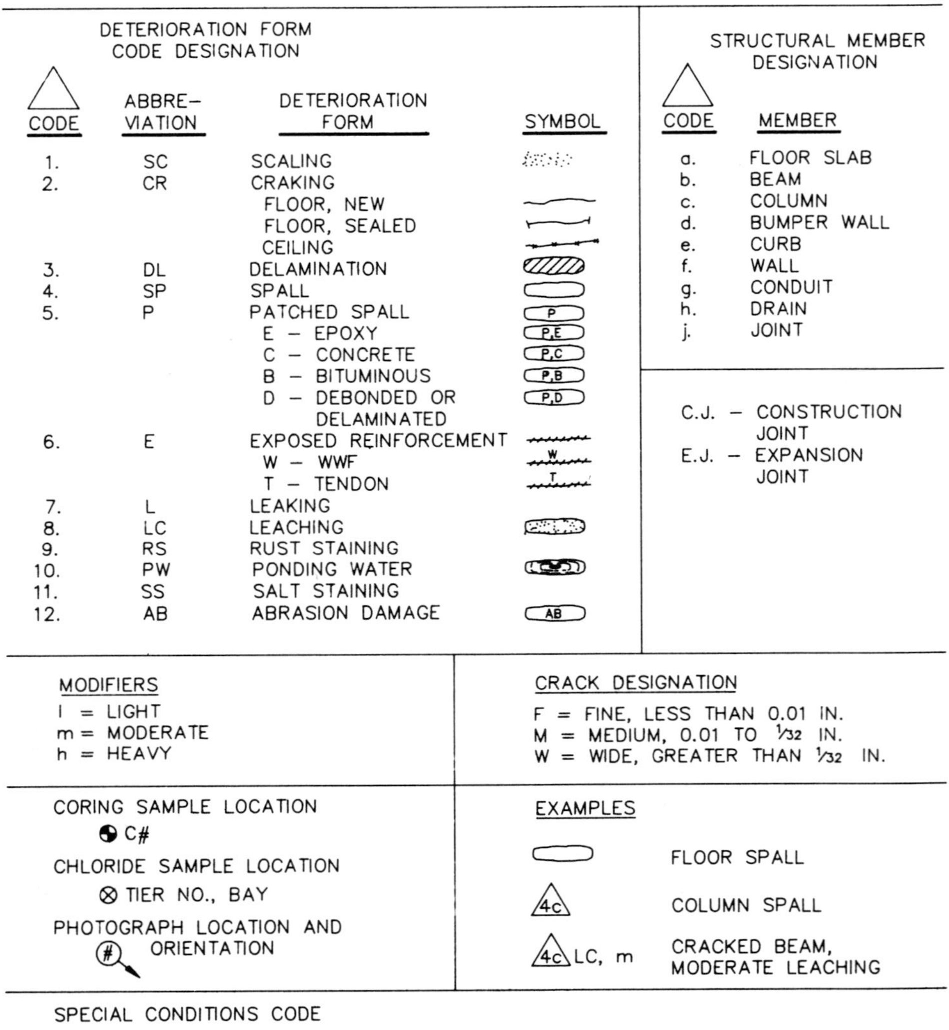

DETERIORATION FORM CODE DESIGNATION

CODE	ABBRE-VIATION	DETERIORATION FORM	SYMBOL
1.	SC	SCALING	
2.	CR	CRAKING	
		FLOOR, NEW	
		FLOOR, SEALED	
		CEILING	
3.	DL	DELAMINATION	
4.	SP	SPALL	
5.	P	PATCHED SPALL	P
		E − EPOXY	P,E
		C − CONCRETE	P,C
		B − BITUMINOUS	P,B
		D − DEBONDED OR DELAMINATED	P,D
6.	E	EXPOSED REINFORCEMENT	
		W − WWF	W
		T − TENDON	T
7.	L	LEAKING	
8.	LC	LEACHING	
9.	RS	RUST STAINING	
10.	PW	PONDING WATER	
11.	SS	SALT STAINING	
12.	AB	ABRASION DAMAGE	AB

STRUCTURAL MEMBER DESIGNATION

CODE	MEMBER
a.	FLOOR SLAB
b.	BEAM
c.	COLUMN
d.	BUMPER WALL
e.	CURB
f.	WALL
g.	CONDUIT
h.	DRAIN
j.	JOINT

C.J. − CONSTRUCTION JOINT

E.J. − EXPANSION JOINT

MODIFIERS

l = LIGHT
m = MODERATE
h = HEAVY

CRACK DESIGNATION

F = FINE, LESS THAN 0.01 IN.
M = MEDIUM, 0.01 TO 1/32 IN.
W = WIDE, GREATER THAN 1/32 IN.

CORING SAMPLE LOCATION

● C#

CHLORIDE SAMPLE LOCATION

⊗ TIER NO., BAY

PHOTOGRAPH LOCATION AND ORIENTATION

EXAMPLES

FLOOR SPALL

COLUMN SPALL

CRACKED BEAM, MODERATE LEACHING

SPECIAL CONDITIONS CODE

Figure A-4.   Field survey legend.

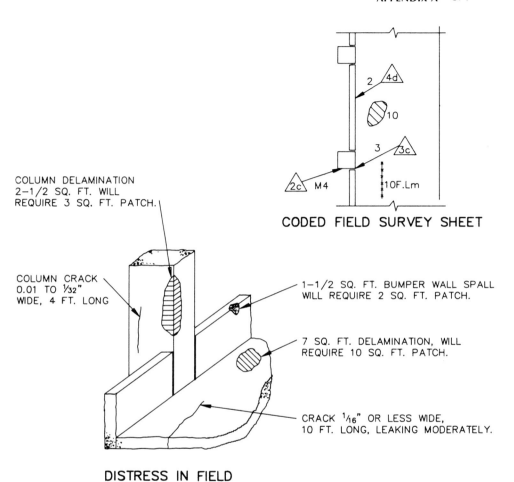

CODED FIELD SURVEY SHEET

COLUMN DELAMINATION
2-1/2 SQ. FT. WILL
REQUIRE 3 SQ. FT. PATCH.

COLUMN CRACK
0.01 TO 1/32"
WIDE, 4 FT. LONG

1-1/2 SQ. FT. BUMPER WALL SPALL
WILL REQUIRE 2 SQ. FT. PATCH.

7 SQ. FT. DELAMINATION, WILL
REQUIRE 10 SQ. FT. PATCH.

CRACK 1/16" OR LESS WIDE,
10 FT. LONG, LEAKING MODERATELY.

DISTRESS IN FIELD

Figure A-5.   Use of field survey legend to record data systematically.

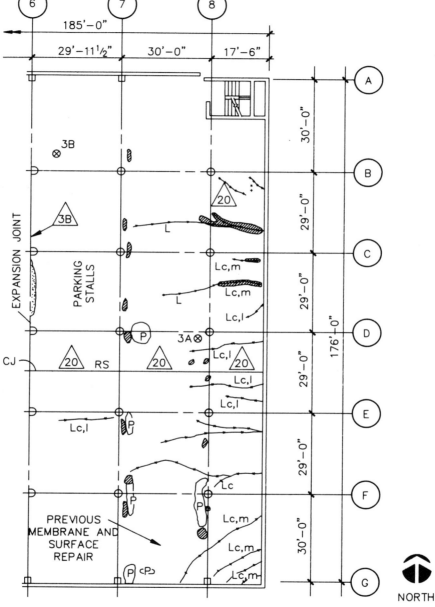

Figure A-6. Sample of data recorded on survey sheets.

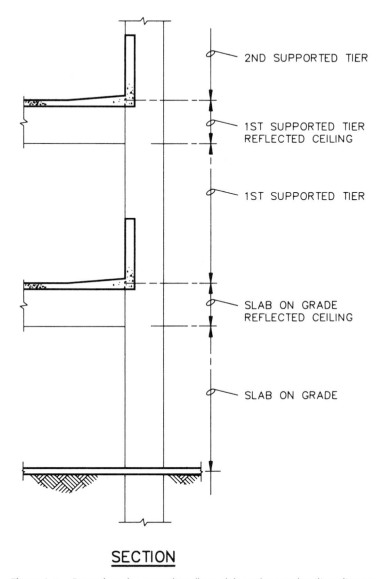

2ND SUPPORTED TIER

1ST SUPPORTED TIER
REFLECTED CEILING

1ST SUPPORTED TIER

SLAB ON GRADE
REFLECTED CEILING

SLAB ON GRADE

SECTION

Figure A-7. Procedure for recording floor slab surface and ceiling distress.

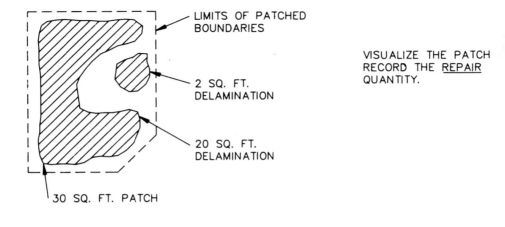

LIMITS OF PATCHED
BOUNDARIES

VISUALIZE THE PATCH
RECORD THE REPAIR
QUANTITY.

2 SQ. FT.
DELAMINATION

20 SQ. FT.
DELAMINATION

30 SQ. FT. PATCH

200 SQ. FT. OF DELAMINATED
CONCRETE FLOOR SLAB CAN
REQUIRE 300 SQ. FT. OF
PATCHING.

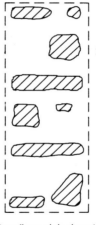

Figure A-8.   Procedure for estimating and recording floor slab deterioration.

Glossary

ABRASION RESISTANCE   Ability to resist being worn away by rubbing and friction.

ABRASIVE NOSING   A strip with a roughened surface cast in, or adhesive applied to, the edge of a stair step to provide traction and minimize slipping.

ACCESS DESIGN   The design of the connection points between the parking facility and the adjacent roadways or streets.

ACTIVE SECURITY   Measures or systems which promote good security by providing a means of alerting the management/employees of the facility to an incident in progress.

ADHESIVES   The group of materials used to join or bond similar or dissimilar materials; for example, in concrete work, the epoxy resins.

ADMIXTURE   Material other than water, aggregate, or cement, used as an ingredient of concrete and added to concrete during mixing to modify the concrete properties.

AGGREGATE   Granular material, such as sand, gravel, or crushed stone, used with cement, water, and admixtures to form concrete.

AIR ENTRAINING   The capability of a material to develop a system of minute bubbles of air in concrete during mixing.

AIR-ENTRAINING AGENT   An admixture for concrete which causes entrained air to be incorporated into the concrete during mixing, usually to increase its workability and frost resistance.

AIR ENTRAINMENT   The inclusion of air in the form of minute bubbles (generally smaller than 1 mm) during the mixing of concrete.

ALKALI-AGGREGATE REACTION   Chemical reaction in mortar or concrete between hydroxyl associated with alkalies (sodium and potassium) from Portland cement or other sources and certain constituents of some aggregates; under certain conditions, deleterious expansion of the concrete or mortar may result.

ALKALI REACTIVITY (OF AGGREGATE)   Susceptibility of aggregate to alkali-aggregate reaction.

ANCHORAGE   In post-tensioning, a device used to anchor a tendon to a concrete member; in pretensioning, a device used to anchor a tendon during hardening of the concrete.

ANGLED   Parking stalls not perpendicular to the driving aisle.

ANTI-PASSBACK   Type of control which prevents several users from sharing one card; it can be either *firm*, which rejects the card outright when it is "out of mode," or *soft*, which opens the gate but prints out an error message for later action by management.

AUDIT TRAIL   Data in the memory of the fee computer or on the journal tape of a cash register that are used for auditing the cash revenues of the system.

AUTOGENOUS HEALING   A natural process of closing and filling of cracks in concrete or mortar when the concrete or mortar is kept damp.

BREAK-OVER EFFECT   The tendency for vehicles to "bottom out" when a change in slope is too abrupt. (*See* TRANSITION SLOPE.)

BUFFER   *See* COMPUTER TERMS at end of Glossary.

CAMBER   Upward deflection intentionally built into a structural element or form to improve appearance or to nullify the deflection of the element under loads, shrinkage, and creep; camber may also be produced by prestressing.

CAPACITY   Used as a short form for either flow capacity or static capacity. The distinction will be determined by the context.

CARD READER   A device which reads coded cards similar to credit cards which, if valid, sends a signal to open the gate. See ANTI-PASSBACK.

CAST-IN-PLACE CONCRETE   Concrete which is deposited in the place where it is required to harden as part of the structure, as opposed to precast concrete.

CCTV   Closed-Circuit Television. A system of television cameras, monitors, and other devices, connected by cable, carrying the signals only within the circuit.

CENTRAL FACILITY COMPUTER (CFC)   A computer that processes an FMS for one facility or a group of facilities.

CHALKING   Disintegration of coatings, such as cement paint, manifested by the presence of loose powder evolved from the paint at or just beneath the surface.

CHARGE-COUPLED DEVICE (CCD)   A solid-state imaging sensor in which the television-scanning function is accomplished by moving the video signal on a silicon chip.

CHECKING   Development of shallow cracks at closely spaced but irregular intervals on the surface of mortar or concrete.

CIRCULAR HELIX   A ramp wound in a tight circle; also called an express helix, it has no parking and provides an extremely fast route up or down.

CLEARANCE BAR   A device set at the posted clearance height to warn drivers whose vehicles are too tall to drive through the parking facility. Due to bouncing and the *breakover* effect, the posted clearance must generally be 2–4 in. greater than the minimum straight vertical clearance at any point.

COMPUTERIZED COUNT CONTROLLER (CCC)   Generally a small microprocessor that replaces the electromechanical CM.

CONCRETE   Mixture of Portland cement, fine aggregate, coarse aggregate, and water, with or without admixtures.

CONTRACT DOCUMENTS   Project drawings and specifications.

CONTROLLER/MONITOR (CM)   A board that indicates the status of various peripherals (gate up, ticket supply low, etc.), the current occupancy (*see* DIFFERENTIAL COUNTER) and the various NON-RESETTABLE COUNTERS.

CORROSION   Disintegration or deterioration of concrete or reinforcement by electrolysis or by chemical attack.

CPU   *See* Computer Terms at end of Glossary.

CRACK COMPARATOR   A viewer capable of magnifying surface cracks, having a graduated linear scale in the viewer which assists in crack width measurements.

CRAZE CRACKS   Fine, random cracks or fissures caused by shrinkage which may appear in a surface of plaster, cement paste, mortar, or concrete.

CRAZING   The development of craze cracks; the pattern of craze cracks existing in a surface. (*See also* CHECKING.)

CREDIT-CARD OPTION   On a machine-readable system, either card or ticket, a system to track use and issue a monthly bill for parking charges for the prior period.

CROSSOVER   A route used to go from an up thread of a helix to a down thread or vice versa.

DAILY FEE PARKER   A parker who pays the appropriate parking fee each time he/she visits the facility. In most cases, a new fee is charged with each visit, even with multiple visits on the same day; also called *transient parker;* daily fee parkers can be either short- or long-term parkers.

DATA MANAGEMENT or DATA-BASE SOFTWARE   *See* Computer Terms at end of Glossary.

DECLINATING TICKETS   A prepaid ticket that tracks usage and associated fees until the prepaid amount is used up; it eliminates the need for cashiers for these users and can replace the card readers.

DEFLECTION   A variation in position or shape of a structure or element due to effects of loads or volume change, usually measured as a linear deviation from an established plane.

DEGRADATION   The deterioration or damage to the coding on a card or ticket that naturally occurs in everyday use.

DELAMINATION   A separation along a plane parallel to a surface, as in the separation of a coating from a substrate or the layers of a coating from each other, or, in the case of a concrete slab, a horizontal splitting, cracking, or separation of a slab in a plane roughly parallel to, and generally near, the upper surface. It is found most frequently in bridge decks and caused by corrosion of reinforcing steel or freezing and thawing; similar to spalling, scaling, or peeling, except that delamination affects large areas and can often only be detected by tapping.

DESIGN VEHICLE   The 85th-percentile vehicle among the mix of expected vehicle sizes.

DETERIORATION   Disintegration or chemical decomposition of a material during test or service exposure. (*See also* DISINTEGRATION.)

DIAGONAL CRACK   An inclined crack caused by shear stress, usually at about 45 degrees to the neutral axis of a concrete member, or a crack in a slab, not parallel to the lateral or longitudinal dimensions.

DIFFERENTIAL COUNTERS   A resettable counter that is increased one digit with each entry and decreased with each exit to show current occupancy of the facility.

DISCOLORATION   Change of color from that which is normal or desired.

DISINTEGRATION   Deterioration into small fragments or particles.

DISKS   *See* Computer Terms at end of Glossary.

DOWN LOAD   *See* Computer Terms at end of Glossary.

DROP SAFE   A safe with a special slot that allows a cashier to deposit revenues without having the combination/key to open the safe.

DRY-MIX SHOTCRETE   Pneumatically conveyed shotcrete in which most of the mixing water is added at the nozzle.

DRYPACKING   Placing of zero slump or near-zero slump concrete, mortar, or grout by forcing it into a confined space.

DURABILITY   The ability of concrete to resist weathering action, chemical attack, abrasion, and other conditions of service.

DURESS ALARM   An alarm, usually silent, that a cashier can depress to summon assistance in the event of a problem at an exit lane.

DUSTING   The development of a powdered material at the surface of hardened concrete.

EFFICIENCY   The GROSS PARKING AREA divided by the STATIC CAPACITY.

EFFLORESCENCE   A deposit of salts, usually white in color, formed on a surface, the substance having emerged in solution from within concrete or masonry and deposited by evaporation.

EMERGENCY TELEPHONE   A special telephone, without a dial, that connects directly to a manned security station or police force. Although emergency telephones are generally "hard wired," microwave-signal transmission can also be used to eliminate the need to run wires to every unit.

END-BAY PARKING   Stalls placed along the TURNING BAY, using that area for access into and out of the stalls.

ENTRAINED AIR   Microscopic air bubbles intentionally incorporated in concrete during mixing, typically between 10 and 1000 micrometers in diameter and spherical or nearly so.

ENTRAPPED AIR   Air voids in concrete which are not purposely entrained and which are significantly larger and less useful than those of entrained air, 1 mm or larger in size.

EPOXY CONCRETE   A mixture of epoxy resin, catalyst, fine aggregate, and coarse aggregate. (*See also* EPOXY MORTAR, EPOXY RESIN, and POLYMER CONCRETE.)

EPOXY MORTAR   A mixture of epoxy resin, catalyst, and fine aggregate. (*See also* EPOXY RESIN.)

EPOXY RESIN   A class of organic chemical-bonding systems used in the preparation of special coatings or adhesives for concrete, or as binders in epoxy-resin mortars and concretes.

EVALUATION   Determining the condition, degree of damage or deterioration, or serviceability and, when appropriate, indicating the need for repair, maintenance, or rehabilitation. (See also REPAIR, MAINTENANCE, and REHABILITATION.)

EXCEPTION TRANSACTION   A transaction that does not follow the normal procedure. It includes lost tickets, validation for parking without an account number, or other transactions where the cashier has circumvented procedures.

EXPOSED CONSTRUCTION   Portions of a building exposed to public view.

EXPOSED TO PUBLIC VIEW   Situated so that it can be seen from eye level from a public location after completion of the building. A public location is accessible to persons not responsible for operation or maintenance of the building.

EXPRESS RAMP   A straight or gently curved ramp between two tiers.

FACILITY MANAGEMENT SYSTEM (FMS)   A *software* package that analyzes and reports activity and runs on a computer of appropriate size and capability. With an FMS, all information about transactions and card uses is transmitted periodically to a central host computer.

FEATHER EDGE   Edge of a concrete or mortar placement such as a patch or topping that is beveled at an acute angle.

FEE COMPUTER   An electronic cash register which calculates the fee due from an input of the ''in'' time and date; the internal clock provides the ''out'' time and date.

FEE INDICATOR   A device designed to display the fee as entered into the fee computer to the driver of the exiting vehicle.

FLOOR   The surface at the bottom of any volume of space, such as a room or an ocean. (See also TIER.) Because all parking and drive areas must be designed for proper drainage (Chapter 6), a floor area in a parking facility will be either sloped only as required for drainage, in which case it is nominally *level*, or sloped for floor-to-floor circulation, in which case it is nominally *sloped*.

FLOOR-TO-FLOOR HEIGHT   The vertical dimension from the top of one floor surface to the top of the floor surface on the tier above or below; varies from 9–12 ft, but 10 ft is most common.

FLOW CAPACITY   The ability to accommodate expected traffic VOLUMES without excessive congestion and delay.

FOOTPRINT   The outline of the structure viewed from directly above, such as would be seen in an aerial photograph.

FULL SIGN   An illuminated sign at each entry lane that is automatically turned on when the occupancy of the facility reaches the preset full level.

FUNCTIONAL DESIGN   The consideration of pedestrian and vehicular flow through a parking facility.

GENERATOR   Land use which creates a demand for parking.

GEOMETRICS   The dimensions of various components of a parking facility, including aisles, stalls, and MODULES.

GRAPHICS   The way in which a message is presented on a SIGN.

GROOVE JOINT   A joint created by forming a groove in the surface of a pavement, floor slab, or wall to control random cracking.

GROSS FLOOR AREA (GFA)   The sum of the floor area on each tier using out-to-out dimensions.

GROSS PARKING AREA (GPA)   The sum of the floor area on each tier, calculated from inside to inside of exterior walls, less enclosed areas devoted to auxiliary uses, such as stair towers, elevator shafts and lobbies, and storage and equipment rooms. Any other uses, such as retail or office space, should be excluded from GPA.

HAIRLINE CRACKING   Small cracks of random pattern in an exposed-concrete surface.

HANDSHAKING   See Computer Terms at end of Glossary.

HELIX   A coiled shape or a spiral. In a parking structure, a series of PARKING BAYS and/or ramps which provide floor-to-floor circulation.

HONEYCOMBING   Voids left in concrete due to failure of the mortar to effectively fill the spaces among coarse aggregate particles.

INBOARD   Locating a stair/elevator tower in a row of parking stalls, displacing those stalls, or tucking it into an unused corner. Towers placed *outboard* are outside the perimeter of the parking area.

INTERFACES   See Computer Terms at end of Glossary.

JOINT SEALANT   Compressible material used to exclude water and solid foreign material from joints.

JOURNAL TAPE   A paper tape in the fee computer that records each step of each transaction.

LAITANCE   A layer of weak and nondurable materials containing cement and fines from aggregates brought by bleeding water to the top of overly wet concrete. The amount is generally increased by overworking or overmanipulating concrete at the surface by improper finishing or by job traffic.

LATEX   A water emulsion of a synthetic rubber or plastic obtained by polymerization and used in coatings and adhesives.

LEVEL OF SERVICE   A qualitative measure of the conditions in a particular traffic carrying component, applied to geometrics, other design parameters, flow capacity, and queueing at entry/exit points herein.

LICENSE PLATE INVENTORY (LPI)   A listing of the license plates of all vehicles parked in a facility at a certain time, usually conducted at midnight or in the very early morning hours by entering the plate numbers of every car into a hand-held calculator. The numbers are later off-loaded into a central computer. The cashier enters the license plate number of the exiting vehicle and the fee computer determines if the ''in'' date on the ticket is correct.

LOAD, DEAD   Dead weight supported by a member, as defined by a building code.

LOAD, LIVE   Live load specified by a building code.

LOAD, SELF   Weight of the member itself.

LONG SPAN   A structural system which spans the full width of the MODULE.

LONG-TERM PARKER   A parker who stays in a facility more than 3 hr; may be either a DAILY-FEE or MONTHLY PARKER.

LOOP DETECTOR   Loops of wire placed in floor slabs connected to a device that magnetically detects vehicle presence.

MACHINE-READ TICKETS   Information on the ticket is coded by the ticket dispenser to be read directly by the fee computer instead of the cashier entering the ''in'' time.

MAINTENANCE   Taking periodic actions that will either prevent or delay damage or deterioration or both. (*See also* REPAIR.)

MAP CRACKING   *See* CRAZING.

MATCHLINES   Markings on floor plans depicting where one tier stops and another begins. When owners and other construction laymen are to view the plans, it helps if the matchlines are located near the center of the sloping bay or ramp. However, on drawings to be used in construction, matchlines should be placed at natural breaks in construction, such as the end of a concrete placement.

MEAN INHIBITING PERIOD   The average spacing of vehicles at capacity flow in time units.

MEMORY   *See* Computer Terms at end of Glossary.

METER   A device that accepts a parking fee and displays how much purchased time remains. One meter is placed at each stall. Originally mechanical, electronic meters are now available that have microprocessors to calculate the remaining fee and keep track of how much money is in the cash box.

METER BOX   A device that accepts parking fees for a number of spaces and has a means of letting the enforcement official know which stalls are paid and/or not paid. Originally, was as simple as a box with a slot for each stall; now, electronic and computerized online meters are available.

MICROCRACKS   Microscopic cracks within concrete.

MICROPROCESSOR   *See* Computer Terms at end of Glossary.

MINICOMPUTER   *See* Computer Terms at end of Glossary.

MODEM   *See* Computer Terms at end of Glossary.

MODULE   The wall-to-wall dimension of a PARKING BAY. The combination of one or two rows of parked vehicles and the driving aisle providing access thereto. A module may be *single loaded*, with parking on only one side of the aisle, or *double loaded*, with parking on both sides.

MONOMER   An organic liquid, of relatively low molecular weight, that creates a solid polymer by reacting with itself, with other compounds of low molecular weight, or with both.

MONTHLY PARKER   A parker who pays, usually in advance, for a month or more of parking. A monthly parker is usually allowed to come or go as often as he/she pleases within the paid period.

MOTION DETECTOR   A device that detects the presence of a person in a normally ''quiet'' area.

MULTITASKING   *See* Computer Terms at end of Glossary.

MULTIUSER   *See* Computer Terms at end of Glossary.

NESTING   A sequence of gates which must be passed through in the proper order for continued authorization.

NETWORK   *See* Computer Terms at end of Glossary.

NON-RESETTABLE COUNTER   A digital or electronic counter that increases one digit with each transaction. When it reaches the maximum number of the counter (usually 99,999), it starts back over at zero (similar to the odometer on a car which records total miles driven).

OFF-LINE   Opposite of ON-LINE; off-line systems can still be computerized by other means of data communication.

ON-LINE   Generally, a device which is hard-wired to a CPU of some sort.

OPERATING SYSTEM   *See* Computer Terms at end of Glossary.

OVERBOOKING   The practice of either selling 10–20% more monthly passes than the number of spaces allocated, or allowing transient parkers to use spaces allocated to monthly parkers when the latter are absent.

OVERHEAD CLEARANCE   The straight vertical clearance encountered by pedestrians and vehicles moving through the structure.

OVERLAY   A layer of concrete or mortar, seldom thinner than 1 in., placed on and usually bonded to the worn or cracked surface of a concrete slab to either restore or improve the function of the previous surface.

PACHOMETER   Instrument for nondestructively locating and estimating concrete cover and/or diameter of embedded reinforcement.

PAN AND TILT   A device which allows a camera to rotate vertically and horizontally to change the view of an area.

PANIC BUTTON   A button placed on an intercom unit that sounds an alarm at the master intercom station.

PARATRANSIT   Enlarged vans or small buses designed to transport a number of individuals for dial-a-ride or shuttle purposes.

PARKING ACCESS AND REVENUE CONTROL (PARC)   Any device or combination of devices that control access, utilization and/or revenues for a parking space.

PARKING BAY   Rows of parking with an aisle in between. A parking bay may be *single-loaded* (parking on one side only) or *double-loaded* (on both sides). (*See* Figure 2-1.)

PARKING GARAGE   A parking facility that does not meet code requirements for openness and therefore must have mechanical ventilation.

PARKING STRUCTURE   A multistory parking facility that meets building code requirements for natural ventilation. Also may be called a parking deck or parking ramp.

PASSIVE SECURITY   Physical features of a parking facility which promote good security, usually by providing increased visibility.

PATTERN CRACKING   Fine openings on concrete surfaces in the form of a pattern, resulting from a decrease in volume of the material near the surface, or increase in volume of the material below the surface, or both.

PAY-ON-FOOT   A system where the patron pays "on foot" upon returning to the facility but before retrieving his car. An exit ticket or token is issued that is surrendered at the gate. There are two types of pay-on-foot systems: *manned* and *automated.* In the manned system, cashier(s) using the same equipment as at a conventional exit lane are stationed at strategic locations, such as the grade level elevator lobby. In the automated system, a machine accepts the patron ticket, and displays and accepts the fee.

PEAK HOUR   The 60 consecutive minutes that together have the highest total volume of traffic.

PEAK-HOUR FACTOR (PHF)   The ratio of the total hourly volume to the maximum 15-min RATE OF FLOW within the hour.

PEELING   A process in which thin flakes of mortar are broken away from a concrete surface, such as by deterioration or by adherence of surface mortar to forms as they are removed.

PERIPHERAL   *See* Computer Terms at end of Glossary.

PERSONAL COMPUTER   *See* Computer Terms at end of Glossary.

PETROGRAPHY   Chemical and microscopic examination mainly in laboratory of concrete samples.

PITTING   Development of relatively small cavities in a surface, due to phenomena such as corrosion or cavitation, or, in concrete, localized disintegration. (*See also* POPOUT.)

PLASTIC CRACKING   Cracking that occurs in the surface of fresh concrete soon after it is placed and while it is still plastic.

PLASTIC SHRINKAGE CRACKING   *See* PLASTIC CRACKING.

POLYETHYLENE   A thermoplastic, high molecular-weight organic compound used in formulating protective coatings or, in sheet form, as a protective cover for concrete surfaces during the curing period, or to provide a temporary enclosure for construction operations.

POLYMER   The product of polymerization; more commonly, a rubber or resin consisting of large molecules formed by polymerization.

POLYMER-CEMENT CONCRETE   A mixture of water, hydraulic cement, aggregate, and a monomer or polymer, polymerized in place when a monomer is used.

POLYMER CONCRETE   Concrete in which an organic polymer serves as the binder; also known as resin concrete. Sometimes erroneously employed to designate hydraulic cement mortars or concretes in which part or all of the mixing water is replaced by an aqueous dispersion of a thermoplastic copolymer.

POLYMERIZATION   Reaction in which two or more molecules of the same substance combine to form a compound containing the same elements in the same proportions, but of higher molecular weight, from which the original substance can be generated, in some cases only with extreme difficulty.

POLYURETHANE   Reaction product of an isocyanate with any of a wide variety of other compounds containing an active hydrogen group, used to formulate tough, abrasion-resistant coatings.

POPOUT   The breaking away of small portions of concrete surface due to internal pressure which leaves a shallow, typically conical, depression.

POT LIFE   Time interval after preparation during which a liquid or plastic mixture is usable.

POULTICE   A smooth paste usually made by mixing some essentially inert fine powder with the solvent or solution to be used and applying it to the surface for cleaning concrete stains.

PRECAST CONCRETE   Concrete cast elsewhere than in its final position.

PRESTRESSED CONCRETE   Concrete in which forces of such magnitude and distribution are applied, that the tensile stresses resulting from the service loads are counteracted to the desired degree.

   Pretensioned concrete is prestressed concrete in which the method of prestressing is to tension the tendons before the concrete hardens.

   Post-Tensioned concrete is prestressed concrete in which the method of prestressing is to tension the tendons after the concrete hardens.

PROTOCOL   *See* Computer Terms at end of Glossary.

PVC   Polyvinyl chloride; a stiff, strong plastic that is commonly used for pipe and conduit.

QUEUE   The line of waiting vehicles in the reservoir.

RATE OF FLOW   The equivalent hourly rate at which vehicles pass a given point, determined by observing volumes in a time interval less than 1 hr, such as 15 min.

REACTIVE AGGREGATE   Aggregate containing substances capable of reacting chemically with the products of solution or hydration of the Portland cement in concrete or mortar under ordinary conditions of exposure, resulting, in some cases, in harmful expansion, cracking, or staining.

REDIAL   A software subprogram that overrides the antipassback control or one-card use.

REHABILITATION   Making major repairs or modifications, which if not performed, could result in loss of serviceability; during rehabilitation, a project is normally out of service. (*See also* REPAIR.)

REINFORCEMENT   Bars (smooth or deformed), wires, fibers, strands, and other elements which are embedded in concrete in such a manner that reinforcement and concrete act together in resisting forces.

   Conventional reinforcement is non-prestressed smooth or deformed bar or wire reinforcement with yield strengths in the 40,000–75,000 psi range.

   Prestressed reinforcement is steel bars, wires, or strands with ultimate strengths in the 250,000–270,000 psi range, strong enough to permit effective pre- or post-tensioning.

REPAIR   Restoring damaged or deteriorated elements to serviceable condition; repair work can normally be performed while a structure remains in service. (*See also* REHABILITATION and MAINTENANCE.)

RESERVOIR   The space for waiting behind the vehicle being serviced. Without a vehicle in the reservoir, there is "dead time" when no vehicle is being serviced.

RESIN   A natural or synthetic, solid or semisolid organic material of indefinite and often high molecular weight having a tendency to flow under stress; usually has a softening or melting range and usually fractures conchoidally.

RESIN MORTAR (OR CONCRETE)   See POLYMER CONCRETE.

RESOLUTION   In CCTV systems, the detail and clarity of the picture.

RESTRAINT (OF CONCRETE)   Restriction of free movement of fresh or hardened concrete following completion of placement in formwork or molds or within an otherwise confined space; restraint can be internal or external and may act in one or more directions.

RISE   The dimension of the vertical component of a sloping element.

RUBBER-STUD FLOORING   A flooring system designed for commercial applications that has raised "studs" about 1 in. in diameter closely spaced throughout the pattern.

RUN   The dimension of the horizontal component of a sloping element.

SANDBLASTING   A system of cutting or abrading a surface such as concrete by a stream of sand ejected from a nozzle at high speed by compressed air; often used for cleanup of horizontal construction joints or for exposure of aggregate in architectural concrete.

SCALING   Local flaking or peeling away from the near-surface portion of hardened concrete or mortar. (See also PEELING and SPALLING.) (Note: Light scaling of concrete does not expose coarse aggregate; medium scaling involves loss of surface mortar of 5–10 mm in depth and exposure of coarse aggregate; severe scaling involves loss of surface mortar of 5–10 mm in depth with some loss of mortar surrounding aggregate particles 10–20 mm in depth; very severe scaling involves loss of coarse aggregate particles as well as mortar generally to a depth greater than 20 mm.)

SECURITY AUDIT   The process of assessing the risk of incidents in the parking facility as a whole and in specific locations within the facility.

SERVICE RATE (μ)   The maximum number of vehicles that can be processed through a lane in an hour under a constant heavy load; it is therefore the capacity.

SHEAR WALL   A wall reinforced to carry the lateral loads placed on a building by wind, earthquakes, and other horizontal forces.

SHORT SPAN   A structural system which does not span the full module, resulting in columns between parked vehicles.

SHORT-TERM PARKER   A parker who stays in a facility 3 hr or less.

SHOTCRETE   Mortar or concrete pneumatically projected at high velocity onto a surface; also known as air-blown mortar; also pneumatically applied mortar or concrete, sprayed mortar, and gunned concrete. (See also DRY-MIX SHOTCRETE and WET-MIX SHOTCRETE.)

SHRINKAGE   Volume decrease caused by drying and chemical changes; a function of time, but not temperature or of stress due to external load.

SHRINKAGE CRACK   Crack due to restraint of shrinkage.

SHRINKAGE CRACKING   Cracking of a structure or member due to failure in tension caused by external or internal restraints as reduction in moisture content develops, as carbonation occurs, or both.

SIGN, SIGNAGE   The system of signs providing directions, warnings, and commands to the user.

SOFTWARE   *See* Computer Terms at end of Glossary.

SOLID STATE   *See* Computer Terms at end of Glossary.

SPALL   A fragment, usually in the shape of a flake, detached from a larger mass by a blow, action of weather, pressure, or expansion within the larger mass; a small spall involves a roughly circular depression not greater than 20 mm in depth or 150 mm in any dimension; a large spall may be roughly circular or oval or, in some cases elongated, more than 20 mm in depth and 150 mm in greatest dimension.

SPALLING   The development of spalls.

SPEED RAMP   Connects two PARKING BAYS with a grade differential of 2–5 ft.

STATIC CAPACITY   The number of parking spaces in a parking facility.

STIRRUP   Reinforcement used to resist shear and torsion stresses in a beam, typically bars, wires, or welded wire fabric (smooth or deformed).

STRUCTURAL STEEL   Rolled-steel structural shapes, plates, and assemblies, as opposed to steel reinforcement.

SWITCHER   In CCTV systems, a device that controls which picture is shown on the monitor at any particular time. Switchers can sequence from one camera to another automatically with manual override when the observer wishes to continue to view the scene from one specific camera.

TENDON   A steel element such as a wire, cable, bar, rod, strand, or group of such elements used to impart prestress to concrete when the element is tensioned.

THERMOSETTING   Becoming rigid by chemical reaction and not remeltable.

TICKET DISPENSERS   Located at entry lanes to issue a ticket stamped or coded with date and time. In first-generation equipment, the ticket is imprinted with a *rate ring* which rotates like a clock with the time of issue. An *out clock* (much like an old-fashioned employee time clock) at the exit-lane cashier station punches the ring in the appropriate fee range, indicating the fee to be charged. In the second-generation system, the ticket is merely stamped with the date and time.

TIDAL FLOW CAPACITY   The ability to accommodate traffic flow that is almost all inbound or outbound in a particular hour.

TIE   Loop of reinforcing bar or wire enclosing the longitudinal reinforcement in a column.

TIER   One story. A tier in a parking facility is defined by the way it is depicted on the construction drawings; because of the sloping PARKING BAYS and ramps, it is sometimes difficult to decide where one tier stops and another begins. The most important thing is to ensure that all floor areas are shown once and only once on the plan views.

TOKEN   *See* VALIDATION.

TRANSIENT   A parker who comes in occasionally, not every day.

TRANSITION SLOPE   An area which softens the change in slope where two floor elements are joined. Minimizes BREAK-OVER EFFECT.

TRANSVERSE CRACKS   Cracks that develop at right angles to the long direction of a member.

TURNING BAY   The space used to turn from one PARKING BAY to another parking bay.

TURNING RADII   The dimension from the center to the edge of the circle traced by the outside front wheel of a design vehicle while making the tightest comfortable turn at 10 mph.

TURNOVER   The average number of times a parking space is used by different vehicles over a specified period, calculated by dividing the number of different parkers in a facility over the period by static capacity.

TURNOVER CAPACITY   The ability to accommodate traffic flow that mixes both arriving and departing vehicles in the same hour.

TWO-WAY INTERCOM   A unit which allows someone at any one of the substations to talk to and listen to the operator at the master station. Communication between substations, however, is not possible. Two-way intercoms are hard wired to each other.

TWO-WAY RADIOS   A unit which allows communication between substations and the master unit, using radio-frequency waves for transmission. Usually, all communications are heard on all units.

VALIDATIONS   A system of validating the ticket of a daily fee parker for a reduced charge or a period of free parking. Validations come in many forms: stickers, credit card-like imprints, ink stamps, or an authorized signature. *Tokens* are a form of validation but eliminate the need for tickets and a cashier. Control of tokens, however, generally is much more difficult because there usually is no audit trail. In some situations, merchants or other users purchase books or rolls of validation stamps or tokens, sometimes at a discount to the face value, and distribute them free to customers. Validations by ink stamp or imprint can be tracked by a ''merchant number'' and bills issued monthly. In other cases, such as hospitals, the subsidy for the validation is not directly accounted for in the revenue stream, but tracking by merchant numbers can help management identify fraud or misuse of the system.

VEHICULAR CLEARANCE   The clearance encountered by vehicles driving through the structure.

VOICE-ACTIVATED INTERCOM   An intercom which filters out background noise, but carries conversations and unusual noises to the master unit without depression of a button or other device at the substation. Generally, the voice-activation capability at the master unit can be turned off, which reduces the effectiveness of the system.

VOLUME   The total number of vehicles passing a point in a certain period; unless otherwise noted, volumes discussed herein are for a full hour.

VOLUME CHANGE   In this context, primarily a change in horizontal dimension due to elastic shortening, drying shrinkage, creep, and temperature change.

WET-MIX SHOTCRETE   Shotcrete in which all ingredients, including mixing water, are mixed before introduction into the delivery hose; it may be pneumatically conveyed or moved by displacement.

ZOOM   A device which automatically refocuses a camera lens for the desired distance and detail.

COMPUTER TERMS

BUFFER   A device that holds a task until the computer can process it.

CPU   Central Processing Unit, which is the brain of any computer.

DATA MANAGEMENT or DATA-BASE SOFTWARE   Handles the organization, storage, and retrieval of data. Allows the CPU to search for certain data events and patterns.

DISKS   The permanent storage system for the operating system, software, and data for a personal computer. May be either *floppy*, which are inserted and removed from the computer as needed or *hard*, which are permanently fixed in the computer.

DOWN-LOAD   To send downstream from the CPU to the peripheral.

HANDSHAKING   Coordination of communication between computer and device. Some handshaking is part of the hardware, such as the timing and speed of transmission.

INTEGRATION   Combining peripherals and/or computer systems from one or more manufacturers into a complete, properly functioning system.

INTERFACES   The physical connection and electronic circuits that connect the computer to a peripheral. Several different types of devices connect the CPU to the peripherals: a *bus* (a line that moves data), a *board* (onto which chips are mounted with plug-in connections), or a *port* (which is the computer equivalent of an electrical outlet).

MEMORY   The storage/retrieval system for the operating system, software, and data currently in use on a computer. Memory in microprocessors is provided either in *ROM* or *RAM*; ROM stands for Read Only Memory and is permanently electrically fixed on the *chip*; RAM (Random Access Memory) is temporary and can be modified or overwritten during use.

MICROPROCESSOR   Essentially a computer each on its own *silicon chip*. A computer such as the popular personal computer that uses a microprocessor for its CPU is generically called a microcomputer. (Larger, more powerful computers have labels such as mini- and mainframe.) A microprocessor can also be designed for just one task.

MINICOMPUTER   A computer class that is larger and faster than a personal computer but smaller than a *mainframe* which, in turn is smaller than a *supercomputer*. Most minicomputers are MULTIUSER and MULTITASKING.

MODEM   Changes the computer's language into audible frequencies so that it can be sent over telephone lines.

MULTITASKING   Capable of processing different tasks or programs at the same time. Only the most recent and most powerful microprocessors can perform even limited multitasking.

MULTIUSER   A single computer that can run tasks for several users, each at a different terminal.

NETWORK   Interconnection of several computers without multiuser capabilities so that data processed by one computer may be used by another.

OPERATING SYSTEM   Set of software that controls the reading, writing, and placing of data in the memory on the disks. The most popular operating system for personal computers compatible with the IBM PC is known as MS DOS, and various versions have been developed similar to the Intel family of microprocessors.

PERIPHERAL   Any device connected to the CPU and memory, including the keyboard, video display, and printer. In PARC systems, peripherals also include gates and ticket dispensers.

PERSONAL COMPUTER (PC)   A version of the microcomputer designed to sit on a desk and process everyday assignments for one user.

PROTOCOL   A set of control characters that defines what information is coming next in a stream of data.

SOFTWARE   Any set of instructions that tells the CPU what to do.

SOLID STATE   A device using transistors, which are electronic semi-conductor devices that control current flow.

Select Bibliography

PART A   STANDARDS

Documents of the various standards-producing organizations are listed with their serial designation, including year of adoption or revision. The documents listed were the most current at the time this book was written. Since some of these documents are revised frequently, generally in minor detail only, you should check directly with the sponsoring group if you wish to refer to the latest revision.

American Association of State Highway and Transportation Officials

T259-80	Standard Method of Testing for Resistance of Concrete to Chloride Ion Penetration
T260-82	Standard Method of Sampling and Testing for Total Chloride Ion in Concrete and Concrete Raw Materials

American Concrete Institute

116R-85	Cement and Concrete Terminology (reaffirmed 1982)
201.1R-68	Guide for Making a Condition Survey of Concrete in Service (reaffirmed 1979)
201.2R-77	Guide to Durable Concrete (reaffirmed 1982)
209R-82	Prediction of Creep, Shrinkage, and Temperature Effects in Concrete Structures
211.1-81	Standard Practice for Selecting Proportions for Normal, Heavyweight, and Mass Concrete
212.2R-81	Guide for Use of Admixtures in Concrete (revised 1983)
214-77	Recommended Practice for Evaluation of Strength Test Results of Concrete
222R-85	Corrosion of Metals in Concrete
224R-80	Control of Cracking in Concrete Structures
224.1R-84	Causes, Evaluation, and Repair of Cracks in Concrete Structures
302.1R-80	Guide for Concrete Floor and Slab Construction
303R-74	Guide to Cast-in-Place Architectural Concrete Practice (revised 1982)

304-73(83)	Recommended Practice for Measuring, Mixing, Transporting, and Placing Concrete (reaffirmed 1983)
304.2R-71	Placing Concrete by Pumping Methods
305R-77	Hot Weather Concreting
306.1-87	Standard Specification for Cold Weather Concreting
306R-78	Cold Weather Concreting
308-81	Standard Practice for Curing Concrete
309-72	Recommended Practice for Consolidation of Concrete (revised 1982)
311.4R-80	Guide for Concrete Inspection
315-80	Details and Detailing of Concrete Reinforcement
316-74	Recommended Practice for Construction of Concrete Pavements and Concrete Bases
318-83	Building Code Requirements for Reinforced Concrete
318R-83	Commentary for Building Code Requirements for Reinforced Concrete
345-82	Standard Practice for Concrete Highway Bridge Deck Construction
347-78	Recommended Practice for Concrete Formwork
352R-76	Recommendations for Design of Beam-Column Joints in Monolithic Reinforced Concrete Structures (reaffirmed 1981)
362R-85	State-of-the-Art Report on Parking Structures
426R-74	Shear Strength of Reinforced-Concrete Members (reaffirmed 1980)
503R-80	Use of Epoxy Compounds with Concrete
503.1-79	Standard Specification for Bonding Hardened Concrete, Steel, Wood, Brick, and Other Materials to Hardened Concrete with Multi-Component Epoxy Adhesive
504R-77	Guide to Joint Sealants for Concrete Structures
515-1R-79	Guide to the Use of Waterproofing, Dampproofing, Protective, and Decorative Barrier Systems for Concrete
546.1R-80	Guide for Repair of Concrete Bridge Superstructures
548R-77	Polymers in Concrete (reaffirmed 1981)
SP-15 (84)	Field Reference Manual: Specifications for Structural Concrete for Buildings (ACI 301-84) with Selected ACI and ASTM References (1984)
SP-49	Corrosion of Metals in Concrete
SP-60	Vibrations of Concrete Structures
SP-70	Joint-Sealing and Bearing Systems for Concrete Structures

American Society for Testing Materials

A 82-79	Standard Specification for Cold-Drawn Steel Wire for Concrete Reinforcement
A 82-85	Standard Specification for Steel Wire, Plain, for Concrete Reinforcement
A 184-84a	Standard Specification for Fabricated Deformed Steel Bar Mats for Concrete Reinforcement

A 185-79	Standard Specification for Welded Steel Wire Fabric for Concrete Reinforcement
A 185-85	Standard Specification for Steel Welded Wire Fabric, Plain, for Concrete Reinforcement
A 416-85	Standard Specification for Uncoated Seven-Wire Stress-Relieved Steel Strand for Prestressed Concrete
A 421-80 (1985)	Standard Specification for Uncoated Stress-Relieved Steel Wire for Prestressed Concrete
A 496-85	Standard Specification for Deformed Steel Wire for Concrete Reinforcement
A 497-79	Standard Specification for Welded Deformed Steel Wire Fabric for Concrete Reinforcement
A 615-85	Standard Specification for Deformed and Plain Billet-Steel Bars for Concrete Reinforcement
A 616-85	Standard Specification for Rail-Steel Deformed and Plain Bars for Concrete Reinforcement
A 617-84	Standard Specification for Axle-Steel Deformed and Plain Bars for Concrete Reinforcement
A 706-84a	Standard Specification for Low-Alloy Steel Deformed Bars for Concrete Reinforcement
A 722-75	Standard Specification for Uncoated High-Strength Steel Bar for Prestressing Concrete (1981)
A 767-85	Standard Specification for Zinc-Coated (Galvanized) Steel Bars for Concrete Reinforcement
A 775-84	Standard Specification for Epoxy-Coated Reinforcing Steel Bars
C 31-84	Standard Method of Making and Curing Concrete Test Specimens in the Field
C 33-84	Standard Specification for Concrete Aggregates
C 39-84	Standard Test Method for Compressive Strength of Cylindrical Concrete Specimens
C 42-84a	Standard Methods of Obtaining and Testing Drilled Cores and Sawed Beams of Concrete
C 94-83	Standard Specification for Ready-Mixed Concrete
C 109-84	Standard Test Method for Compressive Strength of Hydraulic Cement Mortars (Using 2-in or 50-mm Cube Specimens)
C 114-83b	Standard Methods for Chemical Analysis of Hydraulic Cement
C 138-81	Standard Test Method for Unit Weight, Yield, and Air Content (Gravimetric) of Concrete
C 143-78	Standard Test Method for Slump of Portland Cement Concrete
C 144-84	Standard Specification for Aggregate for Masonry Mortar
C 150-85	Standard Specification for Portland Cement
C 171-69 (1980)	Standard Specification for Sheet Materials for Curing Concrete
C 172-82	Standard Method of Sampling Freshly Mixed Concrete
C 173-78	Standard Test Method for Air Content of Freshly Mixed Concrete by the Volumetric Method

C 192-81   Standard Method of Making and Curing Concrete Test Specimens in the Laboratory

C 231-82   Standard Test Method for Air Content of Freshly Mixed Concrete by the Pressure Method

C 260-77   Standard Specification for Air-Entraining Admixtures for Concrete

C 309-81   Standard Specification for Liquid Membrane-Forming Compounds for Curing Concrete

C 311-77   Fly Ash or Natural Pozzolans for Use as a Mineral Admixture in Portland Cement Concrete, Sampling and Testing

C 330-85   Standard Specification for Lightweight Aggregates for Structural Concrete

C 387-83   Standard Specification for Packaged, Dry, Combined Materials for Mortar and Concrete

C 457-82   Standard Practice for Microscopical Determination of the Air-Void Content and Parameters of the Air-Void System in Hardened Concrete

C 494-82   Standard Specification for Chemical Admixtures for Concrete

C 567-85   Standard Test Method for Unit Weight of Structural Lightweight Concrete

C 595-85   Standard Specification for Blended Hydraulic Cements

C 597-71   Test for Pulse Velocity through Concrete

C 618-85   Standard Specification for Fly Ash and Raw or Calcined Natural Pozzolan for Use as a Mineral Admixture in Portland Cement Concrete

C 666-84   Standard Test Method for Resistance of Concrete to Rapid Freezing and Thawing

C 672-76   Test for Scaling Resistance of Concrete Surfaces Exposed to Deicing Chemicals

C 685-85   Standard Specification for Concrete Made by Volumetric Batching and Continuous Mixing

C 845-80   Standard Specification for Expansive Hydraulic Cement

C 856-71   Recommended Practice for Petrographic Examination of Hardened Concrete

C 876-80   Standard Test Method for Half-Cell Potentials of Reinforcing Steel in Concrete

C 898-84   Guide for Use of High Solids Content, Cold Liquid Applied Elastomeric Waterproofing Membrane with Separate Wearing Course

D 994-71   Standard Specification for Preformed Expansion Joint Filler for Concrete (Bituminous Type) (1982)

D 1751-83   Standard Specification for Preformed Expansion Joint Fillers for Concrete Paving and Structural Construction (Nonextruding and Resilient Bituminous Types)

D 1752-84   Standard Specification for Preformed Sponge Rubber and Cork

Expansion Joint Fillers for Concrete Paving and Structural Construction

E 329-77   Standard Recommended Practice for Inspection and Testing
(1983)   Agencies for Concrete, Steel, and Bituminous Materials as Used
            in Construction

American Welding Society

AWS D1.4-79   Structural Welding Code—Reinforcing Steel

Canadian Standards Association

CAN/CSA-S413-87   Parking Structures

Federal Highway Administration

FHWA/RD-86/193   Protective Systems for New Prestressed and Substructure
                  Concrete

National Cooperative Highway Research Program

4-70   Concrete Bridge Deck Durability
57-79   Durability of Concrete Bridge Decks
99-82   Resurfacing with Portland Cement Concrete
165-76   Waterproof Membranes for Protection of Concrete Bridge
           Decks
244-81   Concrete Sealers for Protection of Bridge Structures

National Ready Mixed Concrete Association

Checklist for Certification of Ready-Mixed Concrete Production Facilities (1967)

PART B   REFERENCES

Bridge Deck Rehabilitation Manual, Parts 1 and 2, Manning, D. G. and Bye, D. H.
    Reports SP-016 and SP-017, Ontario Ministry of Transportation, 1984.
Engineer Manual, "Evaluation and Repair of Concrete Structures," EM 1110-2-2002, US
    Army Corps of Engineers.
Expansion Joints in Buildings, Technical Report No. 65, Building Research Advisory
    Board/Federal Construction Council, National Academy of Sciences, Washington,
    DC, 1974.
Parking Garage Maintenance Manual, National Parking Association, Washington, DC,
    1982, 46 pp.
PCI Design Handbook, 3rd edition, Prestressed Concrete Institute, Chicago, 1985.

Post-Tensioning Manual, 4th edition, Post-Tensioning Institute, Phoenix, AZ, 1985.

Raths, C. H., "Spandrel Beam Behavior and Design," *PCI Journal,* vol. 29, N9.2, Mar./ Apr. 1984, pp. 62–131.

Recommended Lateral Force Requirements and Commentary, Structural Engineers Association of California, San Francisco, CA, 1980.

"Removing Stains from Concrete," *Concrete Construction Magazine,* Concrete Publication, Inc., Addison, IL, May 1987, p. 27.

Schupack, Morris, "A Survey of the Durability Performance of Post-Tensioning Tendons," *ACI Journal,* Proceedings, vol. 75, no. 10, Oct. 1978, pp. 501–510.

Index

Index

Durability protection (internal (*Continued*)
finishing, 170; finishing and water,
171; finishing for membranes, 171; fly
ash, 164; formwork for concrete, 170;
impact of, 161; latex, 165; methyl
methacrylate, 166; mixing concrete,
169; placing concrete, 169; reinforcement, 166–169; silica fume, 165; transporting concrete, 169; water, 162;
water reducing additives, 163

Electrical conduit: cast-in, 119; exposed,
118–119; rewiring, 118
Electrical system, maintenance of, 242,
247, 248
Electrochemical reaction, 263. *See also*
Corrosion
Electrolyte, 263
Elevators and elevator lobbies: maintenance, 241; security in, 95
Encroachments, into parking stalls, 28
End bay parking: definition of, 9, 12; efficiency and, 30; site dimensions and,
14; speed ramps and, 12–14
Entry/exit designs: auxiliary spaces,
85–86; CCTV and, 66; considerations
in, 51; importance of, 51; intercoms
and, 66; lane requirements, 75–77; layout, 82–85; level of service criteria, 8;
level of service for lane widths, 13;
level of service for queueing, 81–82;
queueing analysis, 78–81; reversible
lanes, 84; safety and, 105; vehicular/
pedestrian conflicts, 108; vehicular restraints, 107
Epoxy: bonding medium, 332; crack injection, 281, 325; patching mortar material, 333, 335
Expansion joint sealant specifications,
199; installation, 209; materials, 206;
preparation, 208; quality assurance,
201; submittals, 203; warranty, 203.
See also Expansion joint seals
Expansion joint seals, 119; armored strip
seal, 121; compression seal, 121; locating, 121; requirements for, 119; safety
and, 105; tee joint, 119; vibrations and,
124. *See also* Expansion joint sealant
specifications
Exploratory excavation (also test wells),
310, 311

Exposed steel, 241; painting of, 252,
253
Express ramps: to achieve level floors,
20; in circulation system, 9; definition
of, 9; slope and level of service, 13

Federal Highway Administration
(FHWA), 299, 301, 317
Fee computers. *See* Revenue control
Field observation checklist: beams, 225;
for cast-in-place, guidelines, 222; columns, 225; post-tensioned structure,
222, 224; post-tensioned tendons, 224;
precast connections, 226; precast members, 226; for precast structure, 223,
226; problem areas, 226; slabs, 224
Field survey, 256, 257; data evaluation,
312–316; preparation, 287–289; purpose of, 286; recording, 290–296, 367–
374. *See also* Concrete cracking, Deterioration
Fire safety, conflicts with other safety
concerns, 104
Floor finish, safety and, 105
Floor slab: cleaning, 242, 243; flushdown, 243; maintenance, 238–240;
ponding, 240, 272, 310; sweeping,
243. *See also* Deterioration, Floor slab
repair
Floor slab repair: alternatives, 340, 344;
approach, 337–340; cost, 254, 325,
337, 344; continuing corrosion, 340;
economics, 344; life expectancy,
341–344; strategy, 340. *See also* Repair, Sealing
Floor slope: advantages of "level" floors,
20; impact on structure length, 12, 14;
importance in functional design, 7, 10;
level of service classification, 13; security and, 20
Floor-to-floor height: importance in functional design, 6, 7, 10; influences on,
11; typical, 9, 11
Flow capacity: benefits of analysis,
40–41; functional design and, 6; level
of service classification, 13, 35, 40;
level of service criteria, 8; non-parking
circulation components, 35–37; office
uses and, 41, 44; parking angles and,
37–38; parking bays, 37–40; peak-hour

factors, 35; peak-hour volumes, 33–34; retail uses and, 42, 44; rule of thumb, 32; special events and, 42, 44; tidal flow capacity, 37; turnover flow capacity, 38

Fly ash, post-tensioned concrete and, 164 *See under* Durability protection

Freeze-thaw deterioration: influencing factors, 276–278; mechanism, 273–276. *See also* Deterioration

Functional design: flow capacity restraints, 41–43; general implications, 43–44; influences on, 6; rules of thumb, 7, 8, 32, 43

Geometrics, 21–32; flow capacity and, 37; functional design and, 6, 21; level of service classifications, 28–29; level of service criteria and, 8. *See also* Module

Graphics, 249

Grease removal, 243

Half-cell corrosion potential testing, 299, 306, 307. *See also* Corrosion

Handrails, code requirements, 108

Helix: definition of, 9; forms of, 9–10. *See also* Circular helix, Double-threaded helix, Single-threaded helix

High-pressure water: blasting, 319, 330; cleaning, 243

Ice control, measures, 242, 245, 246

Illuminating Engineering Society (IES) Subcommittee on Off-Roadway Facilities, 94

Infrared thermography, 289

Inspection, annual: slab-on-grade, 294; of structure, 250; walk-through, 230, 289

Institute of Transportation (formerly Traffic) Engineers (ITE): geometrics standards, 26; peak-hour volumes, 33; trip generation, 33

Intercoms: coordination with CCTV, 99–100; PARC systems and, 66; parking office and, 85; for security, 99–100; in stair/elevator towers, 92

Iowa method. *See under* Overlays

Isolation joints. *See* Expansion joint seals, Expansion joint sealant specifications; *see under* Design measures for volume change

Joint: distress, 282, 284; expansion joint seals, 284; maintenance, 237, 240, 241; purpose of, 240, 279, 282, 284; sawing, 138; sealant life expectancy, 241, 282; sealing and repairing, 325; tooling, 137; topping, 137

Landscaping, 251

Lateral load resistance: shear walls, 127; shear walls and passive security, 128; truss action, 128

Latex: bonding medium, 332; modified overlay, 336, 351–357; patching, 335, 351–357

Latex modified concrete. *See under* Overlays

Leaching, 261, 285

Level of service: criteria for parking applications, 8; definition of, 7; entry/exit design and, 8; flow capacity and, 8; geometrics and, 8, 28–29; regional impacts, 29

Life expectancy: full-depth slab removal and replacement, 344; joint sealant systems, 241; patching, 341, 343; repair methods 341; sealer, 237, 319. *See also* Service life

Light trucks, vans, utility vehicles (LTVU): clearance for, 12; downsizing of, 25

Lighting: CCTV and, 94; closed areas and, 94; glare, 94–95; passive security and, 93; recommended levels, 94; stain and, 95; uniformity, 94

Live loads, 153; combination with snow, 156

Load testing, 286, 315

Loads. *See* Design loads

Maintenance: aesthetic features, 235, 251; categories, 235; CCTV and, 103; of facility, 233–253; floor slab, 238–240; landscaping and, 97; mechanical systems, 246; operational items, 235, 242–250; safety and, 104; schedules, 237, 242, 251; structural system, 235–241

Maintenance program, 234, 235, 347

Materials testing, 298, 299; core compressive strength, 304; petrographic examination, 305; sampling for chlo-

Quality: concrete control joint sealant specification, 209; concrete sealer specification, 200; control, field, concrete, 184; expansion joint sealant specification, 201; post-tensioning, field control, 198; post-tensioning, specification, 187; precast plant, control, 230; topping, 139; traffic topping specifications, 201; waterproofing specifications, 199

Queueing analysis, 78–81; level of service classification, 81–82

Radar, examination, 289, 299, 309

Ramps. See Driving aisles

Reinforcement: cover for, 166, 298, 299, 302; epoxy coated, 167, 323, 341; epoxy coated and bond, 168; galvanized, 167; locating, 296, 299, 308, 309; post-tensioned, 168; pre-tensioned, 168; stressing pockets, 169. See under Durability protection

Repair: basic requirements, 329; contractor selection, 346; contracts, 345; cost estimate, 344, 346; costs, 346; documents, 256, 288, 345–347; material application, 336; materials, 239, 332–336; methods, 317; objective, 255, 315, 337; priorities, 345; scheduling, 346; specification, 318, 322, 331, 332, 345; types, 236; work items, 290, 345. See also Concrete cracking; Floor slab repair

Replacement, floor slab, 323, 334, 337, 338, 339, 340, 341. See also Repair

Restoration: program, 255, 258, 286; services, 255–258

Revenue control: cost, 75; count systems, 60, 65, 71, 73, 85; facility management systems and, 69–75; fee computers, 58, 59; first generation, 57–58; machine read, 63, 68; pay-on-foot, 63–65; second generation, 58–60; service rates, 77; technology, 68

Safety: accidents versus other life safety concerns, 104; checks, 250; concerns of, in parking facilities, 90–91; head knockers and projectiles, 105–106; maintenance and, 104; operational,

239, 240, 245, 246, 250, 256; tripping and slipping, 104–105; vehicular and pedestrain barriers, 106–108; vehicular and pedestrian conflicts, 108; See also Handrails, Security

Sand-cement, bonding grout, 332

Scabbler, 330, 331

Scaling, 272, 278, 292, 293. See also Freeze-thaw deterioration

Scanning microscope, 299

Scarifier, 330

Schedule: construction, 115; design, 114

Sealants, joint: installation, 172; one part, 172; two part, 172. See also Concrete control joint sealant installation

Sealing, 317, 325. See also Waterproofing specifications

Security: cash, 98–99; concerns of, in parking facilities, 90, 248, 249; landscaping and, 97; negligence versus omission, 90; patrols, 99; perimeter, 96; psychology and, 90; restrooms and, 95–96; risk levels, 91; signs and, 91–98; stair towers and elevators and, 95. See also Active Security, CCTV systems, Passive security, Security alarms, Security audit, Security management

Security alarms: cash, 99; coordination with CCTV, 99–100; types of 99–100

Security audit: definition of, 91; criteria for risk levels, 92; results of, 93

Security management: booth attendants' role in, 103; consultants' role in, 104; importance of, 103; patrols, 99

Security office: location of, 96, 103; manning of, 103

Seismic loads, 157

Service life: cathodic protection system, 325; concrete overlay, 343; parking structure, 254, 337, 341, 343, 344; traffic topping, 335, 343. See also Life expectancy

Service rates, of PARC systems, 77

Shear bond. See Bond

Short span design, impact on flow capacity, 37

Shotblasting, 330

Shotcrete, 335

Silica fume concrete. See under Overlays; *see also* Microsilica and sealer; *see under* Durability protection

Single-threaded helix: advantages and disadvantages of, 17, 20; combinations of, 16; definition of, 9–10; efficiency of, 32; flow capacity, 32, 38, 39, 42–46; with level bays, 19–20; site dimensions for, 12, 14; traffic flow in, 16–17

Site visits, 220; for cast-in-place concrete floor placement, 221–222

Slabs: steel construction, with, 139; structural slabs, 136–137. *See also* Concrete finishing specifications, Drainage, Waterproofing specifications

Slope. *See* Floor slope

Small car only stalls: design vehicle for, 28; introduction of, 22, 26; Parking Consultants Council recommendation, 26; problems with, 26; size of, 22

Smith, Roger, 25

Snow loads, 156

Snow removal, guidelines, 117, 243–245, 321

Spalling, 266. *See also* Corrosion, Deterioration

Specifications: communication in, 178; computer aided specifying, 180; copyright, 180; exceptions in, 180; keeping current, 183; latex modified mortar or concrete, 351–357; local conditions, and, 179; low slump dense concrete, 357–364; performance, 179; prescription, 179; production of, 180; review of, 314; suppliers and, 178; surface preparation, 318, 331, 347–351. *See also* individual headings

Speed Ramps: in circulation systems, 9; definition of, 9; length of, 14; level of service classification, 13; parking on, 12–14; site dimensions and, 12; structure length and, 14

Stair towers: CCTV in, 100; door swings, 106; efficiency and, 30; maintenance, 241; problems, 146; security in, 95

Stall dimensions: design vehicle and, 27; determinants of, 21–22; impact of downsizing, 26; impact on flow capacity, 37. *See also* Geometrics

Striping, 250

Structural design: conflicts with safety concerns, 104; structural maintenance, 235–241

Structural slabs: cast-in-voids, with, 136; hollow core units, 136; one-way slabs, 137; pan joists, 136; precast joists, with, 136; protection for, 136; thin cast-in-place slabs, 136; two-way slabs, 137; waffle slabs, 136

Structural steel components: beams, 140; cover plates, 140; floors, 139; steel forms for floors, 139; studs, 140

Structural steel superstructure systems: advantages, 135; cost, 135. *See also* Structural steel components

Structural system selection: engineer's role, 125; factors, 125; foundations, 126; summary, 140; Table 5-2, 142–143. *See also* Cast-in-place concrete superstructure systems, Precast concrete superstructure systems, Structural steel superstructure systems, Volume change

Tees. *See* Double tees

Tensile stress control: bending, 124; columns, 124; walls, 124

Thermal coefficient of expansion, 333. *See also* Repair

Thermal compatibility, 332–333. *See also* Repair

Ticket dispenser. *See* Revenue control

Tidal flow capacity. *See* Flow capacity

Topping, concrete. *See under* Precast concrete structural components; *see also* Traffic topping

Torsion, 151

Traffic flow, 15–20. *See also* One-way traffic flow, Two-way traffic flow

Traffic topping: application, 238–239, 320, 340; evaluation and selection, 301, 335; protected, 173; traffic bearing, 173; vapor and, 174; wear grades, 174. *See also* Traffic topping specifications, Waterproofing specifications

Traffic topping specifications, 199: installation, 209; materials, 205; preparation, 207; quality assurance, 201; submittals, 202; warranty, 203. *See also* Waterproofing specifications, Traffic topping